Climatology

Physical and Dynamic Climatology

CHAPTER 1: The Basis of Modern Climatology
CHAPTER 2: Energy and the Atmosphere
CHAPTER 3: Temperature and Temperature Distribution
CHAPTER 4: The Physical Basis of the Hydrologic Cycle
CHAPTER 5: Precipitation and the Water Balance
CHAPTER 6: The Primary Circulation System
CHAPTER 7: The Climatology of Severe Storms

1 The Basis of Modern Climatology

CLIMATOLOGY: AN ATMOSPHERIC SCIENCE

A BRIEF CHRONOLOGY

THE CURRENT VIEW
Atmospheric Composition
The Atmosphere as a Gas
The Layered Atmosphere

SUMMARY

APPLIED STUDY 1
Applications of Climatology

The earth's atmosphere is a fundamental component of the planet's environment. Together with the *geosphere* (the solid earth), the *hydrosphere* (the waters of the ocean), and the *biosphere* (living parts of our world), it makes earth the habitable, hospitable place that it now is. The atmosphere serves a number of major functions on the planet. Beyond providing the reservoir of air required for life, it also plays a critical role in the distribution or redistribution of energy over the planet. The atmosphere provides an insulating layer around the earth, which raises the mean temperature of the surface from $-23°C$ ($-9°F$) to near 15°C (59°F) and, at the same time, shields the surface from the large doses of ultraviolet radiation that would destroy most life forms. It also serves as a major transporter of heat horizontally across the earth's surface. Tropical regions receive far more energy than polar regions, and atmospheric circulation helps to equalize this imbalance by moving heat from the warmer to the colder areas. In addition to transferring energy, the atmosphere transfers water from low to high latitudes and from ocean to land.

CLIMATOLOGY: AN ATMOSPHERIC SCIENCE

Because of the complexity of the gaseous envelope that surrounds the planet, atmospheric scientists often subdivide its study into specific areas of interest. One such division identifies the three fields of aerology, meteorology, and climatology.

Aerology (or *aeronomy*) is essentially the study of the free atmosphere through its vertical extent. Initially, the major concern of the aerologist was the identification of atmospheric structure and the amount and distribution of its component parts. Today aerology deals mostly with the chemistry and physical reactions that occur within the various layers. The word *aerology* is less widely used now than it once was, and its content is frequently considered as part of meteorology.

Meteorology is the science that deals with motion and phenomena of the atmosphere with the view to both forecasting weather and explaining the processes involved. It deals largely with the status of the atmosphere over a short period of time and utilizes physical principles to attain its goals.

Climatology is the study of atmospheric conditions over a longer period of time. It includes the study of different kinds of weather that occur at a place. It concerns not only the most frequently occurring types, the average weather, but the infrequent and unusual types as well. Dynamic change in the atmosphere brings about variation and occasionally great extremes that must be treated on the long-term as well as the short-term basis. As a result, climatology may be defined as the aggregate of weather at a place over a given time period.

Given this definition climate is partly meteorological; but since it is also concerned with specific climatic conditions at locations on the earth's surface, it is also geographical. The British climatologist, E. T. Stringer (1975), has written

> The variations in the Earth's surface have profound effects on the interchange of heat, moisture, and momentum between land, water, and atmosphere, and are vital in determining specific climatic conditions;

ONE

CONTENTS

Climatic Change: The Human Impact   324

    The Role of Surface Changes □ Atmospheric Changes

The Climate of Cities   329

    The Modified Processes □ The Observed Results

Summary   334
Applied Study 15   336

    Acid Rainfall: A Change in Quality

16   Future Climates   *339*

The Need to Know   340

    Climate and Energy □ World Food Production □ Global Ecology □ Other Impacts

Modeling Future Climates   344

    Analog and Digital Models □ Atmosphere-Ocean Relationships □ Scenarios

The Current View of Future Climates   351
Summary   351
Applied Study 16   352

    Climatic Change and Food Production

Appendix 1   Background Notes   *355*

1. The Gas Laws   356
2. Radiation Laws   357
3. Energy Flow Representation   358
4. The Heat Budget   359
5. Lapse Rates and Stability   360
6. Thornthwaite's Water Balance   362
7. Vorticity and Angular Momentum   364

Appendix 2   Glossary   *367*

References Cited   *372*

Index   *373*

The Climate of the West Coast   262

    Summer Weather □ Winter Weather □ The Effects of the Mountains on Local Climate

Summary   266

Applied Study 12   267

    Setting Records—A Case of Probability

13 Polar and Highland Climates   271

The Arctic Basin   273

    The Polar Wet and Dry Climate

The Antarctic Basin   277

    The Polar Dry Climate □ The Polar Wet Climate

Highland Climates   283

Summary   286

Applied Study 13   286

    Climate and Human Physiology

THREE   Climates of the Past and Future

14 Reconstructing the Past   293

Surface Features   294

    Evidence from Ice □ Periglacial Evidence □ Sediments □ Sea Level Changes

Past Life   298

    The Faunal Evidence □ The Floral Evidence

Evidence from the Historical Period   303

    Records of Floods □ Records of Drought □ Migrations □ Contemporary Literature □ Evidence of Agriculture and Settlements □ The Instrumental Period

The Reconstructed Climate   307

Summary   311

Applied Study 14   311

    The Little Ice Age

15 Causes of Climatic Change   315

The Natural Changes   316

    Earth-Sun Relationships □ Solar Output Variation □ Atmospheric Modification □ Distribution of Continents □ Variation in the Oceans □ Other Theories

CONTENTS

Regionalization of the Climatic Environment   181

9

The Rationale   182
Approaches to Classification   183
Empiric Systems   185

    The Köppen System □ The Thornthwaite Classification

Genetic Systems   192

    The Model

Climatic Regions and Environmental Characteristics   196

    Climate and the Distribution of Vegetation □ Soil Classification and Climate

Summary   202
Applied Study 9   203

    Climatic Regions and Development: The Transamazon Highway

Tropical Climates   207

10

General Characteristics   208

    Radiation and Temperature □ Precipitation

Climatic Types   210

    The Tropical Wet Climate □ The Tropical Wet and Dry Climate □ The Tropical Deserts □ Coastal Deserts □ Desert Storms

Summary   221
Applied Study 10   221

    Desertification

The Mid-Latitude Climates   227

11

General Characteristics   228

    Middle Latitude Circulation □ Summer Season □ Winter Season □ Water Balance

Climatic Types   236

    Mid-Latitude Wet Climates □ Mid-Latitude Wet and Dry Climates □ Summer-Dry Climates □ Mid-Latitude Deserts

Summary   247
Applied Study 11   247

    Supplemental Irrigation in the Upper Mississippi Valley

Climates of North America   251

12

Climate East of the Continental Divide   254

    Summer Weather □ Winter Weather

6 Atmospheric Circulation — 111

Factors Affecting Air Movement   112
  Pressure Gradient □ Rotation of the Earth □ Friction
Global Patterns of Wind   115
  Tropical Circulation □ Mid-latitude Circulation □ Polar Circulation □ Surface Winds of the World □ Seasonal Changes in the Global Pattern
Diurnal Wind Systems   129
  Land and Sea Breeze □ Mountain and Valley Breeze
Summary   131
Applied Study 6   132
  Energy From the Wind

7 The Climatology of Severe Storms — 135

Thunderstorms   136
  Thunderstorm Formation □ Associated Hazards □ Hail Regions
Tornadoes   143
  Formation and Characteristics □ Distribution and Severity
Tropical Cyclones   147
Summary   154
Applied Study 7   155
  Storms and Planning

TWO   Regional Climatology

8 Regional Climates: Scales of Study — 161

Definitions   162
Microclimates   164
  General Characteristics □ The Role of Surface Cover □ The Role of Topography
Local Climatology   169
Meso- and Macroclimates   170
  Synoptic Climatology □ Regional Climatology
Summary   176
Applied Study 8   177
  Donora: An Air Pollution Episode

xi

The Heat Balance   38
Summary   40
Applied Study 2   40
   Direct Use of Solar Energy

3 Temperature and Temperature Distribution   45

The Concept of Temperature   46
The Temperature Lag   47
Temperature Data   48
Factors Influencing the Vertical Distribution of Temperature   49
Factors Influencing the Horizontal Distribution of Temperature   52
   Locational Factors □ Dynamic Factors
The Distribution of World Surface Temperatures   59
Summary   60
Applied Study 3   64
   Temperature Thresholds

4 The Physical Basis of the Hydrologic Cycle   67

Physical Properties of Water   68
The Hydrologic Cycle   69
Atmospheric Humidity   72
Air Masses   73
Evaporation and Transpiration   74
Condensation   79
   Dew and Frost □ Fog and Clouds
Summary   84
Applied Study 4   84
   The Hydrologic Response to Winter Conditions

5 Precipitation and the Water Balance   87

Vertical Transport of Water Vapor   89
   Convective Lifting □ Cyclonic Lifting □ Orographic Lifting
Spatial Variation in Precipitation   94
Temporal Distribution of Precipitation   98
   Floods □ Drought
Regional and Local Water Balances   104
Summary   105
Applied Study 5   106
   Impacts of Snowfall

Contents

ONE   Physical and Dynamic Climatology

1   The Basis of Modern Climatology   3

Climatology: An Atmospheric Science   4
A Brief Chronology   6
The Current View   9

    Atmospheric Composition ☐ The Atmosphere as a Gas ☐ The Layered Atmosphere

Summary   18
Applied Study 1   18

    Applications of Climatology

2   Energy and the Atmosphere   21

The Nature of Radiation   23
The Solar Source   25
The Role of the Atmosphere   26

    Scattering ☐ Absorption ☐ Reflectance ☐ Transmissivity

Earth-Sun Relationships   29
Global Patterns of Energy Receipts   31
Terrestrial Radiation and the Greenhouse Effect   33
The Planetary Energy Balance   35

PREFACE

For those readers who do require further explanation of some of the concepts, Appendix 1 provides a brief mathematical account. For those to whom many of the terms used are new, a glossary is provided in Appendix 2. References and footnotes are kept to a minimum. For further readings, the *Instructor's Manual* that accompanies the text provides an appropriate listing.

The ultimate aim of this book is to provide those interested in climatology with an up-to-date yet readily comprehensible account of the causes and effects of climate. In attaining this goal, the authors, of necessity, have been selective in the material presented. It is their hope, however, that in making this selection, they have provided concepts and ideas that will influence readers to continue their study of climatology.

The authors would like to thank the following persons who read the manuscript and made many valuable suggestions that improved the final version: Anthony Brazel of Arizona State University, David Greenland of the University of Colorado, Glen A. Marotz of the University of Kansas, and Daniel L. Wise of Western Illinois University. Helpful comments and suggestions were also provided by David Marczely of Southern Connecticut State College, Richard Skaggs of the University of Minnesota, and Stephen J. Stadler of Oklahoma State University.

Preface

Current events are creating an unprecedented growth in interest in climatology. Media coverage of droughts, floods, and very cold or exceptionally mild winters expose people to many of the basic concepts of climate and climatic anomalies. Added to this are the many articles on changing climate that appear in the popular press.

As a partial response to the heightened interest in climatology, this text provides a basic yet thorough introduction to a number of aspects of climate. Part One of the text deals with the major processes that ultimately produce the climate at any given location. This coverage is followed, in Part Two, by a systematic treatment of regional climates. The climatic types considered are discussed in terms of both the basic characteristics of the climate type and the processes that produce them.

Part Three considers the topic of climatic change. To enable the student to grasp the meaning and significance of this topic, considerable attention is given to the nature of climates of the past and the ways in which they are interpreted. Thereafter, the problem of attempting to predict future climates is presented.

Climatic data and knowledge are used today to help solve a host of related environmental and societal problems. This applied aspect of climatology is exemplified in this text by presenting an "Applied Study" at the end of each chapter. In each case, a topic covered within the chapter is selected to illustrate how the climatic knowledge is applied to a related topic. Obviously, the coverage cannot be complete, but this approach demonstrates many applications of climatology without greatly extending the length of the book.

Many potential users of this text will not have had rigorous mathematical training; as a result, mathematics and the use of physical formulas are minimized.

The atmosphere, together with the surface of the earth, and the sea, is the laboratory in which all our different sorts of weather, our winds, and our rainstorms and blizzards and our dry periods, are manufactured. As these influence our happiness and well-being, every hour of our lives, we ought to discuss them in considerable detail.

—Hendrik Willem van Loon

Published by Charles E. Merrill Publishing Company
A Bell & Howell Company
Columbus, Ohio 43216

This book was set in Optima.
The production coordinator was Rebecca Money.
Text design by Cynthia Brunk
Cover design by Cathy Watterson
Line illustrations by Cartographics, Department of Geography, Texas A&M University

Front cover:
Wheat Field with Cypresses by Vincent van Gogh
Reproduced by courtesy of the Trustees, the National Gallery, London.
Back cover:
Vincent van Gogh, *The Starry Night* (1889), oil on canvas, 29" by 36¼". Collection, The Museum of Modern Art, New York. Acquired through the Lillie P. Bliss Bequest.
Page xvi:
Winter in Flanders by Pieter Brueghel the Elder
Reproduced by permission of Musées royaux des Beaux-Arts de Belgique.
Page 158:
Boy and Calf—Coming Storm by George Bellows
Columbus Museum of Art, Ohio: Gift of Mr. and Mrs. Everett D. Reese.
Page 290:
Along the Shore by Maurice Prendergast
Columbus Museum of Art, Ohio: Gift of Ferdinand Howald.

Copyright © 1984 by Bell & Howell Company. All rights reserved. No part of this book may be reproduced in any form, electronic or mechanical, including photocopy, recording, or any information storage and retrieval system, without permission in writing from the publisher.

Library of Congress Catalog Card Number: 83-63079
International Standard Book Number: 0-675-20144-6
Printed in the United States of America
1 2 3 4 5 6 7 8 9—88 87 86 85 84

John E. Oliver
Indiana State University

John J. Hidore
The University of North Carolina
at Greensboro

Climatology
An Introduction

Charles E. Merrill Publishing Company
A Bell & Howell Company
Columbus Toronto London Sydney

PHYSICAL AND DYNAMIC CLIMATOLOGY

SUMMARY

Information about earth's atmosphere represents results of many years of inquiry into the nature of the atmosphere. The study of atmospheric science has evolved into three main subgroups, with climatology representing the aggregate of meteorological conditions for a given time and place. A knowledge of the structure and composition of the atmosphere is essential to understanding its processes. In dealing with structure, it is useful to consider the atmosphere as consisting of a series of shells. The division of these shells depends upon the nature of the investigation. Climatologists find the subdivisions based upon vertical temperature gradient to be an appropriate classification procedure. In terms of composition, it is necessary to identify both permanent and variable constituents and to note that the proportions of the permanent gases are constant up to an elevation of some 88 km.

APPLIED STUDY 1

Applications of Climatology

Climatology is a fascinating area of study. Not only does it explore the workings and nature of the atmosphere but it is also an area of study that relates directly to the way in which the environment functions and the everyday lives of people.

The relationship of climate to other disciplines and activities comes under the heading of applied climatology, the basic purpose of which is ultimately to aid society in attaining a better adjustment to and understanding of the climatic environment. Applied climatology is used to—

1. improve efficiency of various economic activities that are influenced by climate,
2. aid in the needs of societal activities,
3. reduce the losses incurred from climatic hazards.

Consider the following examples:

Energy Climate plays a part in the use and development of both fossil fuels and the renewable energy resources. The amount of heating and cooling needed for buildings depends upon the climate of the area in which the building is located. The use of fossil fuels to generate energy is so climate-related that the concept of heating and cooling degree-days is used by engineers to estimate energy demands (Applied Study 3).

The potential an area has for the development of renewable resources is directly related to its atmospheric conditions. To evaluate the potential for the direct use of solar energy (Applied Study 2) and wind power (Applied Study 6) requires detailed knowledge of the climatic conditions that prevail.

Food The success of agriculture is determined by how well farmers adapt their crops and activities to climate. To be sure, in modern agriculture the use of good

THE BASIS OF MODERN CLIMATOLOGY

The composition of air can also be used to classify the atmosphere. In the lower portions of the atmosphere, to a height of about 88 km, a uniform mixture of gases is found. That is, at sea level, 78 percent of the gases would be nitrogen, 21 percent oxygen; similar proportions of the same gases would be found at 30, 50, and 80 km. This layer is termed the *homosphere*. Above the homosphere the composition of the air is no longer uniform and is thus called the *heterosphere*. Four layers are identified in the heterosphere (figure 1–4). From the top of the homosphere to a height of 220 km, the atmosphere is largely made of molecules of nitrogen. Above this, to perhaps 1125 km, atomic oxygen prevails. The final two layers are composed of helium (to 3540 km) and hydrogen atoms. Note that while figure 1–4 shows distinctive boundaries between these layers, they are in reality ill-defined.

One final method used to identify atmospheric layers draws upon the various chemical reactions that occur and the regions in which solar radiation produces ionization. These are called the *chemosphere* and the *ionosphere*, respectively. A summary of the major features of these, together with other methods of atmospheric layering, is outlined in figure 1–5.

FIGURE 1–5 Systems of nomenclature of atmospheric shells (From U.S. Air Force, *Handbook of Geophysics*).

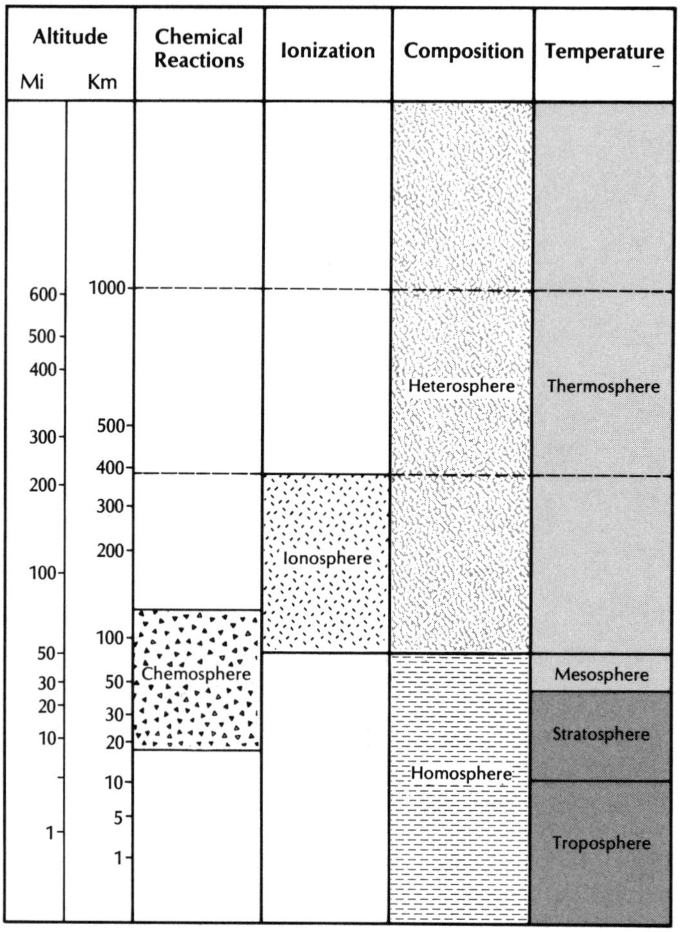

above any location decreases. There is a corresponding decrease of pressure with height. The pressure data in figure 1-3 indicate that surface pressure is in the order of 10^3 (1000) millibars (mb). In the lower stratosphere, some 15 km above the surface, the pressure is only 10^2 (100) mb. The rapid decrease with height means that about one-half of the total mass of the atmosphere is found below 5.6 km, while 99 percent of the mass occurs below 30 km.

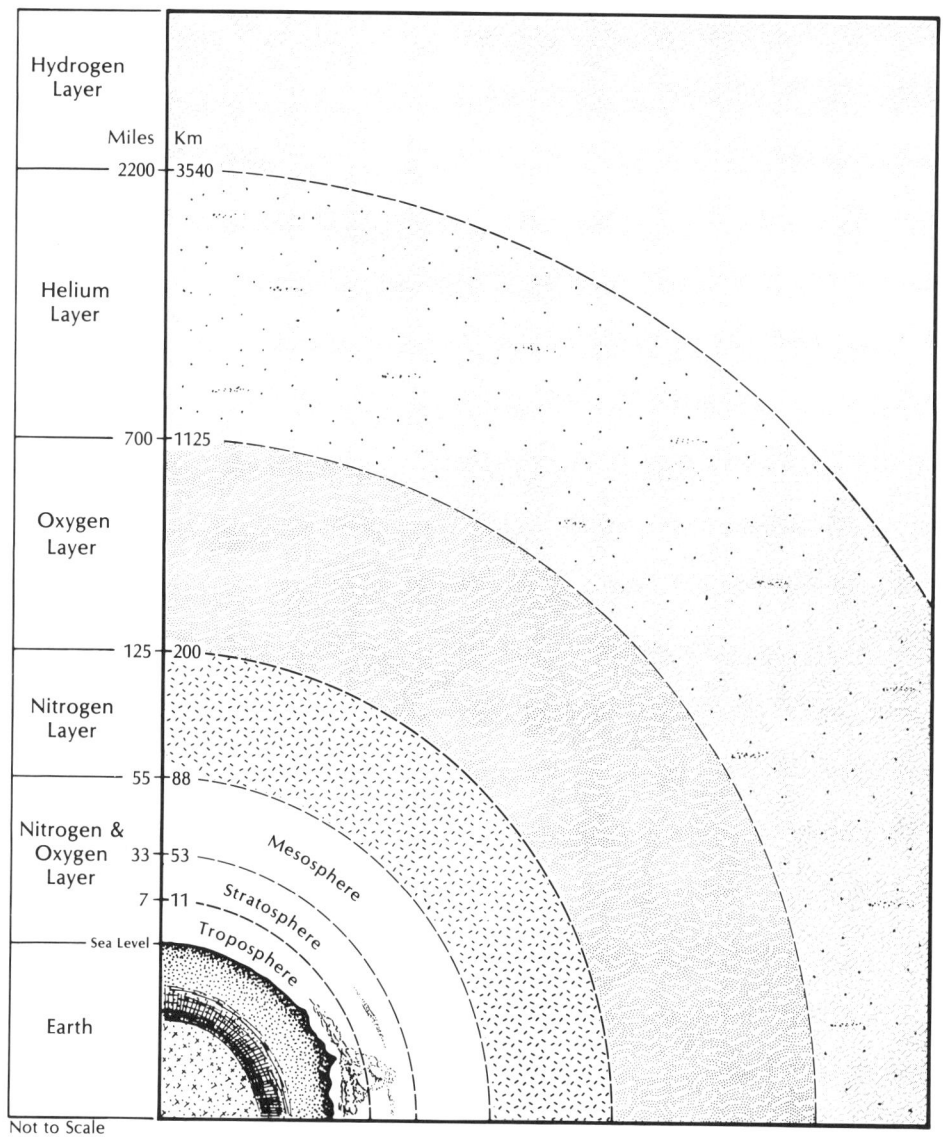

FIGURE 1-4 Schematic cross-section of the earth's atmosphere, showing broad chemical divisions (From John J. Fagan, *The Earth Environment*, © 1974, p. 13. Reprinted by permission of Prentice-Hall, Inc., Englewood Cliffs, New Jersey).

THE BASIS OF MODERN CLIMATOLOGY

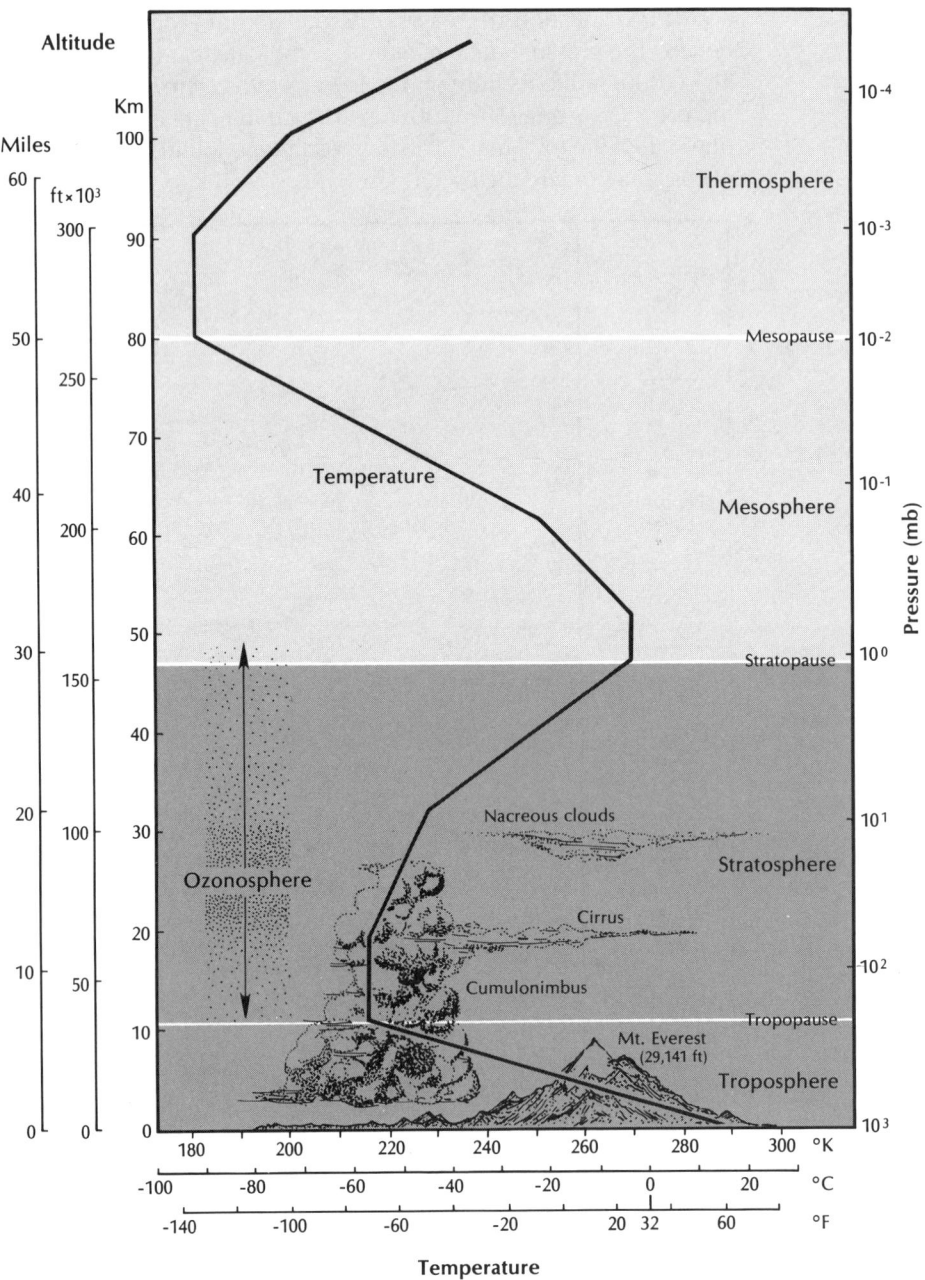

FIGURE 1-3 Temperature structure of the atmosphere.

thus capable of flowing. Water is a fluid; and in some ways, water and the atmosphere act much alike. Motion in fluids can be in any direction; depending on the direction of the force applied, the flow regime generally can be classified as laminar or turbulent. *Laminar flow* takes place when each individual particle in the fluid travels in a straight line, and the path taken by each particle is parallel to the path taken by other particles. *Turbulent flow* occurs when the individual particles are traveling in any direction with respect to the mean direction of travel of the fluid. The eddies in streams and the circulation in dust devils are visible examples of turbulent flow.

The Layered Atmosphere

In general terms, the atmosphere can be considered as a series of concentric layers or shells surrounding the earth. Various names are given to the identified layers, depending upon the criteria for the divisions.

The most commonly used method to describe atmospheric layering uses temperature as the variable. This is illustrated in figure 1–3, which shows temperature changes with height and the manner in which the layers can be identified.

The *troposphere* is the lowest level, the part in which our day-to-day lives are carried out and in which the events that we call "weather" occur. In this layer, there is a generally uniform decrease of temperature with height. The lowest part of the troposphere, up to 1.5 km or 2 km (5000–7000 ft), is called the *friction layer*. Winds in this zone are modified through friction caused by the surface and surface irregularities. The friction layer represents the area in which most day-to-night variability occurs.

The upper limit of the troposphere is the *tropopause*. It is a zone in which the general decrease of tropospheric temperature ceases; temperature remains fairly constant with height through the tropopause, which is an *isothermal* (equal temperature) layer. The tropopause also represents the upper limit of large-scale turbulence and mixing of the atmosphere.

The *stratosphere* extends from about 10 km to 45 km above the surface. In its lower section, temperatures are fairly constant, but at an elevation of about 30 km, they increase toward the upper limit of this zone, the *stratopause*. Air circulation in the stratosphere is characteristically persistent, with winds blowing at high velocities.

The *mesosphere,* above the stratopause, is identified by a marked temperature decrease with altitude. Beginning at an elevation of about 80 km, the decline continues until the mesopause is reached. The *thermosphere* lying above the mesopause has no defined upper limit. It is so named because of very high thermodynamic temperatures attained. These temperatures represent a concept of heat that differs from the notion of *sensible heat,* the idea of hot and cold as sensed by our bodies. This important differentiation factor is considered in chapter 2.

The temperature pattern shown in figure 1–3 is supplemented by data showing pressure of the atmosphere at the various altitudes. The *atmospheric pressure* at any location may be considered as the force exerted by the weight of the atmosphere above that point. On ascending in the atmosphere, the amount of air atmosphere until the level reached 0.1 PAL. The land surfaces as well as the oceans were now partially screened from the ultraviolet radiation, and life moved from sea to land. An acceleration in the evolution of land life began at the end of the Silurian period, when primitive land plants became established. Increased photosynthesis added large amounts of oxygen to the atmosphere, and by the Carboniferous period, environments indicate that forests with abundant trees and ferns were widely spread over the earth. Since Carboniferous times, the amount of oxygen fluctuated, until today's levels were attained.

The Atmosphere as a Gas

Although the atmosphere is a mixture of gases, it behaves in many ways as if it were a single gas. Accordingly, the general laws that apply to all gases describe the behavior of atmosphere. These laws state the relationships among pressure, density, volume, and temperature. They were initially formulated in the seventeenth and eighteenth centuries and based upon the study of gases in closed containers. The equations representing the relationships are given in appendix 1 (note 1). They use the concepts of *proportionality* (if two variables are proportional, then as one increases so does the other), and *inverse proportionality* (the increase in one quantity results in the decrease of the other). The following relationships result:

1. If temperature is held constant, the density of a gas is proportional to pressure, while volume is inversely proportional to pressure. Thus an increase in pressure results in an increase in density and a decrease in volume.

2. If volume is kept constant, the pressure of a gas is proportional to temperature. Pressure will increase as temperature increases, assuming no change in volume.

3. If pressure is kept constant, the volume of a gas is proportional to temperature and inversely proportional to density. Accordingly, raising temperature causes volume to increase and density to decrease if there is no change in pressure.

These laws can be combined into a single law, the Ideal (General) Gas Law, so named to indicate that it is for ideal conditions and to remind users that gases do not obey the law precisely. It is given as

$$PV = RT$$

or

Pressure × Volume = Constant × Temperature

The constant R is referred to as the gas constant; it is a particular value for each gas. In using this formula, all variables must use the same system of units and temperatures must be given in absolute values.

The atmosphere has not only the characteristics of a gas but also those of a fluid. A *fluid* is a substance having particles that move easily and change their relative position without separation of the mass. They easily yield to pressure and are

TABLE 1-4 Geologic time and oxygen levels.

Years Before Present	Geologic Time Scale			Events Influencing Oxygen Level		Oxygen Levels
	Era	Period	Lithosphere (Geologic Events)	Biosphere (Biologic Events)	
2 million	Cenozoic	Quaternary	Glaciation	Man Mammals diversify	Oxygen approaches PAL
50 million		Tertiary			
	63 million	Cretaceous	Mountain building, e.g. Alps, Rockies Extensive sea deposits —chalk beds	Development of cereals and grass Dinosaurs extinct Development of flowering plants	Oxygen levels fluctuate
100 million	Mesozoic	Jurassic			
200 million		Triassic			
	230 million	Permian Carboniferous Devonian Silurian Ordovician Cambrian	Mountain building, e.g. Appalachians Coal formation Extensive limestones Thick marine deposits	Conifers spread Widespread forest of spore-bearing plants First land plants Abundant marine life	Oxygen may have been 10 × PAL Oxygen 3–10% PAL
500 million	Paleozoic				
	570 million	Pre-Cambrian	Glaciation Many mountain building episodes Volcanoes	Sea life develops	Oxygen 1% PAL ozone screen effective Oxygen increases, carbon dioxide decreases
1 billion					
2 billion			First red beds (sediments) containing iron oxides) Oldest sediments	First oxygen-bearing photosynthetic cells Abiogen (life from matter) evolution	Oxygen in atmosphere
5 billion					No free oxygen
			Origin of Earth		

Source: From J. E. Oliver (1977, p. 80). Used by permission of Wadsworth Publishing Company.

vapor, oxides of nitrogen, methane, and ozone are typical variable gases. The ratio of the constant gases to one another applies essentially to the lower levels of the atmosphere. At higher altitudes, the proportions change as the form and occurrence of gases are modified.

The present composition is the result of the long evolution of the atmosphere. The primitive initial gaseous envelope of the earth was dependent upon the nature of earth formation. There is no single theory accepted by all to explain the origin of the earth, but of the theories available, two are considered viable alternatives. One suggests that primordial gases and solids coalesced to form a hot, molten earth; the other that a great upheaval or cataclysm in space produced debris that eventually became the earth. Irrespective of which is correct, it can be assumed that the earth began as a molten body with temperatures at the surface in excess of 8000°C. The gases present, predominantly hydrogen and helium, were so hot that the molecules overcame the gravitational pull of earth and escaped to space.

As the earth cooled and a solid crust formed, gases such as carbon dioxide, nitrogen, and water vapor were released. These formed the secondary primitive atmosphere, which may have had a composition of 60–70 percent water vapor, 10–15 percent carbon dioxide, 8–10 percent nitrogen, with sulfur compounds making up the remainder. Continued cooling caused the water vapor to condense, clouds to form, and rain to occur. Initially, the rain did not reach the surface; for at earth's still-high temperature, the water evaporated well above the surface. Eventually, the surface temperature was low enough to allow rain to reach the surface. The vast amount of water vapor in the atmosphere must have caused rains that went on continuously for thousands of years. The rain washed out large amounts of carbon dioxide, which was then joined with surface materials to form carbonate rocks.

Thus far there has been no mention of oxygen, a gas that now forms 21 percent by volume of the atmosphere. Oxygen was a latecomer to the scene; it probably originated when water vapor dissociated to form hydrogen (H_2) and oxygen (O_2). The light hydrogen escaped to space, but the small traces of oxygen remained to begin formation of an ozone (O_3) layer some 30 km above the surface. Ozone, triatomic oxygen, plays a fundamental role in screening short-wave ultraviolet radiation from the earth's surface. Energy from the sun that penetrates the high atmosphere is composed of a variety of wavelengths (see chapter 2), including ultraviolet. This radiation, in shorter wavelengths, is lethal to most forms of life on earth, and it is fortunate that ozone absorbs energy at this wavelength. So important is its presence that both the development of atmospheric oxygen to present levels and the evolution of life have been related to formation of the ozone layer.

The low oxygen content of the secondary primitive atmosphere, which lacked an ozone screen, allowed ultraviolet radiation to reach the surface and penetrate the seas. Photodissociation of water vapor contributed a small amount of oxygen to the atmosphere, but the levels were only 0.001 that of the present atmospheric level (PAL). By this process the amount of oxygen increased slowly; when the 0.01 PAL level was attained, sufficient screening from ultraviolet radiation permitted the development of life that occurred during Cambrian times. The rocks of this age bear many fossils of marine life. (See table 1–4 for an outline of geologic time and oxygen levels.) Photosynthesis in the oceans released more oxygen into the surprising since, as compared to nitrogen and oxygen, there is only a small amount of atmospheric carbon dioxide. Atmospheric nitrogen was discovered by Rutherford in 1772, and was at first called "Mephitic air." Shortly after, Joseph Priestley isolated oxygen, which he called "dephlogistated air."

Over time other gases were discovered, the last of which was argon, isolated in 1894. The gases were given their modern names and their relative volumes in the atmosphere determined. As shown in figure 1-2, nitrogen and oxygen make up the bulk of the atmosphere, with other gases totalling less than 1 percent of the whole.

The gases shown in figure 1-2 are the permanent or constant constituents. At any given time, traces of other gases are also present in small amounts; water

FIGURE 1-2 Gaseous composition of the atmosphere: (a) gases making up the lower atmosphere in percent by volume; (b) the variation of gaseous components with altitude (From the U.S. Navy Research Facility).

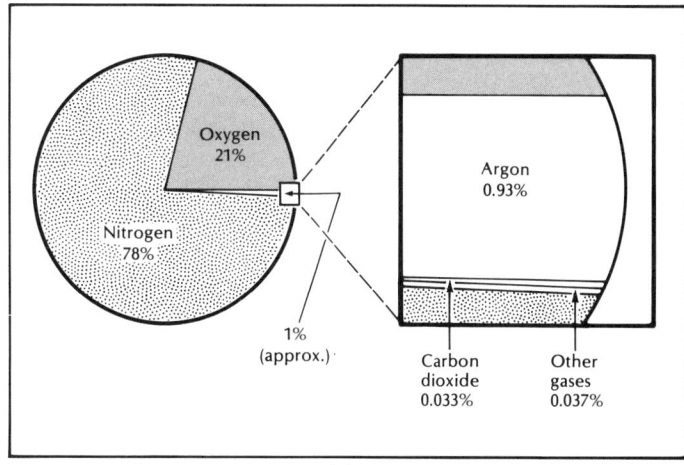

(a)

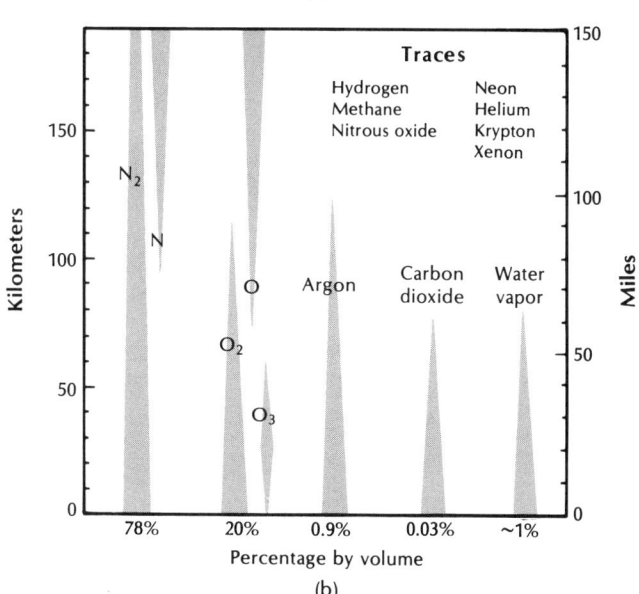

(b)

TABLE 1-3 Elements of the National Climate Program Act of 1978.

The Programs shall include, but not be limited to, the following elements:
1. assessment of the effect of climate on the natural environment, agricultural production, energy supply and demand, land and water resources, transportation, human health and national security;
2. basic and applied research to improve the understanding of climatic processes, natural and man-induced and the social, economic and political implications of climatic change;
3. methods for improving climate forecasts;
4. global data collection on a continuing basis;
5. systems for the dissemination of climatological data and information;
6. measures for increasing international cooperation in climatology;
7. mechanisms for climate-related studies;
8. experimental climate forecast centers;
9. submission of 5 year plans.

(Abbreviated from the National Climate Program Act—Conference Report, Section 5 (d). This act was passed by the 95th Congress and signed into law by President Carter on 17 September 1978.)

program seeks to evaluate the practical limits of weather forecasting and to determine the statistical properties of the general circulation of the atmosphere, which would lead to a better understanding of the physical basis of climate.

Conceived in the 1960s, the first large-scale field project of GARP was the GARP Atlantic Tropical Experiment (GATE), which began in 1974. So many data were gathered in this experiment, whose aim was to relate tropical cloud clusters to the general circulation, that they are still being analyzed.

Of great significance in the recent development of climatology is the creation of the National Climate Program Act. Signed into law in September 1978, this act has the stated purpose "to establish a national climate program that will assist the nation and the world to understand and respond to natural and man-induced climate processes and their implications." The elements of the programs represent an assessment of the role and understanding of climate in many ways. Table 1-3 lists these elements: their content points to both the importance of climatology and the focus of climatological research after 1978.

THE CURRENT VIEW

The knowledge accumulated over time provides the modern climatologist with a basic understanding of the structure and composition of the atmosphere. Such information provides a necessary background to the understanding of climate.

Atmospheric Composition

Knowledge that the atmosphere comprises a number of gases began with the work of chemists in the eighteenth century. The first gas to be isolated and studied in detail was carbon dioxide (CO_2). Its discovery by Black in 1752 is somewhat

PHYSICAL AND DYNAMIC CLIMATOLOGY

TABLE 1-2 Significant events in the development of climatology.

ca. 400 B.C.	The influence of climate and health are discussed by Hippocrates in *Airs, Waters, and Places*.
ca. 350 B.C.	Weather science is discussed in Aristotle's *Meteorologica*.
ca. 300 B.C.	The text *De Ventis* by Theophrastus describes winds and offers a critique of Aristotle's ideas.
ca. 1593	The thermoscope is described by Galileo. (The first thermometer is most likely attributed to Santorre, 1612).
1622	A significant treatise on the wind is written by Francis Bacon.
1643	The barometer is invented by Torricelli.
1661	Boyle's law on gases is propounded.
1664	Weather observations begin in Paris; although often described as the longest continuous sequence of weather data available, the records are not homogeneous or complete.
1668	Edmund Halley constructs a map of the trade winds.
1714	The Fahrenheit scale is introduced.
1735	George Hadley's treatise on trade winds and effects of earth rotation is written.
1736	The Centigrade scale is introduced. (It was first formally proposed by du Crest in 1641).
1779	Weather observations begin at New Haven, Conn., the longest continuous sequence of records in the United States.
1783	The hair hygrometer is invented.
1802	Lamark and Howard propose the first cloud classification system.
1817	Alexander von Humboldt constructs the first map showing mean annual temperature over the globe.
1825	The psychrometer is devised by August.
1827	Beginning of the period during which H. W. Dove developed the laws of storms.
1831	William Redfield produces the first weather map of the U.S.
1837	Pyrheliometer for measuring insolation is constructed.
1841	Movement and development of storms are described by Espy.
1844	Gaspard de Coriolis formulates the "Coriolis force."
1845	First world map of precipitation is constructed by Berghaus.
1848	First of M. F. Maury's publications on winds and currents at sea is written.
1848	Dove publishes the first maps of mean monthly temperatures.
1862	First map (showing western Europe) of mean pressure is drafted by Renou.
1879	Supan publishes a map showing world temperature regions.
1892	Beginning of the systematic use of balloons to monitor free air.
1900	The term *classification of climate* is first used by Köppen.
1902	Existence of the stratosphere is discovered.
1913	The ozone layer is discovered.
1918	Beginning of the development of the polar front theory by V. Bjerknes.
1925	Beginning of systematic data collection using aircraft.
1928	Radiosondes are first used.
1940	Nature of jet streams is first investigated.
1960	The first meteorological satellite, Tiros I, is launched by the U.S.

Other modes of development in climatology concern international programs for weather and climate studies. Of particular note is the Global Atmospheric Research Program, GARP, (also called the Global Weather Experiment), a concerted research effort by more than one hundred forty countries; the World Meteorological Organization (WMO); and the International Council of Scientific Unions. This

slope—that is, the location of a place in relation to parallels of latitude. This mathematical derivation represents only one of the numerous contributions to mathematical geography by such philosophers as Eratosthenes and Aristarchis. The Greek quest for knowledge about the world resulted in treatises on the atmosphere. The first climatography is attributed to Hippocrates, who wrote *Airs, Waters, and Places* in 400 B.C.; in 350 B.C., Aristotle wrote the first meteorological treatise, *Meteorologica*.

The Greek interest in the nature of the atmosphere was not replicated thereafter for many hundreds of years and only acquired new importance in the middle of the fifteenth century, with the Age of Discovery. With the extended sea voyages and development of new trading areas, descriptive reports of non-European climates became available. Many of these descriptions were quite fanciful and provided the basis for long-held misconceptions.

Scientific analysis of the atmosphere began in the seventeenth century, when instruments to measure the weather were developed. These provided data from which laws applicable to the atmosphere could be derived. The thermometer was invented by Galileo in 1593, the barometer by Torricelli in 1643; and, in 1662, Boyle discovered the fundamental relationship between pressure and volume in a gas.

The eighteenth century was essentially a time when instruments were improved and standardized and when extensive data collection and description of regional climates were undertaken. Explanation of the observed phenomena through the study of the physical processes causing them began in the following century. In 1817, von Humboldt constructed what appears to be the first map that showed temperatures using isotherms; soon after, in 1827, Dove explained local climates in terms of polar and equatorial air currents. Thereafter, contributions became more frequent and ideas essential to understanding the atmosphere slowly evolved. The significant contributions of individuals since the beginning of the nineteenth century are too numerous to detail; instead, table 1–2 provides a partial listing of some of the more important research findings that occurred up to the launching of the first satellite specifically designed for study of the atmosphere. Since that event, major developments in satellite monitoring have had substantial impact on climatological science.

A new dimension in the study of the atmosphere was created in 1966 with the launching of the geostationary ATS-1 (Applications Technology Satellite) series. A *geostationary* (or earth-synchronous) satellite orbits the earth at a height of approximately 35,900 km (22,300 mi). At this elevation, the velocity necessary to maintain orbit is equal to the rotational velocity of the earth. Even though the earth and satellite are both in motion, the effect is that the satellite appears to remain in place over a given point on earth. The geostationary operational satellites are in the GOES/SMS (Geostationary Operational Environmental Satellite/Synchronous Meteorological Satellite) series. As part of a global network, five such satellites are in orbit, two of which provide data for the United States. The satellite images seen on many television weather programs are derived from this source. Their significance is exemplified by the effect of the loss of signal in early 1983, when the U.S. satellite over the Pacific Ocean malfunctioned; its loss precluded accurate prediction of extensive storms that struck California at that time.

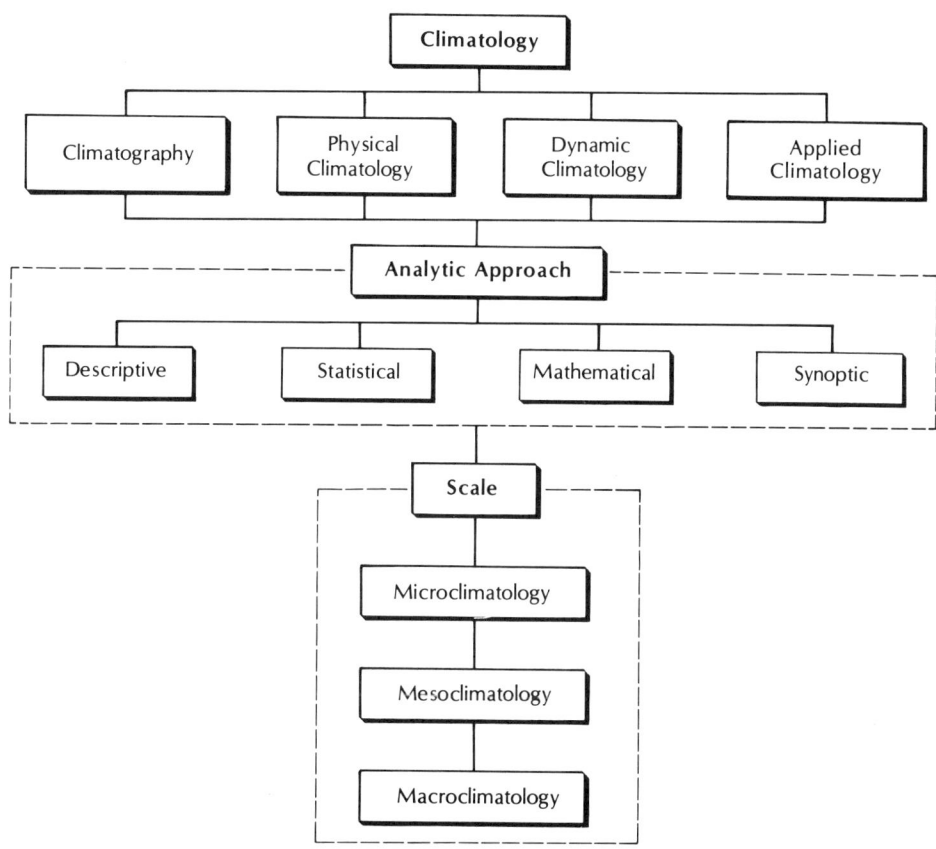

FIGURE 1-1 Subgroups, analytic methods, and scales of climatic study (From J. E. Oliver [1981, p. 4]. Used by permission of V. H. Winston and Sons.).

The analytic approaches suggested in figure 1-1 are largely self-explanatory, with the possible exception of the *synoptic approach*, an analytic method that combines each of the others. The object of synoptic climatology is to relate local or regional climates to atmospheric circulation. Chapter 8 examines this component of climatology in greater detail. As indicated in figure 1-1, the division of climates from large (*macroclimates*) to small (*microclimates*) appears logical; but there are definitional problems involved in their identification, which will be considered in chapter 8.

A BRIEF CHRONOLOGY

Climatology originated in ancient Greece; in fact, the word *climate* is derived from a Greek word meaning "slope." In such a context, it refers to the slope, or inclination, of the earth's axis and is applied to an earth region at a particular elevation on that

here local empirical observation as well as meteorological theory is absolutely necessary. Climatology thus does not belong wholly within the fields of meteorology or geography. It is a science—really an applied science—whose methods are strictly meteorological but whose aims and results are geographical.

As part of this geographic aspect, maps play a significant role in the depiction of climates and climatic data. Many of the maps use *isolines*—lines joining locations of equal value to show distribution of the elements. The name given to the isoline depends upon the climatic element being shown on the map; table 1–1 provides some of the more commonly used terms.

There is a diversity of approaches available in climatic studies. Figure 1–1 illustrates the major subgroups of climatology, the approaches that can be used in their implementation, and the scales at which the work can be completed. *Climatography* consists of the basic presentation of data and its verbal or cartographic description. As the names imply, physical and dynamic climatology are related to the physics and dynamics of the atmosphere: *Physical climatology* deals largely with energy exchanges and physical components, while *dynamic climatology* is more concerned with atmospheric motion and exchanges that lead to and result from that motion. *Applied climatology* is the scientific application of climatic data to specific problems within such areas as forestry, agriculture, and industry. It can involve the application of climatic data and theory to other disciplines, such as geomorphology and soil science. Applied Study 1 examines some examples of the nature of such studies.

TABLE 1–1 Equal value lines (isolines) used in climatology.

	Lines of
Isallobar	Equal pressure tendency showing similar changes over a given time
Isanomaly	Equal anomalies or departures from normal
Isamplitude	Equal amplitude of variation
Isobar	Equal barometric pressure
Isocryme	Equal lowest mean temperature for specified period, e.g., coldest month
Isohel	Equal sunshine
Isohyets	Equal amounts of rainfall
Isokeraun	Equal thunderstorm incidence
Isomer	Equal average monthly rainfall expressed as percentages of the annual average
Isoneph	Equal degree of cloudiness
Isonif	Equal snowfall
Isophene	Equal seasonal phenomena, e.g., flowering of plants
Isoryme	Equal frost incidence
Isoterp	Equal physiological comfort
Isotherm	Equal temperature

land practices and application of nutrients has resulted in improved yields; but a year in which there is a deviation from what is considered the climatic normal can result in agricultural disaster. In Applied Study 16, the topic of the potential impact of climatic change (as opposed to year-to-year variability) is examined. It is shown that world food production could be adversely affected with but a small change in the climate that currently prevails.

Water Many human problems are related to excesses or deficits of water. The excesses, as represented by flooding, are again a result of variations from the expected climatic normal. The applied climatologist supplies important information on everything from the chances of floods occurring to the relative severity of that flood.

A shortage of water, as represented by drought, has extended consequences that can ultimately lead to starvation of animals and humans. Climatologists have investigated the many aspects of droughts, not only in the widely publicized episodes (such as the drought in the Sahel of Africa) but also in areas where a water shortage is less severe but water planning is essential to successful development (Applied Study 11). Methodologies have been derived wherein the water balance that relates incoming to outgoing water can be deduced, and the needs of agriculture, for example, determined with some accuracy. The applied climatologist's concern with drought also includes the environmental degradation that can result when a location experiences slowly increasing aridity. The account of desertification in Applied Study 10 illustrates this problem.

Precipitation in forms other than water, such as snow and hail, also creates problems. In Applied Studies 4 and 5, some of the impacts of high snowfalls are outlined. They are shown to have an extended effect upon many aspects of everyday life.

Health Studies of the relationships between health, disease, mortality, and climate show some remarkable results. When mortality from two diseases in Great Britain and Australia are compared, the graphs are a mirror image of one another. Because the two places are in different hemispheres, the summer and winter seasons are reversed. Figure 1–6 clearly shows that in widely separated places, deaths from respiratory problems and heart diseases both peak in the winter months.

The human body reacts to the climatic environment, and extreme conditions can have dire effects. The use of the temperature-humidity and wind-chill indices are reminders of the vulnerability of the body to atmospheric conditions (Applied Study 13). Similarly, the impact of air pollution is shown to rest not only upon the amount of pollutants in the air, but also on the state of the atmosphere at the time of emission (Applied Study 8).

Climate influences the relative merits of a location as a health resort or as a tourist center. There are some regions that impose great stress upon the human body; in Applied Study 13 the effects of climates at high altitudes provides an apt example of the hazard.

FIGURE 1-6 Comparisons of seasonal mortality from two causes in England/Wales and Australia (From J. E. Oliver [1981, p. 194]. Used by permission of V. H. Winston and Sons.).

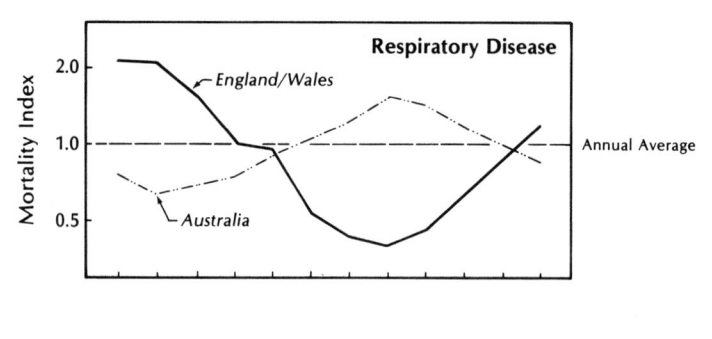

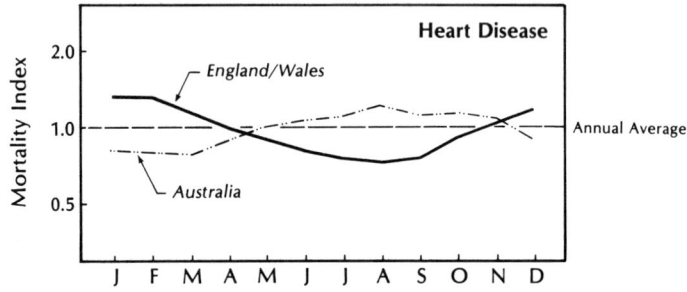

Industry and Trade Applied climatologists have a significant input into the industrial and trade sectors of an economy (for an example of planning and climate, see Applied Study 7). Consider the following examples, which name both the type of activity and the areas in which climatic data and analysis are needed:

  A. Building and Construction: Site conditions, design in relation to climate, energy costs, construction conditions
  B. Commerce: Storage of materials, accidents, plant operation, sales planning
  C. Communication: Construction, maintenance of systems, appropriate design
  D. Services: Insurance, environmental law, disaster control
  E. Forestry: Productivity, hazards, regeneration, biological hazards

At first glance, it may be difficult to accurately assess the relationship of climatology to many of these activities. However, a little thought will clearly indicate the significance of applied climatology.

2 Energy and the Atmosphere

THE NATURE OF RADIATION
THE SOLAR SOURCE
The Role of the Atmosphere
Scattering
Absorption
Reflectance
Transmissivity
EARTH-SUN RELATIONSHIPS
GLOBAL PATTERNS OF ENERGY RECEIPTS
TERRESTRIAL RADIATION AND THE GREENHOUSE EFFECT
THE PLANETARY ENERGY BALANCE
THE HEAT BALANCE
SUMMARY
APPLIED STUDY 2
Direct Use of Solar Energy

Energy is defined as the capacity for doing work; it can exist in a variety of forms (table 2-1) that can be changed from one to another. Of major importance in climatological studies is the way that energy can be transferred from place to place. There are basically three ways in which this transfer can take place: conduction, convection, and radiation. *Conduction* consists of energy transfer directly from molecule to molecule and represents the flow of energy along a temperature gradient. *Convection* involves the transfer of energy by means of mass motions of the medium through which the heat is transferred. Both conduction and convection depend upon the existence of a medium in which to operate. This medium may be solid, liquid, or gaseous. *Radiation* is the only means of energy transfer through space without the aid of a material medium. The major source of energy in the environment is the sun; between the sun and the earth, where a minimum of matter exists, radiation is the only important means of energy transfer. So important is this solar input that all climatological and meteorological phenomena can ultimately be traced to it.

To comprehend fully the role of solar energy in the functioning of the atmosphere requires an assessment of the nature of energy emitted by the sun, the way it enters the earth-atmosphere system, and the transformations that it undergoes in the system.

TABLE 2-1 Energy forms and transformations.

Forms

Radiation	The emission and propagation of energy in the form of waves
Kinetic energy	The energy due to motion; one-half the product of the mass of a body and the square of its velocity
Potential energy	Energy that a body possesses by virtue of its position and that is potentially converted to another form, usually kinetic energy
Chemical energy	Energy used or released in chemical reactions
Atomic energy	Energy released from an atomic nucleus at the expense of its mass
Electrical energy	Energy resulting from the force between two objects having the physical property of charge
Heat energy	A form of energy representing aggregate internal energy of motions of atoms and molecules in a body

Examples of Transformations

Atomic energy ⟶ Radiation ⟶ Heat ⟶ Radiation
(Sun)   (Sunlight)   (Earth surface) (terrestrial)

Radiation ⟶ Chemical energy ⟶ Food chain
(Sunlight)   (Photosynthesis)

Potential energy ⟶ Kinetic energy ⟶ Heat
(Water vapor)   (Raindrop)   (Friction)

THE NATURE OF RADIATION

Any object above the temperature of absolute zero ($-273°C$) radiates energy to its surrounding environment in the form of electromagnetic waves that travel at the speed of light. The fact that energy is transferred in the form of waves means that its characteristics will depend upon wave properties: wavelength, amplitude, and frequency (figure 2-1). Thus, using wavelength as a criterion, radiant energy along an entire spectrum from very short to very long wavelengths can be identified. Figure 2-2 shows the spectrum and illustrates the ways wavelength bands are characterized. In a range between 10^{-4} and 10^{-5} cm is a highly significant band, for at that wavelength the energy is characterized as light, a form visible by the human eye. Wavelengths in the range of visible light are often given in micrometers (μm), where 1 μm = 10^{-4} cm. As the inset in figure 2-2 shows, the visible spectrum ranges from violet (0.40 μm) through the spectrum of the rainbow colors to red, at a wavelength of 0.7 μm.

The type of radiant energy that a body emits is described by radiation laws. The mathematical expression of these laws is provided in Appendix 1 (Note 2). Essentially, these equations provide the following information:

1. The energy emitted by a body is a direct function of its absolute temperature. The hotter the body, the greater the flux (or flow) of energy from it.

2. The maximum wavelength of energy emitted by a body is inversely proportional to its absolute temperature. The higher the temperature, the shorter the wavelength.

The sun provides the energy input for the earth-atmosphere system, and radiation laws can be used to explain the nature of its energy output.

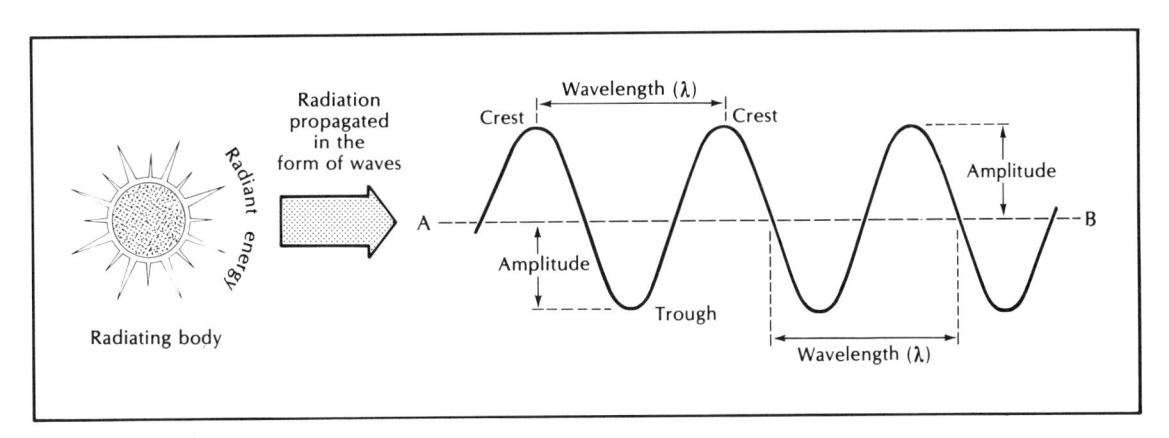

FIGURE 2-1 Radiant energy is transferred in the form of waves, which are characterized by wavelength, amplitude, and frequency.

PHYSICAL AND DYNAMIC CLIMATOLOGY

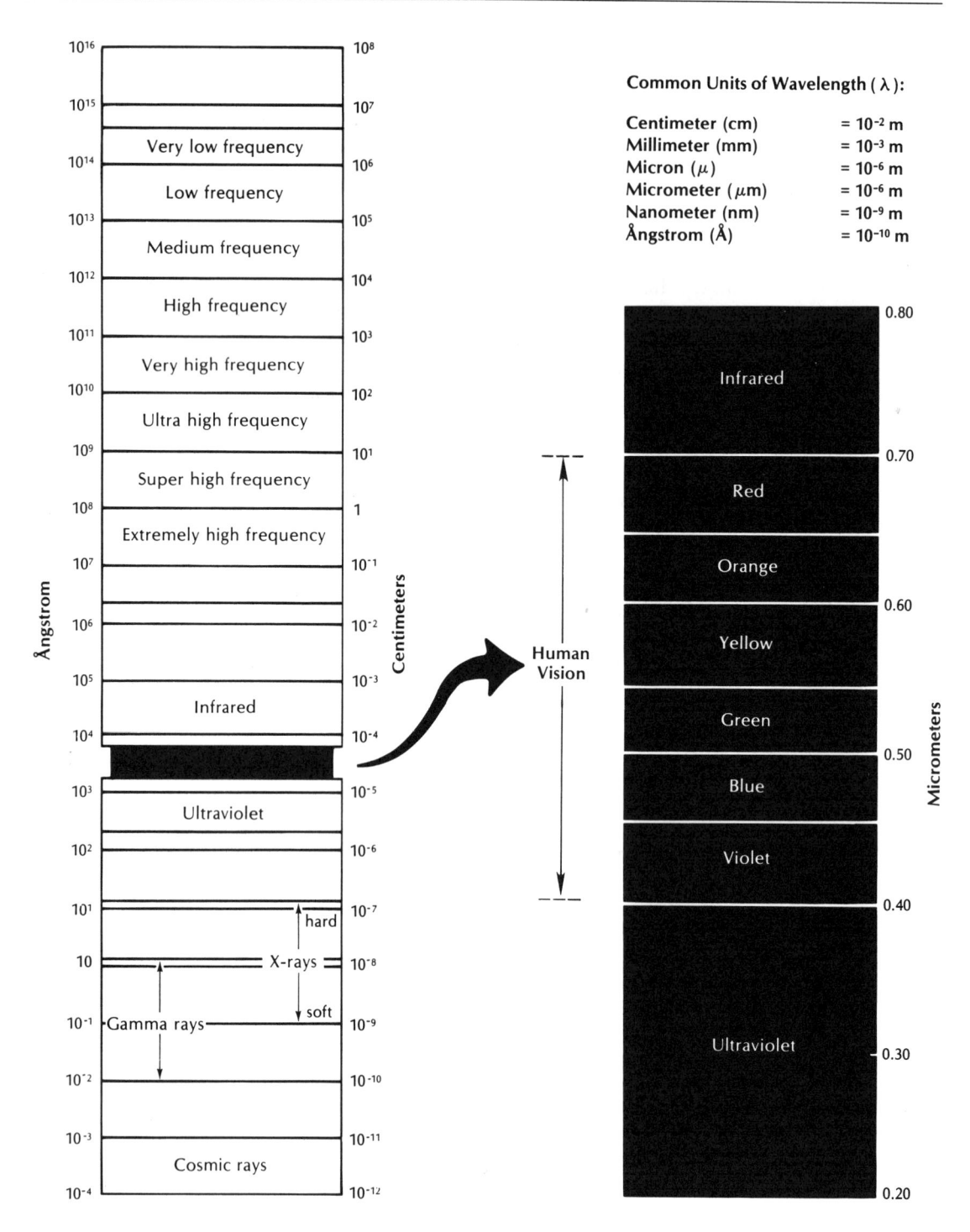

FIGURE 2-2 The electromagnetic spectrum.

ENERGY AND THE ATMOSPHERE

THE SOLAR SOURCE

In the core of the sun, hydrogen atoms are converted to helium in a nuclear reaction. This reaction produces energy that radiates outward from the core to the visible surface of the sun, the *photosphere*. In the core, very high temperatures prevail; at the radiative surface, the effective temperature is on the order of 5800°C. This temperature determines both the amount of energy emitted and the maximum wavelength at which it occurs. The flux of energy from the sun is some 100,000 cal/cm²/min. The bulk of solar energy lies in the visible band centered at 0.5 μm, decreasing in either direction to the ultraviolet and infrared portions of the spectrum.

For general purposes, the energy output of the sun is considered constant. There are, however, a number of variable solar activities that may influence weather and climate on earth. Above the photosphere, the sun's atmosphere is divided into two parts: the lower *chromosphere* and the upper (seen only during eclipses) *corona*. From the corona, ionized gases acquire enough velocity to escape the sun's gravitational pull, producing a solar wind, a stream of hot protons and electrons travelling through space at high speeds.

The intensity of solar winds is influenced by the best-known of solar features, *sunspots*, which are dark areas of the surface that are about 1,500°C cooler than the surrounding chromosphere. Because records of sunspots have been kept for many years and sunspots are relatively easy to observe, they are used as an indicator of solar activity. It has been found that the number of sunspots observed on the sun appears to follow an 11-year cycle. First, the sunspots increase to a maximum, with one hundred or so visible at a given time (the individual spot may have a lifetime ranging from a few days to a few months). Then, over a period of years, the number diminishes until only a few, or none at all, are seen. Change from the maximum to the minimum number takes about 5.5 years, and the complete cycle is thus 11 years.

Just how the sunspot cycle influences weather and climate is a matter of controversy. Some researchers claim to have found that weather patterns in a selected area follow the sunspot cycle closely; in other cases, no relationships can be shown. Even the physical process by which sunspots affect weather is not totally clear because the wavelength of their emissions, which corresponds to the wavelength of X rays, has different effects on different parts of the atmosphere. A further complication is that sunspot activity coincides with solar wind intensity, and it is difficult to separate the effects of each.

The energy emitted by the sun passes through space until it is intercepted by another body. The intensity of solar radiation striking that body is determined by a basic physical law known as the *inverse square law*. This law merely states that the area illuminated (and hence the intensity) varies with the squared distance from the light or energy source. Thus if the intensity of radiation at a given distance is one unit, at twice the distance, the intensity is only one-fourth. This law has obvious impact in relation to the intensity of solar energy intercepted by planets in the solar system.

Given the amount of energy radiated by the sun (100,000 cal/cm²/min) and a mean earth-sun distance of 149.5 million km (93 million mi), amount of radiation intercepted by a surface at right angles to the solar beam at the outer limits of the atmosphere can be calculated. This *solar constant* amounts to 2.0 cal/cm²/min.

THE ROLE OF THE ATMOSPHERE

The solar constant is the amount of energy available at the outer limits of earth's atmosphere. As the solar beam passes through the atmosphere, it is modified through a number of processes. These include scattering, absorption, reflectance, and transmittance.

Scattering

Scattering is the process by which small particles and gas molecules in the atmosphere diffuse part of the incident radiation in all directions; such scattering occurs without any transformation of energy. The first adequate scientific explanation of scattering was developed by the English scientist Lord Rayleigh (1842–1919), and the effect is sometimes called *Rayleigh scattering*.

Rayleigh showed that the amount and direction of scatter depends upon the ratio of the radius of the scattering particle to the wavelength of the energy that is scattered. Furthermore, the amount of scatter is inversely proportional to the fourth power of the wavelength. This relationship means that if a given wavelength is scattered, then a wavelength twice as long will only be scattered one-sixteenth times as effectively. Such selective scattering has a number of important results.

Perhaps the most obvious effect of the scattering of the sun's rays is the blue color of our sky (figure 2-3). The sky is blue because most light waves are larger than air molecules and bypass them with no effect; shorter wavelengths of light are more likely to strike the air molecules and to be scattered in all directions. The blue portion of the light spectrum—the shorter wavelengths of light—is thus scattered more effectively, with the result that we perceive the sky as blue.

Scattering also explains the red-colored sky often seen at dawn and sunset. At these times the sun's rays pass obliquely through the atmosphere, and much more scattering occurs, so much, in fact, that most of the shorter wavelengths (violet-green) are extinguished and only the longer wavelengths (orange-red) are perceived.

Rayleigh's explanation does not account for all types of scattering in the atmosphere. According to a theory developed by Gustave Mie in 1908, molecule sizes that have a larger ratio to wavelength than those that give rise to Rayleigh scattering can result in light at all wavelengths being scattered. According to this theory, light is scattered more often in a forward than a backward direction. Such an effect helps explain why the sky, on a cloudless day, may appear a light blue or sometimes almost white (figure 2-3).

Absorption

Absorption may be defined as a process in which incident radiation is retained by a substance and converted to some other form of energy, eventually to be

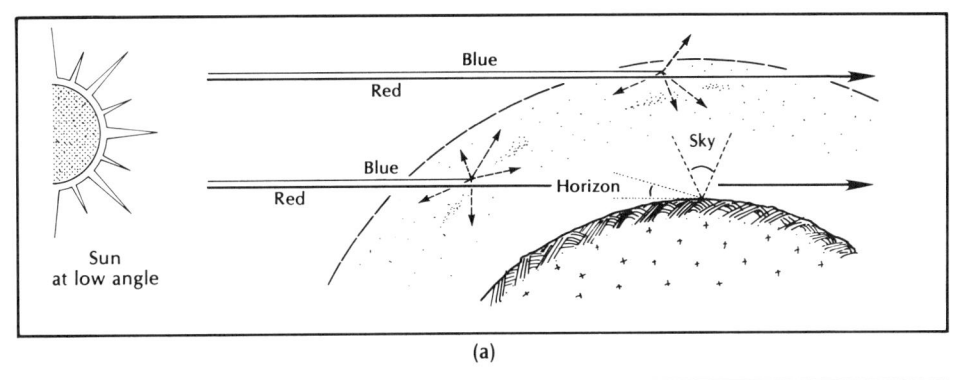

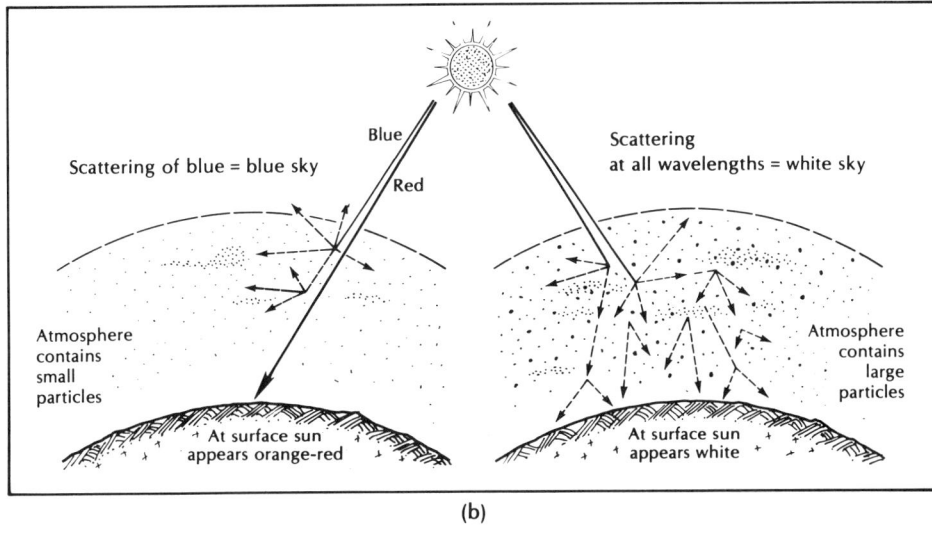

FIGURE 2-3 Scattering of sun's rays produces different sky colors: (a) At sunset on a hazy day, the sky at the horizon is red while the sky above appears blue. (b) Sky color depends upon the number and size of particles in the air and the amount of scattering at various wavelengths.

emitted by the substance. Thus, sunlight striking the side of a house, for example, is absorbed and becomes the energy that heats the exterior materials of the wall.

In the atmosphere, gas molecules, cloud particles, haze, smoke, and dust absorb part of the incoming solar radiation. Such absorption is selective, for gases absorb only in certain wavelengths: they each have an *absorption spectrum characteristic*. Because of this, the types of energy that different gases absorb can be identified. The two most common gases in the atmosphere, nitrogen and oxygen, have absorption bands in the ultraviolet, whereas triatomic oxygen (ozone) absorbs shorter ultraviolet radiation. In recent years much concern has been expressed about the potential depletion of atmospheric ozone, a layer that absorbs the shorter-wave ultraviolet radiation and prevents it from penetrating to the surface. This property is of marked significance, for excessive exposure to ultraviolet radiation is lethal to humans, animals, and plants.

Water vapor and carbon dioxide are strong absorbers of infrared radiation. While they are responsible for absorbing incident radiation falling in their absorption bands, their most significant role is in creating the greenhouse effect, an atmospheric property dealt with in more detail later in this chapter.

Reflectance

Reflectance occurs when part of the incident radiation striking a body is reflected from its surface. There is considerable variation in the amount of reflectivity of natural surfaces, which is expressed as a percentage of the incident radiation reaching a surface. Clouds are by far the most significant atmospheric reflectors, with reflectivity ranging from 40 percent to 90 percent, depending upon thickness. Reflectance from land surfaces varies with the type of surface cover, with snow (80 percent) and dry sand (45 percent) being efficient reflectors. The reflectivity of water surfaces is a function of the angle of the sun in the sky, with high reflectivity corresponding to low sun angle. Table 2–2 provides examples of the reflectivity of various natural surfaces.

Reflectivity is often described by the term *albedo*. Although defined rather loosely, the term may be taken to describe the reflectivity of surfaces—over various wavelengths—when the surfaces are viewed from above. Thus, the entire earth has an albedo that represents reflectance from both atmosphere and surface. This earth-atmosphere albedo is about 35 percent.

Transmissivity

The factors itemized thus far deal with attenuation of the solar beam as it passes through the atmosphere, and thus the amount available at the surface varies with the state of the atmosphere. The fractional part of the incident radiation that penetrates the atmosphere is expressed by *transmissivity*. Clearly, transmissivity will be determined by both the state of the atmosphere and the distance that the solar beam must travel through the atmosphere. This latter effect may be described by the concept of the optical air mass. A value of 1 is given to an optical air mass when the sun is directly overhead, and increases as the angle of the sun above the horizon decreases. Using appropriate geometric ratios, a value is derived for any sun angle; for example, if the sun is 30° above the horizon, the optical air mass has a value 2. This value indicates that the solar beam has twice the distance to travel through the atmosphere.

Each of the foregoing factors modifies the flow of energy to the surface. The amount of radiation striking a unit area at the surface, the *irradiance,* is made up of two components: the energy that is received directly from the sun and the energy that is diffused through scattering. These two components comprise the *global solar radiation* falling on a surface. On days of heavy cloud cover, the direct beam is excluded, and global solar radiation is made up totally of the diffuse component. Knowledge of the relative inputs of global solar radiation is significant in many aspects of applied climatology (See Applied Study 2).

TABLE 2-2 Albedos for short-wave (<4.0 μm) radiation.

Surface	Albedo %	Vegetation cover	Albedo %
Fresh snow	75–95	Sea ice	30–40
Dry (dune) sand	35–45	Tundra	15–20
Dark soil	5–15	Desert	25–30
Blacktop road	5–17	Coniferous forest	5–15
Concrete surface	17–27	Deciduous forest	10–20
Clouds		Savanna	
Cumuliform	70–90	Wet	15–20
Stratus	59–84	Dry	25–30
Cirrostratus	44–50	Green meadows	10–20

Water

Zenith angle of sun	0°	40°	60°	80°	90°
Reflectivity (%)	2	2.5	6	35	100

EARTH-SUN RELATIONSHIPS

The most important variable associated with differences in energy received at various locations on earth result from basic motions of the earth, rotating on its axis and revolving around the sun. Figure 2-4 is a schematic diagram of these earth-sun relationships in the period of a year. In its revolution around the sun, the earth follows an elliptical orbit, with the result that earth-sun distance varies. The earth is closest to the sun (*perihelion*) on 1 January and most distant (*aphelion*) on 1 July. This variable earth-sun distance means that the amount of solar radiation intercepted by the earth at perihelion is about 7 percent higher than at aphelion. This difference, however, is outweighed in its effect on energy receipts by the influences of number of hours of daylight and the angle of the sun about the horizon.

The major periodic changes in radiation that occur at most places are associated with both intensity and duration of solar radiation. The intensity of solar radiation is largely a function of *angle of incidence,* the angle at which the sun's rays strike the earth. The angle of incidence directly affects both the energy received per unit area of surface and the amount of energy absorbed. Since the earth is spherical, the surface exposed to the radiation is curved. In those parallels of latitude where the solar radiation is perpendicular to the surface (the *solar equator*), the intensity of radiation is maximum (see figure 2-5). The intensity of radiation decreases north and south of the solar equator as the angle of incidence decreases.

The seasonal energy regimes are brought about by the phenomenon of the earth's being inclined on its axis by approximately 23°30' (see figure 2-6). As the earth revolves about the sun, the solar equator moves through a range of 47° on the earth. The geographic equator (0° latitude) is the mean location of the solar equator. Thus the solar equator moves north and south through the year from

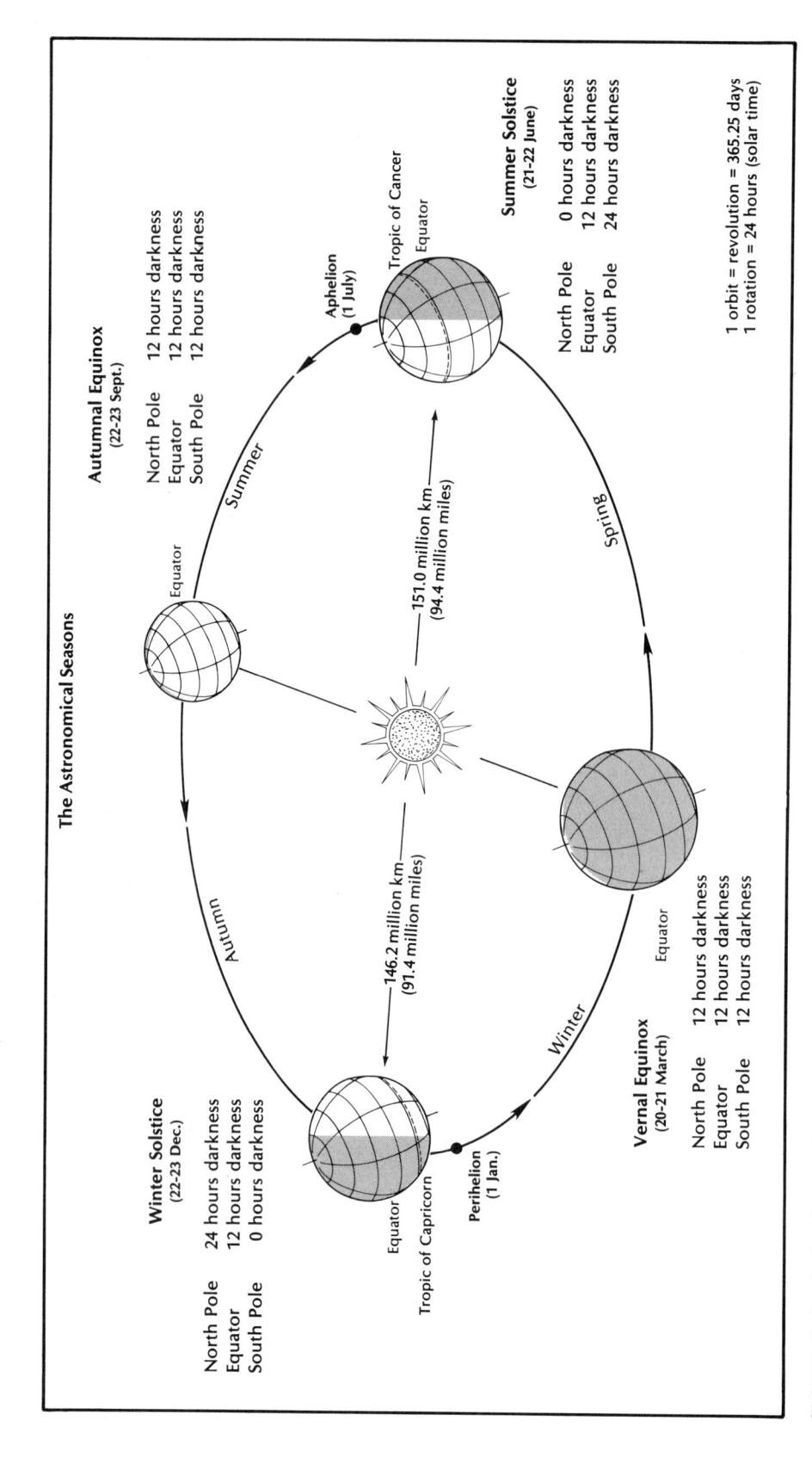

FIGURE 2-4 Orbit of the earth around the sun.

FIGURE 2-5 How sun angle decreases the intensity of radiation. The same amount of energy is contained in both sets of rays, but those that strike the earth obliquely are spread out over a larger area.

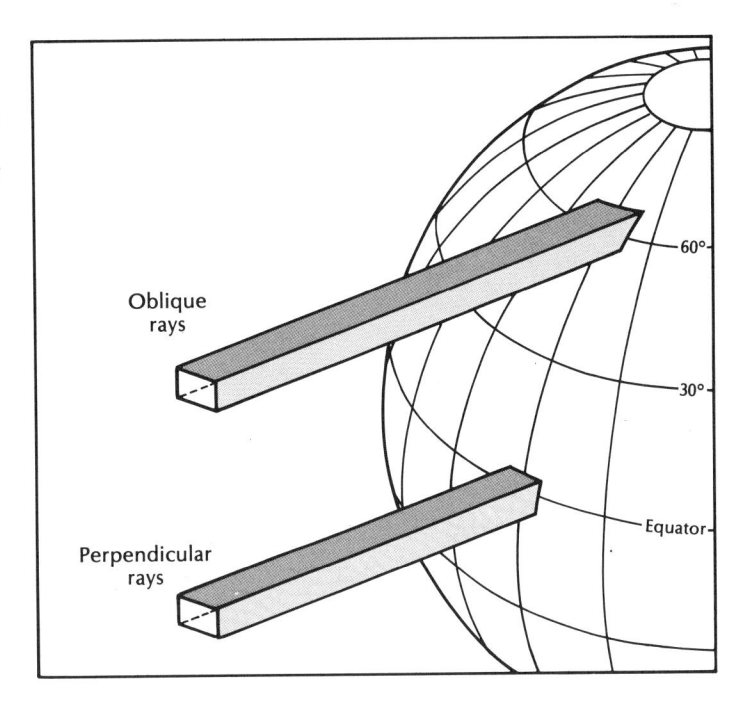

23°30'N (Tropic of Cancer) to 23°30'S (Tropic of Capricorn). This movement brings about changing intensity of solar radiation during the year for all locations.

The geographical equator has the least variation in the angle of incidence of any location. Here the sun is never more than 23°30' from the zenith, while all latitudes between 23°30'N and 23°30'S experience perpendicular rays of the sun sometime during the year. All locations between 23°30' and 66°30' of latitude experience a change of 47° in the angle of the sun's rays. From the Arctic and Antarctic Circles to the poles, the change in the angle of the sun's rays decreases from 47° down to 23'30". The seasonal energy patterns are strengthened by the changes in the length of the daylight period. Only at the time of the spring and fall equinoxes are the length of daylight and darkness everywhere equal over the earth. At all other times the daylight period is longer in the hemisphere where the solar equator is located. Thus, when the sun's vertical rays are north of the equator, the daylight period in the Northern Hemisphere is longer than 12 hours (h). In fact, the length of daylight at the time of the summer solstice increases from 12 h at the equator to 24 h at the Arctic Circle (table 2-3). The imbalance in daylight and darkness results in a greater accentuation of the seasons than the position of the solar equator alone would produce.

GLOBAL PATTERNS OF ENERGY RECEIPTS

The combined effect of earth-sun relationships and modification of the solar beam as it passes through the atmosphere produce energy patterns over the earth's surface. Of

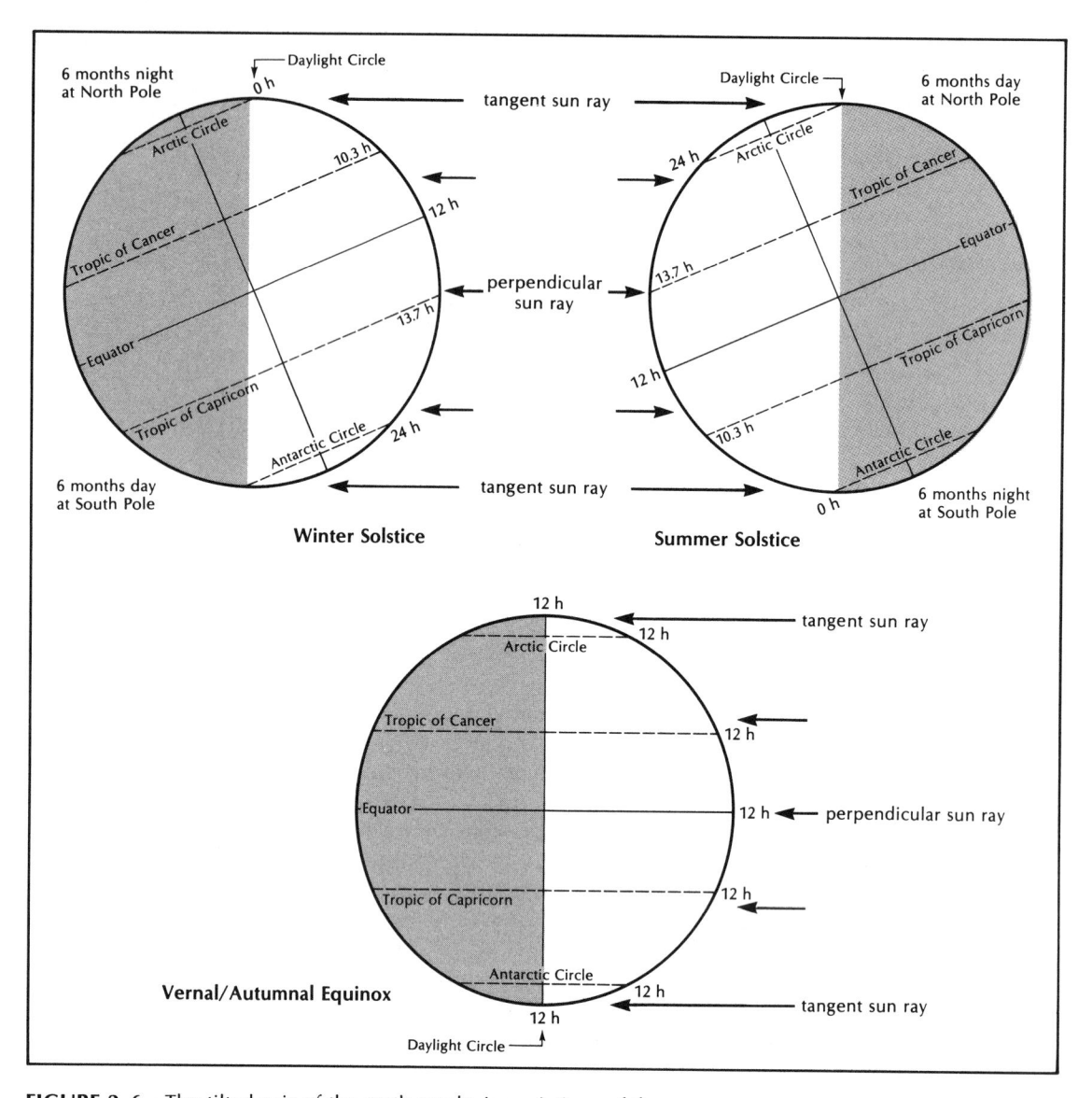

FIGURE 2-6 The tilted axis of the earth results in variations of the angle of overhead sun and length of daylight. Diagram shows conditions at the equinoxes and solstices (From J. J. Hidore [1974, p. 43]. After World Weather, Navaer OO-8ON-24).

prime importance is the distribution of solar radiation receipts. Figure 2–7 reveals a number of significant features.

A. A general zonal (east-west) pattern occurs, with values outside the tropics declining toward the poles. In middle and high latitudes, values over the land, as compared to those at the same latitude over the oceans, are

TABLE 2-3 Extremes of day lengths at various latitudes.

Latitude°	0 h min	10 h min	20 h min	30 h min	40 h min	50 h min	60 h min	66.5 h min
Longest day	12:00	12:35	13:13	13:56	14:51	16:09	18:30	24:00
Shortest day	12:00	11:25	10:47	10:04	9:09	7:51	5:30	0:00

generally higher. The greater cloudiness over the water accounts for this variation.

B. Highest values occur in tropical latitudes, with maxima over the tropical deserts. In the eastern Sahara in North Africa, in excess of 220 kly/year are received, while most other tropical deserts have values in excess of 180.

C. The equatorial lands experience totals similar to those in such middle-latitude locations as the central United States and south central Europe. This phenomenon results from the extensive cloudiness that is associated with lands on either side of the equator.

The annual values provide no indication of the seasonal variations that occur. These can be very large in high and middle latitudes (see chapter 12).

TERRESTRIAL RADIATION AND THE GREENHOUSE EFFECT

If it is not reflected, solar energy arriving at the earth's surface is absorbed or transformed to other energy forms. The absorbed energy causes the surface temperature to rise, and, in turn, the surface becomes an energy radiator. The form of radiation emitted is determined by the radiation law that relates the wavelength of emission to the temperature of the radiating body. The earth's surface is a relatively cool body (compared, for example, to the sun) and thus radiates longer wave energy that falls within the infrared portion of the electromagnetic spectrum. This surface radiation, together with other forms of earth-atmosphere energy exchanges, means that the atmosphere is heated from below rather than directly by the sun's rays.

The infrared energy emitted by the earth does not pass directly back to space. The absorption spectra of gases for infrared radiation are quite different from that for light. Water vapor and carbon dioxide effectively absorb infrared radiation to give rise to the *greenhouse effect*. This term represents an analogy between the heating of air in a greenhouse and the heating of the lower atmosphere (an analogy that is not totally correct, because a large part of the temperature rise in greenhouses is due to exclusion of moving air and turbulent exchange from its interior). Like the glass of a greenhouse the atmosphere allows light energy to pass through. This energy is absorbed by the surface on which it falls and is reemitted as infrared energy. In the analogy, both the glass of the greenhouse and the absorbing gases in the lower atmosphere prevent direct outflow of most infrared energy. In turn, the

FIGURE 2-7 Average annual solar radiation on a horizontal surface expressed in kilo-langleys per year (From W. D. Sellers, *Physical Climatology* [Chicago: The University of Chicago Press, 1965], p. 25. Copyright © 1965. The University of Chicago Press. Reproduced by permission.).

interior of the greenhouse and its atmospheric equivalent retain heat and are warmer than in the case of simple input-output of energy.

The actual process that occurs in the atmosphere is obviously a little more complex than the analogy suggests. Molecules of water vapor and carbon dioxide absorb the terrestrial radiation and then reemit it. Some of the energy will pass to space but a larger proportion will radiate downwards. This counterradiation increases the heat energy available in the lower atmosphere and raises its temperature significantly. In fact, if there were no counterradiation, the mean global temperature would be in the order of between $-22°C$ and $-26°C$, while the average is actually a comfortable 15°C.

Not all of the energy radiated by the earth contributes to the greenhouse effect. Selected wavelengths of the radiant energy are not absorbed by gases and pass freely through the atmosphere to space. The "gaps" occur around wavelengths 8 and 11 μm and are appropriately called *atmospheric windows*.

THE PLANETARY ENERGY BALANCE

The energy flows and transformations dealt with in the preceding pages are continuous processes. Energy from the sun enters the earth-atmosphere system and is ultimately returned to space. A basic necessity of the exchange is that incoming energy and outgoing energy must be in balance. If more energy came in than was returned, the earth would become progressively hotter. Were the reverse true, then the earth would get colder. Thus a basic assumption in the derivation of energy balance is that global net radiation over a suitable time period (in excess of 1 year) must be in balance. In attaining such a balance, many exchanges occur. It is possible to visualize the way that energy is partitioned and to formulate an energy budget. As a first step in this partitioning, the budgeting of incoming solar radiation can be represented (figure 2-8). Data used are expressed as percentage values of the total annual input.

Figure 2-8 shows the relative amounts of solar energy as it is partitioned in its passage through the atmosphere. Some 17 units are absorbed by the atmosphere: short-wave ultraviolet energy by ozone and longer wavelengths by CO_2 and water vapor. Scattering of solar energy ultimately leads to a larger portion's reaching the ground as diffuse light (16 units), while a lesser amount (6 units) is returned to space. Reflection by clouds and the earth's surface adds to outward-going energy to account for some 36 units. This return of energy to space that has not been incorporated into the earth-atmosphere system makes up the planetary albedo.

The amount of solar energy eventually absorbed at the surface (31 units of direct and 16 units of diffuse sunlight) is less than one-half of that which originally entered the atmosphere. It is this energy, together with that already absorbed by the atmosphere, that contributes toward the heating of air. The atmosphere is heated from below rather than directly by solar rays.

The way in which energy from the earth enters the atmosphere is shown in the right-hand portion of figure 2-9. Essentially, three major processes are at work:

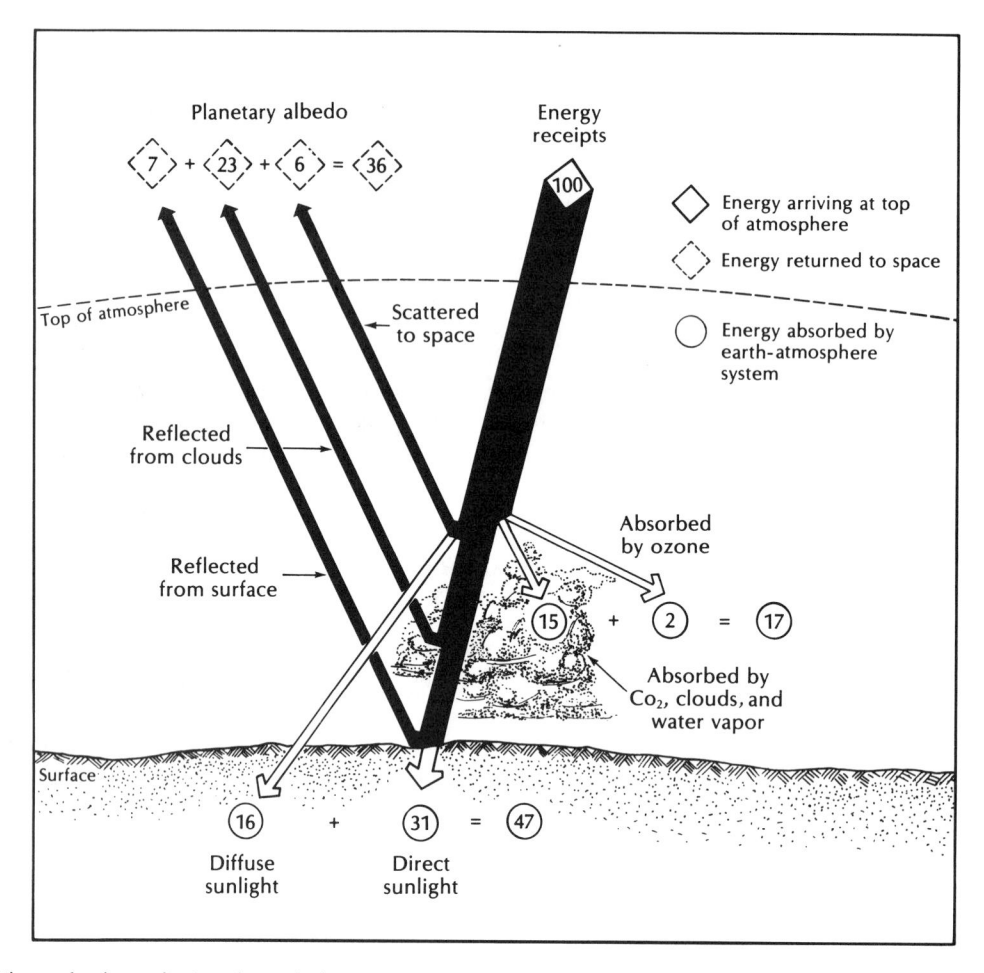

FIGURE 2-8 Flow of solar radiation through the atmosphere; data in percentages of available energy.

Radiation As already noted, radiation occurs in the infrared portion of the electromagnetic spectrum. Of the 98 units radiated, some 7 pass directly to space through atmospheric windows. The major part, 91 units, is absorbed by the atmosphere, reradiated, and gives rise to the greenhouse effect.

Latent heat transfer Energy is required to change liquid water to water vapor; 540–600 cal are needed to vaporize 1 g of water. When the water vapor condenses in the atmosphere, this latent energy is released. This evaporation-condensation process, together with other changes of state of water, redistributes energy throughout the atmosphere (see chapter 4).

Sensible heat transfer Energy absorbed by the earth's surface is transferred by conduction to a very thin layer of air next to the surface. Because air is a very poor

ENERGY AND THE ATMOSPHERE

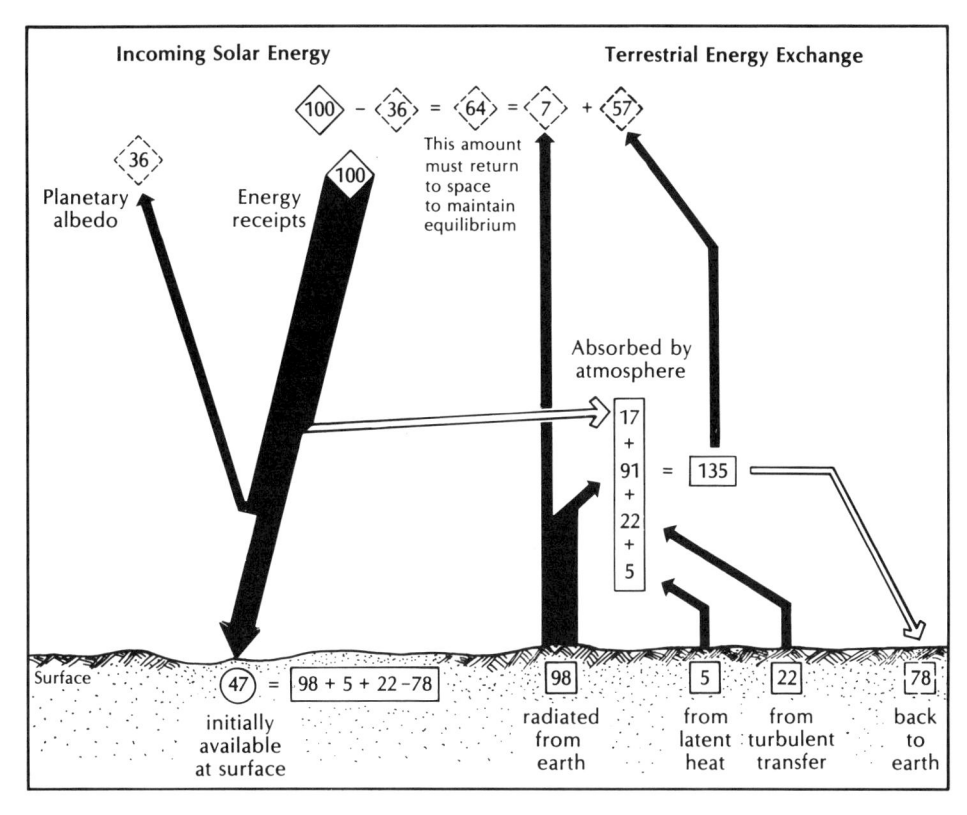

FIGURE 2-9 The terrestrial energy budget: data in percentages of available energy units.

conductor of heat, only a few centimeters are heated in this way: this warm layer of air, however, rises and the heat energy enters the atmosphere through the more efficient convection process. The conduction-convection process is the main heating transfer mechanism in the lower atmosphere. Heat resulting from the transfer is called *sensible heat* (i.e., relating to senses) to differentiate it from other forms of energy transfer.

To describe these energy flows and exchanges, it is useful to draw upon the symbols that have been devised for each of the processes. Appendix 1, Note 3 lists the internationally recognized symbols used to express the variables that form part of the energy budget.

Of major importance in the planetary energy balance of incoming and outgoing energy is the geographical imbalance that occurs. Figure 2-10 shows that equatorward of about 35° latitude, incoming solar radiation exceeds outgoing terrestrial radiation, leading to a surplus of energy. Poleward of these latitudes, the reverse holds true and an energy deficit exists. Surplus energy at low latitudes and a deficit at high latitudes result in energy transfer from one region to another. It is this interchange of energy that is responsible for the major characteristics of atmospheric

PHYSICAL AND DYNAMIC CLIMATOLOGY

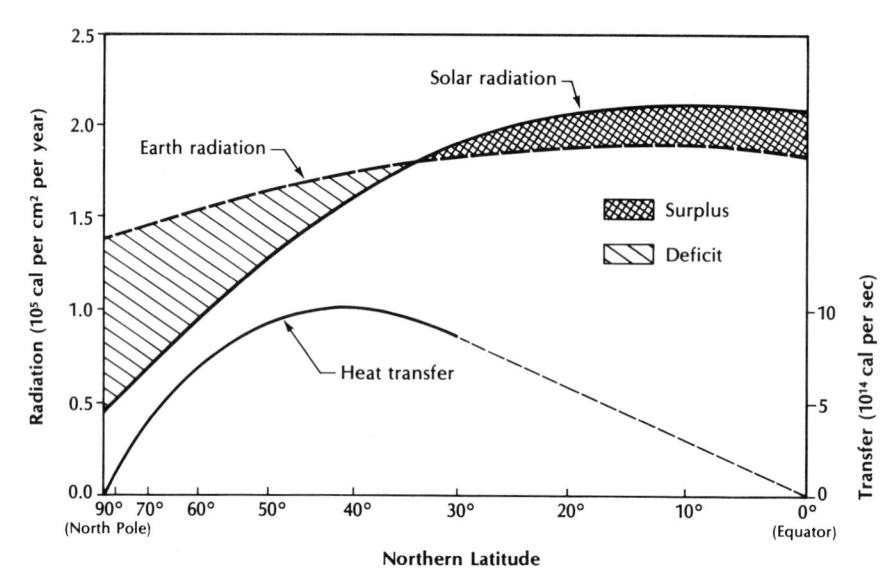

FIGURE 2-10 Latitudinal distribution of earth and solar radiation. A surplus exists in low latitudes and a deficit in high latitudes. (From H. G. Houghton, *Journal of Meteorology*, 11(1): 7, figure 3. Used by permission of the American Meteorological Society.).

circulation and, ultimately, the nature of climates that occur throughout the world. If there were no energy transfer, the energy balance of the atmosphere could only be maintained if the poles were 25°C colder and the equator 14°C warmer than they are now.

THE HEAT BALANCE

Given all of the factors determining the amount of energy available at a location, it is evident that there will be geographical differences between incoming and outgoing radiation. This difference is expressed by the term *net radiation*. When there is more incoming than outgoing energy, net radiation is said to be *positive; negative* values indicate that outgoing exceeds incoming radiation. When a surplus does occur, it transfers heat to the atmosphere by sensible and latent heat transfers and by radiation. If such transfers do not account for all of the surplus energy, some will become stored energy.

Stored energy is that absorbed and retained by the earth's surface to be released at a later time. The amount of absorption and storage on land surfaces is significant on a local basis. Of much greater importance is storage of energy by water in the oceans. The vertical and horizontal mobility of water permit large amounts of energy to be stored and transported to other locations.

ENERGY AND THE ATMOSPHERE

It follows that the heat balance at a place depends upon the net radiation and the amount of energy that is utilized or goes to storage. Appendix 1 (Note 4) provides an account of the symbolic representation of heat balance. It points out that the heat balance at a place is represented as

$$Q^* = LE + H + S$$

where Q^* is net radiation, H is sensible heat transfer, LE is latent heat transfer, and S is the storage factor.

Values for each of these variables have been calculated for latitudinal bands of the earth. Representative values are given in table 2–4. Note that the storage factor S applies only to the oceans; on land masses, the annual mean is zero.

Study of the data in table 2–4 shows that the transfer of energy is of prime importance in maintaining a global balance. As an example, the oceans in the latitudinal zone 60–70°N display a net radiation (23 units) that is less than the sum of energy utilized by LE and H transfers (33 + 16 = 49 units). The difference (26 units) is supplied by the transport of stored energy. Conversely, net radiation in the latitudinal zone 10–20°N is greater than the sum of LE and H by 14 units. As indicated by the negative sign in the table, this excess energy goes into storage for transfer to higher latitudes. Thus, through transport of surplus energy to areas of deficit, energy is redistributed over the earth.

TABLE 2–4 Energy balance at the earth's surface (kcal/cm²/year).

	Oceans				Land				Earth			
	Q^*	LE	H	S	Q^*	LE	H	S	Q^*	LE	H	S
°N 60–70	23	33	16	+26	20	14	6	0	21	20	9	+ 8
50–60	29	39	16	+26	30	19	11	0	30	28	13	+11
40–50	51	53	14	+16	45	24	21	0	48	38	17	+ 7
30–40	82	86	13	+16	60	23	37	0	73	59	23	+ 9
20–30	113	105	9	+ 1	69	20	49	0	96	73	24	+ 1
10–20	119	99	6	−14	71	29	42	0	106	81	15	−10
0–10	115	50	4	−31	72	48	24	0	105	72	9	−24
°S 0–10	115	84	4	−27	72	50	22	0	105	76	8	−21
10–20	113	104	5	− 4	73	41	32	0	104	90	11	− 3
20–30	101	100	7	+ 6	70	28	42	0	94	83	15	+ 4
30–40	82	80	8	+ 6	62	28	34	0	80	74	11	+ 5
40–50	57	55	9	+ 7	41	21	20	0	56	53	9	+ 6
50–60	28	31	10	+13	31	20	11	0	28	31	10	+13
Entire earth	82	74	8	0	49	25	24	0	72	59	13	0

Source: M. I. Budyko, *Climatic Change*, p. 33, 1977, © Copyrighted by the American Geophysical Union.

Q^* = Net radiation
LE = Latent heat of vaporization
H = Sensible heat
S = Storage

SUMMARY

The sun provides the earth with energy. Solar energy is transferred in the form of electromagnetic radiation that is concentrated in the light portion of the spectrum. Earth motions in relation to the sun account for major variations in the receipts of solar energy at locations on the globe. These motions result in differences in intensity and duration of sunlight and provide a general zonal distribution of solar energy.

In passing through the atmosphere, the solar beam is depleted through scattering, absorption, and reflectance. Energy absorbed by the earth is reradiated in the infrared portion of the spectrum. This long-wave energy does not pass freely to space because of absorption by water vapor and carbon dioxide. Reradiation of this energy back to earth gives rise to the greenhouse effect.

To maintain an energy equilibrium, solar energy entering the earth-atmosphere system must be balanced by energy leaving the system. This balance is maintained by the transfer of energy from earth to the atmosphere by radiation, latent heat, and sensible heat exchanges. Ultimately, the amount of energy leaving the system equals that which arrives. There is, however, an imbalance in the distribution of net radiation over the earth, which leads to the movement of energy from surplus to deficit areas, an exchange that ultimately causes the major characteristics of global atmospheric motion.

APPLIED STUDY 2

The Direct Use of Solar Energy

Worldwide price fluctuations in the cost of energy have resulted in the intensive investigation of alternate sources to replace or subsidize the use of fossil fuels. Of the many alternatives possible, the direct use of solar energy is perhaps most frequently mentioned; in recent years, much has been accomplished in implementing systems that take advantage of this natural resource.

There are a number of approaches available in directly using the sun's rays as an energy source (note that with the exception of geothermal energy, all current sources initially derive their energy from the sun, so that the key word in modern developments is *direct*). The most obvious are the so-called active solar systems, in which radiant energy is collected, stored, and transformed to a usable form. In contrast, passive systems rely upon the manipulation of microenvironment so that, depending upon need, solar energy load is maximized or minimized. This approach is essentially summed in the phrase "design with climate."

Examples of active systems range from large government-sponsored schemes to small units in solar homes. A fine example of a large scheme is seen in the solar furnace at Odeillo in the Pyrenees Mountains of France (see figure 2–11).

As shown in the diagram, the system consists of eight tiers of reflecting mirrors, each tier containing seven reflectors. These are designed to follow the

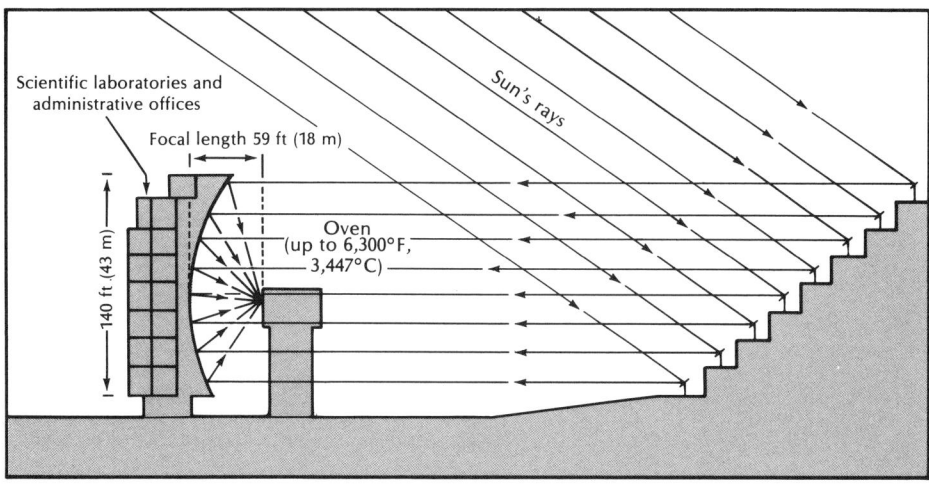

FIGURE 2-11 The solar furnace at Odeillo: (a) a view of the reflecting surface and oven unit (EDS [NOAA]); (b) design of the system (From *Climate and Man's Environment*, John E. Oliver, Copyright © [1973, John Wiley & Sons]. Reprinted by permission of John Wiley & Sons, Inc.).

path of the sun and are reset to an easterly direction once the sun goes down. The mirrors reflect light onto a huge parabolic surface that contains 9000 individual mirrors. From this, the sun's rays are focused into an opening, just 30 cm (12 in.) in diameter in the oven. Temperatures in the oven attain some 3500°C (6300°F), and the heat is used to produce high-quality products, such as ceramics. The high quality is achieved because, unlike in conventional high-temperature furnaces, there are no contaminants that might degrade the product in the solar oven.

Like the solar oven, many solar homes utilize solar energy as a heat source. Typically, the home system consists of a collecting surface and a method whereby the heat energy produces hot water or heats the house. A very simple design is illustrated in figure 2–12.

Solar systems in which the energy generates electricity are more complex. In such systems, the solar collectors are backed by, for example, cadmium sulfide solar cells for conversion to electrical current. Ongoing research is directed toward investigating other ways to generate electrical energy.

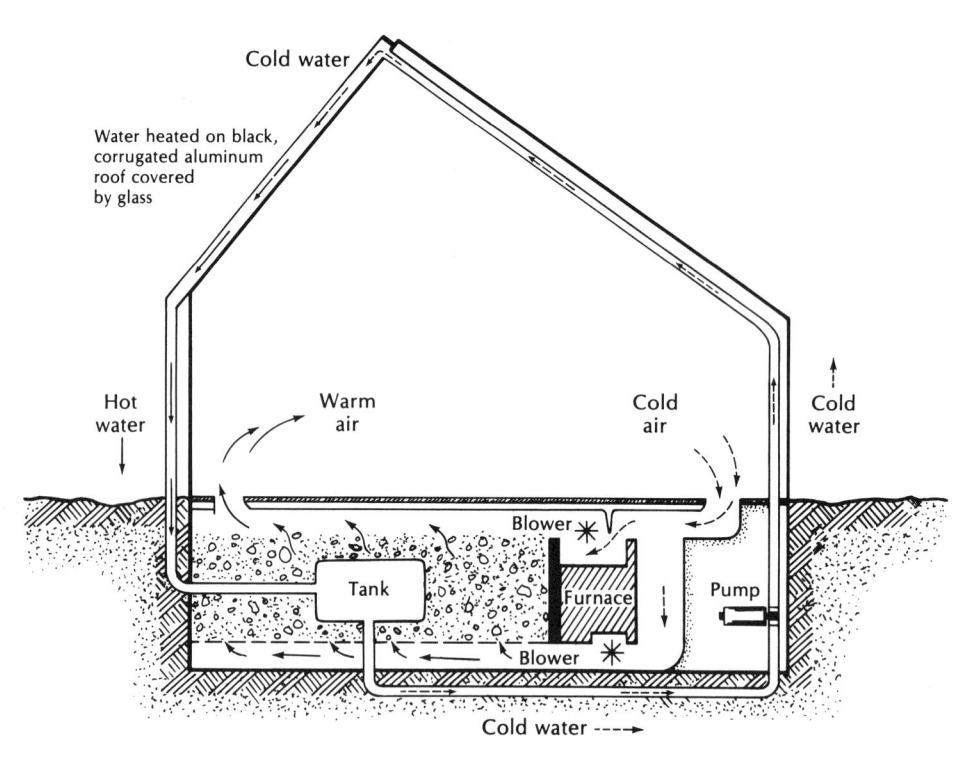

FIGURE 2–12 Basic design of a home heated through direct use of solar energy (From J. E. Oliver, [1977, p. 53]. Used by permission of Wadsworth Publishing Co.).

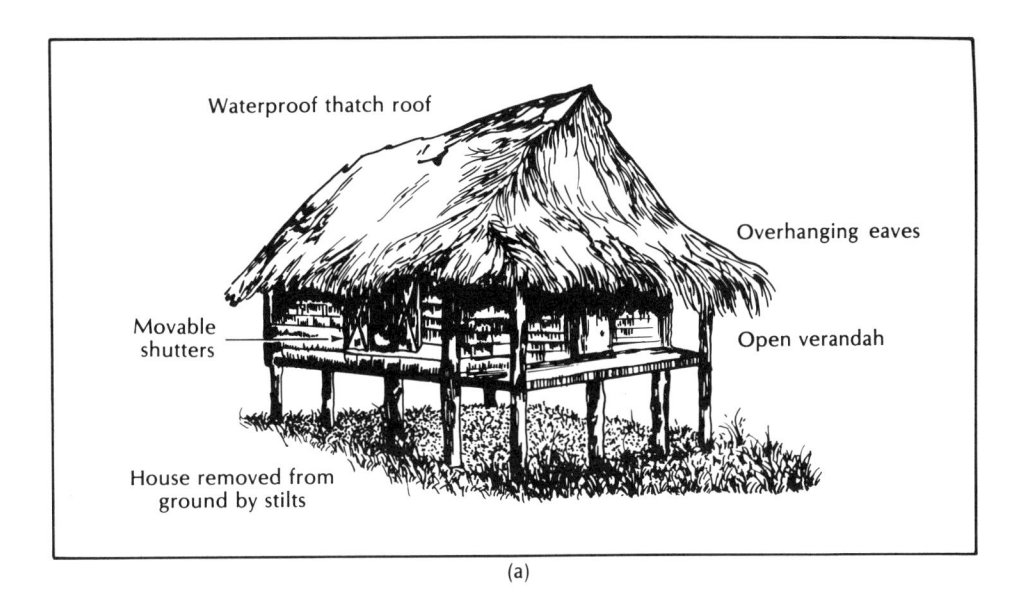

FIGURE 2-13 Use of passive solar design principles in traditional architecture.

Passive systems have a number of approaches to use solar energy in an effective way:

- ☐ Surface color and material of a structure may be designed so that absorption and reflection of solar energy are enhanced or reduced, depending upon local temperatures.
- ☐ The building materials determine the flow of energy through a wall or roof and, by appropriate use of materials and insulation, heat flow can be partially controlled.
- ☐ Wall openings greatly influence the passage of energy into and out of a building. They can be used to limit entry of solar energy, as in hot, dry climates, or to reduce outward flow of heat, as in cold climates. Alternatively, hot wet climates may require maximum wall openings.
- ☐ Building form, the relative dimensions of a building, influences the ease with which buildings can be heated or cooled. A compact, boxlike building is most functional in cold climates. Where a free flow of air is needed, as in warm, middle-latitude climates, an elongated structure is most efficient.
- ☐ Orientation of a structure can be designed so that shadow lengths and shade can be used to shield the building in warm seasons, yet allow access of sun in cold weather. The use of vegetation around homes plays an important part in enhancing this factor.

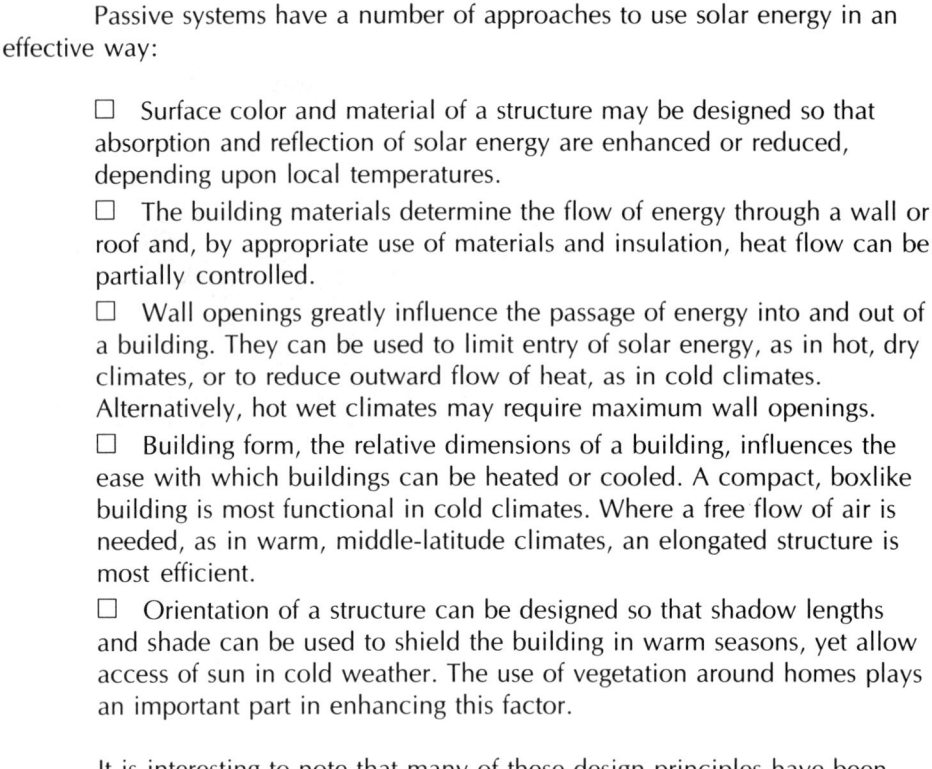

It is interesting to note that many of these design principles have been used in traditional architecture in various world areas. As figure 2-13 (page 43) shows, a traditional home in the hot, wet tropics uses materials from native vegetation growth, has maximum wall openings for access of breezes, and has overhanging eaves to reduce entry of direct solar radiation.

The desert dwelling, the other traditional form shown in the figure 2-13, is made of an earthlike material with a low heat transmissivity. The amount of radiation entering the building is decreased by having few openings. The clustering of such homes—epitomized by narrow, crowded Arab communities—provides shading from direct rays of the sun.

3 Temperature and Temperature Distribution

THE CONCEPT OF TEMPERATURE
THE TEMPERATURE LAG
TEMPERATURE DATA
FACTORS INFLUENCING THE VERTICAL DISTRIBUTION OF TEMPERATURE
FACTORS INFLUENCING THE HORIZONTAL DISTRIBUTION OF TEMPERATURE
Locational Factors
Dynamic Factors
THE DISTRIBUTION OF WORLD SURFACE TEMPERATURES
SUMMARY
APPLIED STUDY 3
Temperature Thresholds

PHYSICAL AND DYNAMIC CLIMATOLOGY

The flows of energy into and out of the earth's atmosphere eventually determine the amount of heat energy at any location. To provide a measure of the heat available in a given body, temperature scales have been devised. It is important to note that temperature is not the same thing as heat, despite their interchangeable use in an everyday context. To examine the relationship further, it is necessary to clearly understand the meaning of temperature.

THE CONCEPT OF TEMPERATURE

Heat is a measure of the quantity of energy present in a body; *temperature* provides a measure of the intensity or degree of hotness of that body. The heat possessed by a

TABLE 3-1 Temperature conversions.

Scale	Steam Point	Ice Point	Absolute Zero
Fahrenheit (°F)	212	32	−459.69
Celsius (°C)	100	0	−273.15
Kelvin (K)	373.15	273.15	0.0

Conversions
  Note: 1 Fahrenheit degree (F°) is 5/9 as large as 1 Kelvin degree (K) and 1 Celsius degree (C°).
  To convert from °C:

$$°F = 9/5 °C + 32$$
$$K = °C - 273.15$$

To convert from °F:

$$°C = 5/9 (°F - 32)$$
$$K = 5/9 (°F - 32) - 273.15$$

To convert from K:

$$°C = K + 273.15$$

Changes in temperature

$$9 \, C° = 5 \, F°$$

where C° is Celsius temperature change and F°, Fahrenheit temperature change, e.g., a change in temperature from 50°F to 60°F = 18 F° in Celsius.

$$9 \times C° = 5 \times 18$$
$$C° = \frac{5 \times 18}{9} = 10$$

substance depends not only on its temperature but also on its mass. Raising the temperature of 25 g of water from 20°C to 25°C would require five times more energy than raising 5 g through the same 5 degrees. Both would measure 25°C, but the heat content of the larger mass would be much greater.

By definition, *temperature* is the condition that determines the flow of heat energy from one substance to another, with the flow always being from high to low temperature. Such a definition indicates that study of the temperature characteristics of the earth-atmosphere is, in fact, a study of heat energy and the way in which it is distributed over the earth's surface.

Temperature is probably the best-known and most widely used climatological measurement; and, as illustrated in Applied Study 3, it is used in many applications. Despite its common use, however, it has important characteristics that are frequently overlooked. One arises from the fact that cited temperatures refer to sensible temperatures as distinguished from kinetic temperature and temperature changes brought about by latent heat transfer (see chapter 4).

Another everyday problem concerning temperature is the units in which it is given; this is a result of historical circumstances. In 1714, Daniel Fahrenheit proposed a scale in which an arbitrary number, 96, representing the approximate equivalent of the temperature of a healthy person, was used as the base. Using the expansion and contraction of mercury about this number, the freezing point of water was set at 32 units, or *degrees,* and the boiling point at 212. Most countries of the world (the United States is a notable exception) have discarded the Fahrenheit scale in favor of the Celsius or Centigrade scale. The latter scale, introduced by Anders Celsius in 1729, has 100 units between the freezing (0°) and boiling points (100°) of water. In 1848 the absolute, or Kelvin, scale, named for the English physicist Lord Kelvin, was introduced. The zero on this scale is theoretically the lowest temperature in the universe, the temperature at which molecular motion ceases. Table 3–1 provides conversion factors needed to pass from one to another of the scales.

THE TEMPERATURE LAG

Solar energy inputs are periodic at most places on earth. The cycles reflect basic earth motions in relation to the sun. Earth rotation gives rise to day and night and results in diurnal temperature variations. As might be anticipated, temperatures are highest during the day, when large amounts of energy are flowing to earth, and lowest at night. There is not, however, a one-to-one relationship in the time of highest solar input and highest temperature because, as already emphasized, temperature largely results from earth radiation following absorption of solar energy.

Highest daytime temperatures usually occur in mid- to late afternoon, several hours after the time of greatest solar input, local noon. Despite the common notion that the hour of highest temperature occurs at the time when solar energy input equals the flow of outgoing energy, such is not true. Measurements show that equilibrium between incoming and outgoing radiation often occurs about an hour and a half before sunset and not at the time of maximum temperature. The lag between the time of maximum input of energy and maximum daily temperature actually occurs because during the early afternoon the earth receives a steady flood of incoming long-wave radiation from the lower atmosphere. It is when this flow, which supplements solar energy inputs, reaches a maximum that highest temperatures occur.

Minimum daily temperatures occur at a time when incoming solar energy and outgoing earth radiation are balanced. Terrestrial radiation decreases through the hours of darkness until, shortly after dawn, incoming solar rays provide sufficient energy to balance outgoing terrestrial rays.

Lag of temperature also occurs on a seasonal scale. Earth revolution causes maximum and minimum solar inputs (outside the tropics) to occur at the time of the summer solstices in each hemisphere. Thus, June and December represent the time of greatest and least solar energy receipts in the Northern Hemisphere; the reverse holds true in the Southern Hemisphere. But these months are not the warmest or coldest. As table 3–2 indicates, in most cases there is a 1-month lag between the highest and lowest energy inputs and the warmest and coldest months. One of the examples given, Naples, Italy, experiences a 2-month lag.

TEMPERATURE DATA

In this and subsequent chapters, temperature data are frequently given as mean values. The mean, or average, represents a value that gives a reasonable approximation of what is the normal, where *normal* can be defined as the value that locates the center of the measurement distribution. It is derived by adding together the individual measurements and dividing them by the number of values added (sum of values/ number of values).

TABLE 3–2 Sample data illustrating temperature lag.

	Average Temperature at Solstice		Average Monthly Temperature		
	June	Dec.	Warmest Month		Coldest Month
Charleston, S.C. (33°N)	26°C(79°F)	11°C(51°F)	July 28°C(82°F)	Jan.	10°C(50°F)
Urbana, Ill. (40°N)	22°C(72°F)	0°C(32°F)	July 25°C(77°F)	Jan.	−1°C(30°F)
Naples, Italy (41°N)	22°C(72°F)	11°C(51°F)	Aug. 25°C(77°F)	Jan.	9°C(48°F)
Moscow, U.S.S.R. (43°N)	19°C(66°F)	−6°C(22°F)	July 21°C(70°F)	Jan.	−8°C(17°F)
Edmonton, Canada (53°N)	14°C(57°F)	8°C(18°F)	July 17°C(63°F)	Jan.	−14°C(7°F)

A whole set of climatological means can be identified, the type largely depending upon the unit of time that the mean measures. For example, the average daily temperature for a given location is the sum of the means of that day [(maximum + minimum)/2] that has occurred every year for which the data are available, divided by the number of years of record. The monthly mean uses the daily means to derive an average temperature for each of the days of a single month and then finds the average of these for many years. In all, the daily values form the base of all average temperatures in whichever form they may be expressed.

The *mean,* however, is a measure of the central tendency of the data set and gives no indication of the way the data are dispersed about that value. The mean of the set 2, 4, 6, 8, and 10 is 6. The same average occurs for 4, 5, 6, 7, and 8. To account for these different distributions about the mean, other statistical measures are used. One that will be presented later in this chapter is the *range*. This is obtained by subtracting the smallest individual value from the largest in the data set. When the range is given in conjunction with the mean, it provides more insight into the data that are being examined. For example, the mean annual temperature at Quito, Ecuador, is 15°C (59°F); at Nashville, Tennessee, it is 15.2°C (59.5°F). However, the range at Quito is only 0.5°C (15.2° − 14.7°C), while at Nashville it is 22.2°C (26° − 3.8°C). Despite the very similar average annual temperatures, the two locations experience different climate regimes.

Another factor that must be considered in analyzing temperature data is the fact that temperature changes over time. The changes can occur through both natural variations and the activities of people. Both of these factors are considered in some detail in Part Three, but their significance at this point can be demonstrated by providing a few examples. Natural changes in temperatures are perhaps best exemplified by merely noting that not long ago, in geologic terms, the earth was in the midst of an Ice Age. Continental glaciers extended far into the United States, and areas that now experience temperate climatic conditions had a climate resembling that which now occurs in Greenland. In the 10,000 years or so that separate the two extremes, great temperature changes have occurred.

The influence of human activities upon temperature is best examined in areas of high population density where the earth's surface has been changed from vegetation to a concrete-brick-macadam covering: it is best seen in cities. So marked is the influence of cities upon temperature regimes that it has given rise to a special area of climatic study called *urban climatology*. In the urban climate, one of the most frequently observed changes is the occurrence of higher temperatures over the city. At night, especially under calm conditions, the city becomes so much warmer than the surrounding countryside that it is identified as an urban heat island. The causes and amount of temperature change that occur in cities are described in chapter 15.

FACTORS INFLUENCING THE VERTICAL DISTRIBUTION OF TEMPERATURE

In chapter 1, the layers of the atmosphere were differentiated on the basis of temperature. It was shown that the upper limit of the troposphere, the tropopause, was distinguished by an isothermal layer separating it from the stratosphere. The atmosphere is densest in the near-earth parts and, as indicated by the significance of energy transformations, is heated from below. There is, as a result, a decrease of temperature with increasing elevation in the troposphere. The amount of decrease is called the *lapse rate*. Temperature decrease with altitude has been calculated for many world areas and the average lapse rate is 6.5°C/km (3.6°F/1000 ft). This is a large temperature difference for relatively small distances and is much greater than that of the existing horizontal gradients. Permanent snow and ice occur at 5 km above the equator but not for many thousands of kilometers in the horizontal toward the poles.

The vertical distribution of temperature is influenced by the nature of the underlying surface. For example, temperature decreases most rapidly with altitude over continental areas in summer. Figure 3-1 shows a cross-section of the atmosphere up to 23 km along the 80th meridian (which passes through eastern Canada, the United States, Cuba, and Panama) from the North Pole to the equator in January

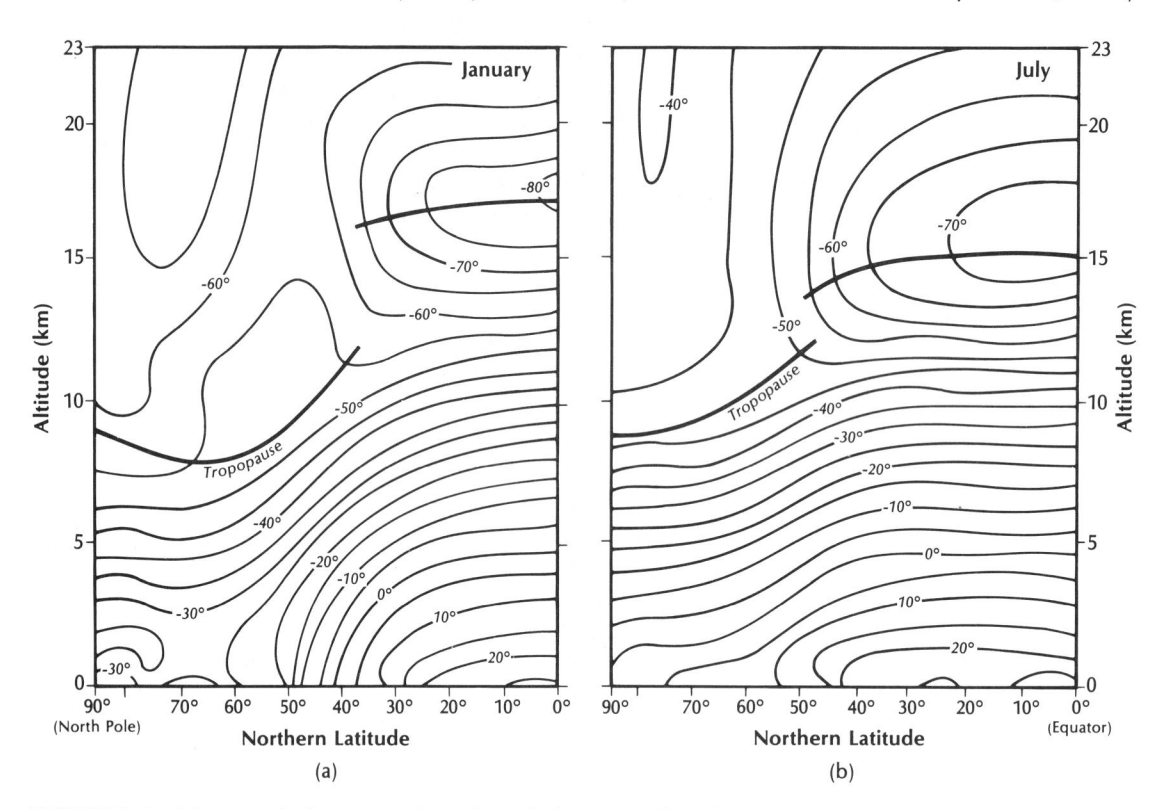

FIGURE 3-1 Mean vertical cross-sections through the atmosphere in (a) January and (b) July. Both sections are along the meridian 80° W in the Northern Hemisphere. (From H. B. Byers, *General Meteorology*, 4th edition, McGraw-Hill Book Company, © 1974. Used by permission.).

TEMPERATURE AND TEMPERATURE DISTRIBUTION

and July. The heavy lines on the graphs show the tropopause. The following features are to be noted:

1. The tropopause is lower over high latitudes than low latitudes. A well-marked break occurs in the middle latitudes.
2. The north-south temperature gradients are much steeper in winter.
3. The strongest horizontal gradients in both summer and winter are in middle latitudes, the area of greatest storm activity.
4. The coldest part of the troposphere occurs over the equator in the region of the tropopause.

Another effect of altitude is that the diurnal range of temperature at higher elevations is greater than that at an equivalent climate at sea level. Figure 3-2 provides a graphic model of this effect. The reason for the greater range is that at higher elevations the atmosphere is rarer than at lower altitudes. This variation will cause maximum daily temperatures to be about the same or slightly less than at a lowland counterpart. The main difference occurs at night when the escape of terrestrial energy occurs much more readily because of the lower density of gases at higher elevations.

Decrease of pressure with altitude modifies the meaning of given values on temperature scales. The gas laws (see Appendix 1, Note 1) indicate the relationships between temperature and pressure; at higher elevations, pressure changes also become important. The reduced pressure, for example, means that molecules of water vapor escape more easily from a water surface. Thus, at sea level, water boils at a temperature of 100°C. At Quito, Ecuador, at an elevation of 2849 m (9350 ft), water will bubble and boil at 90°C; at the top of Mount Everest, the boiling point of water is only 71°C.

The decrease of temperature with increasing elevation is periodically interrupted, and a temperature increase with altitude may occur. Such conditions are described as *temperature inversion*. Inversions can occur at ground level, when

FIGURE 3-2 Diurnal range of temperature in a mountain and a lowland station at places of similar latitude.

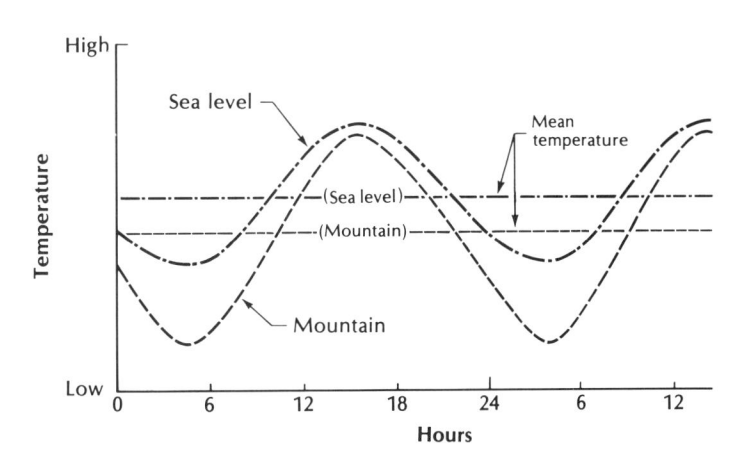

associated with radiational cooling, or above ground level because of subsidence. The significance of both surface and upper air inversions is described in later chapters.

FACTORS INFLUENCING THE HORIZONTAL DISTRIBUTION OF TEMPERATURE

Were the earth a homogenous body without the land-ocean distribution, earth temperature would be latitudinal, with temperature decreasing evenly from the tropics to the poles. It is possible to compare actual temperatures with those hypothetical values and thus identify world areas where thermal anomalies occur, where the anomalies represent deviations from temperature resulting only from solar energy inputs. Isolines called *isanomals* can be drawn through points of equal values to produce world isanomalous maps.

Figures 3-3 and 3-4 show isanomalous temperature maps for January and July, respectively. Study of the maps allows a number of general observations to be made:

a. In winter (in each respective hemisphere), the land masses have large negative anomalies indicating values as much as $-20°C$ below the hypothetical mean. By contrast, ocean areas either have positive or no anomalies.

b. In summer (in each respective hemisphere) the largest anomalies are still over the continents, but now they are positive values. In some areas a small negative anomaly occurs over the oceans.

c. The patterns of the anomalous conditions over the oceans are distinctive in shape and appear to be associated with ocean movement.

d. Smallest anomalies occur in the equatorial/tropical realms and attain highest values in high middle latitudes.

These observations permit the identification of the major factors that modify the zonal solar climate and hence the patterns of temperature over the globe. The controls may be conveniently dealt with as two types. First are those factors that are due primarily to geographical location on the earth's surface, which basically determines the amount of energy received directly from the sun. In contrast to these locational factors are temperature characteristics that result from the transport of energy by the mobile atmosphere and ocean. This dynamic effect can greatly modify the temperature a location experiences.

Locational Factors

Latitude The latitude of a location is of prime importance in determining solar energy receipts. Earth-sun relationships show that both the angle of the sun in the sky and the length of day are major determinants of solar energy receipts. These two

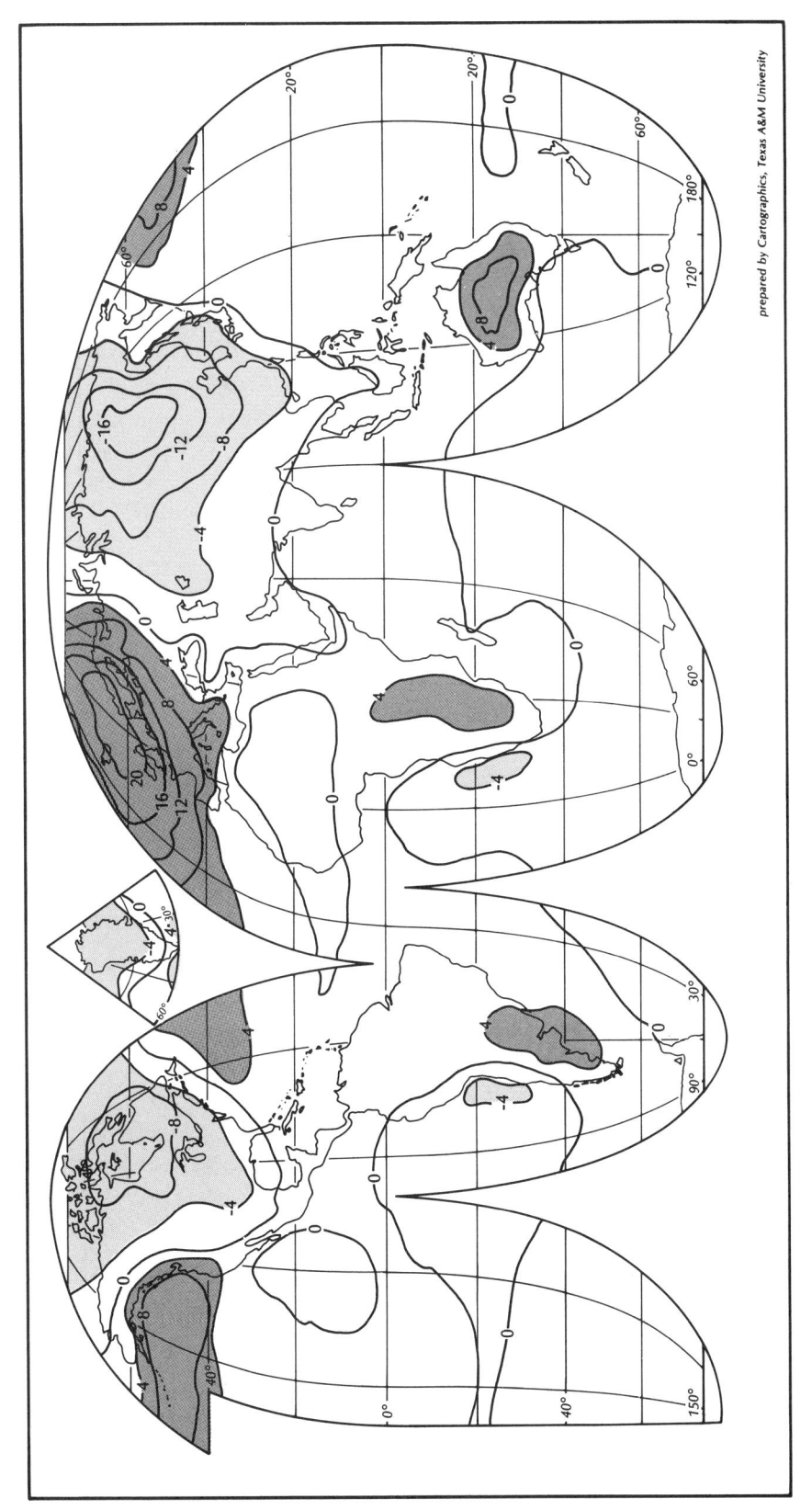

FIGURE 3-3 Isanomalies of temperatures (°C) in January (After Hann-Süring).

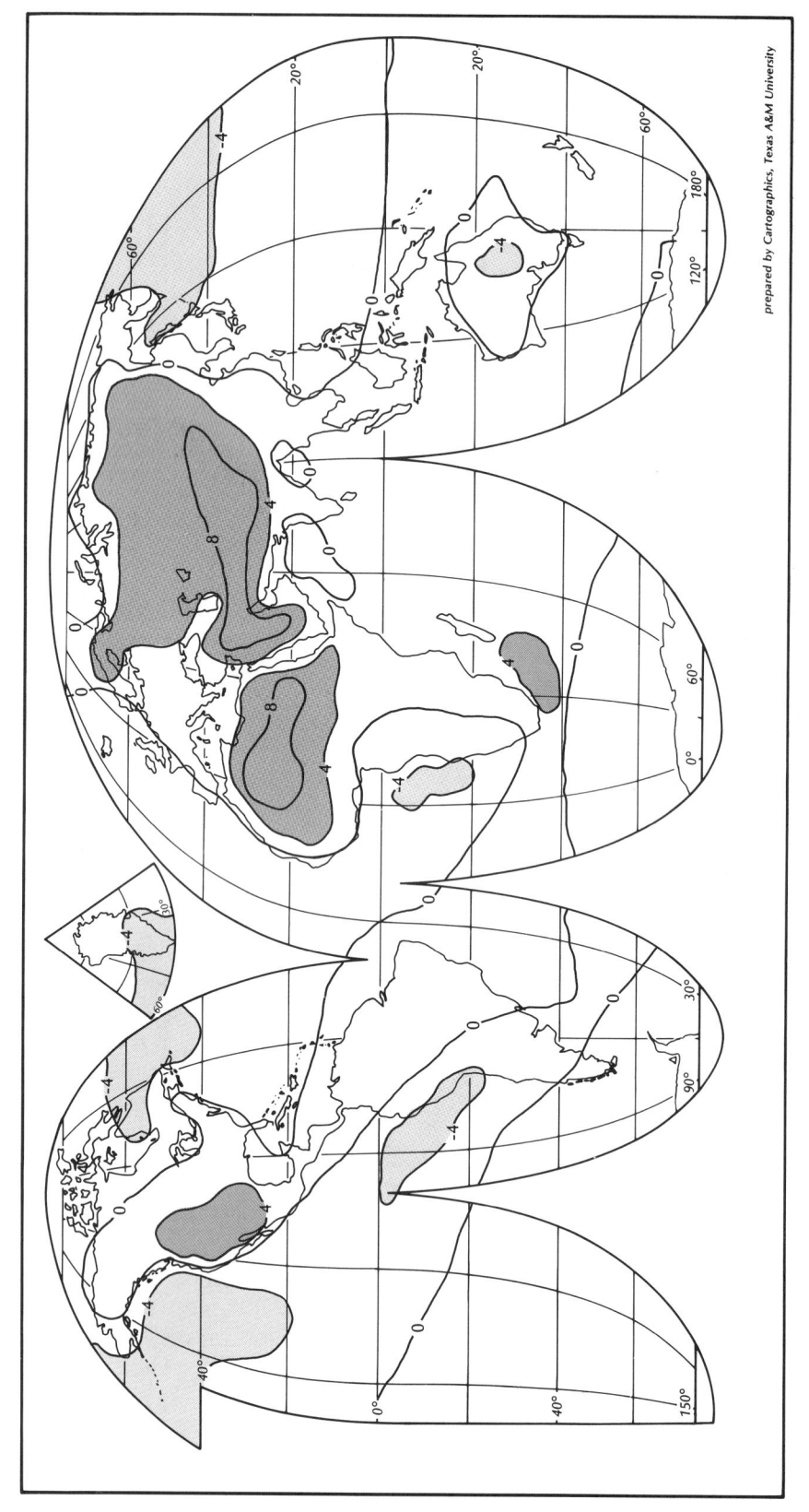

FIGURE 3-4 Isanomalies of temperatures (°C) in July (After Hann-Süring).

factors are determined by latitude. The significance of earth-sun relationships in explaining temperatures over the earth go beyond the obvious equator-pole differences, for they appear in more subtle ways. For example, they relate to the fact that highest temperatures on earth are not found at the equator but near the tropics of Cancer and Capricorn. A partial explanation of this occurrence is in the apparent migration of the sun between 23.5° N to 23.5° S. In its passage the sun seems to move relatively quickly over the equator but slows as it progresses north and south. Thus, between 6° N and 6° S, the sun's rays are vertical for 30 days during the time of the equinoxes. Between 17.5° and 23.5° N and S, the vertical rays occur for 86 days near the solstice. The longer period of high sun and the concurrent longer days allow time for surface heat accumulation and thus give rise to the zone of maximum heating near the tropics. The heating is further enhanced by the clear skies near the tropics, compared to the very cloudy equatorial belt.

Temperature regimes, especially the seasonal cycles, are also related to earth-sun motions. The temperature regime at equatorial stations shows few variations, with two maxima at the period of equinoxes. Stations farther north show a distant summer-winter maximum and minimum, with the range generally increasing with latitude. The actual temperatures are influenced by factors other than latitude, but the relationships indicate the relative control of solar energy inputs.

Surface properties The various pathways taken by solar energy when striking a surface depend largely upon the type of surface. Of particular note is reflectivity, for surfaces with high albedo absorb less incident radiation, with the result that the total energy available is diminished. Thus polar icecaps can be effectively maintained, for as much as 80 percent of the solar radiation falling on them is reflected.

Even if two surfaces have a similar albedo, the incident energy does not always cause them to have similar temperatures, since the heat capacities of the surfaces differ. The *heat capacity* of a substance is the amount of heat required to raise its temperature 1°C. When the substance has a mass of 1 g, the heat capacity is called *specific heat*. Thus, specific heat is defined as the amount of heat (number of calories) required to raise the temperature of 1 g of a substance through 1°C. As table 3-3 shows, the specific heat of substances can vary appreciably. Of particular significance is the fact that the specific heat of water is some five times greater than that of rock and the land surface in general. This means that the amount of heat required to raise the temperature of water 1°C is five times greater than required for the same temperature increase on land. The same amount of energy applied to a land surface and a water surface would result in the land's becoming much hotter than the water.

The difference is also heightened by the different heat conductivity of the two materials (table 3-3). Loose, dry soil is a very poor conductor of heat and only a superficial layer will experience a rise in temperature following energy input. Water has only a fair conductivity, but its general mobility and transparency permit heat to circulate below the surface layer. A natural unisturbedsoil wih a vegetation cover may have daily temperature changes recorded to 1 m; a quiet stand of water has daily temperature variations that can be measured to a depth of perhaps 6 m.

TABLE 3-3 Thermal properties of selected substances.

Specific Heat	
Water	1.0
Ice (at freezing)	0.5
Air	0.24
Aluminum	0.21
Granite	0.19
Sand	0.19
Iron	0.11

Thermal Conductivity	
	cal/cm²/s °C*
Air	0.000054
Snow	0.0011
Water	0.0015
Dry soil	0.0037
Earth's crust	0.004
Ice	0.005
Aluminum	0.49

* The number of calories passing through an area 1 cm² in a second when temperature gradient is 1°C.

When such differences are applied to the continental scale, it is evident that land masses will heat up much more rapidly than oceans in summer. However, since this heat does not penetrate to great depths, it is rapidly radiated to the atmosphere as winter approaches. Hence land masses tend to experience extreme temperatures, while water bodies are more equable and show less change (figure 3-5).

The thermal differences of land and water are heightened by the way in which solar energy is partitioned at the surface. Recall that energy inputs raise the temperature of the absorbing surface and evaporate moisture from that surface *(LE)*. The absorbed energy is either radiated back to the atmosphere or passed to the air in the form of sensible heat *(H)*. In this, the vertical transfer of heat to the lower levels of the atmosphere is completed through turbulent transfer. The amount of sensible heat available thus depends in part upon the fraction of incident energy that is used up in changing the state of water. Over the oceans, evaporation will be continuous; on land, it will depend upon the water available at the surface. Some indication of the significance of this relationship can be obtained by comparing the ratio of sensible heat to latent heat for different parts of this earth. This ratio (the *Bowen ratio*) is given by H/LE, so that the larger the value of the ratio, the more sensible heat (rather than latent heat) is passing to the atmosphere. Some values in kilolangleys (kly) per year follow:

	Net Radiation	LE	H	H/LE
All oceans	+82	−74	−8	0.11
All lands	+49	−25	−24	0.96

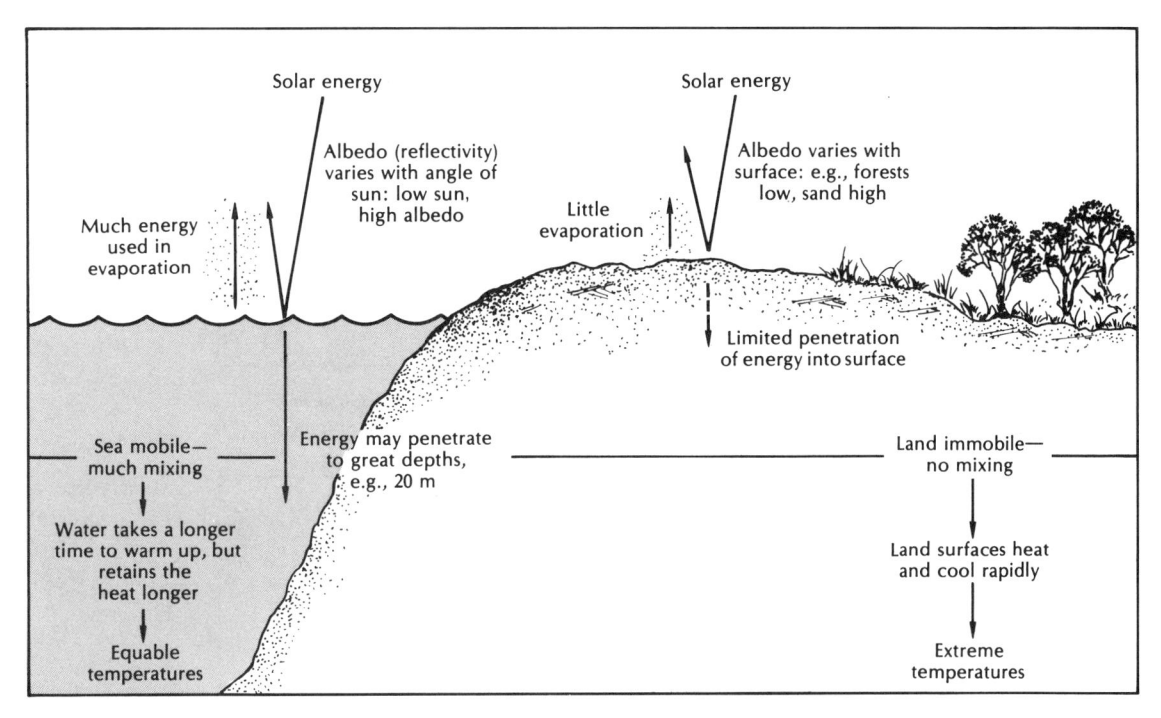

FIGURE 3-5 Solar energy striking land and sea surfaces is partitioned in different ways. The water surfaces tend to be conservative, warming and cooling slowly; the land masses experience large temperature differences.

The much larger value for land further modifies the already observed land-sea temperature differences. In fact, so distinctive is this effect that climatologists use the concept of *continentality,* a measure of the continental influence on climate (see chapter 14).

The relative partitioning of *LE* and *H* also varies at individual stations. Figure 3-6 provides examples of a tropical moist station (West Palm Beach, Florida), a desert station (Yuma, Arizona) and a station on the west coast of middle latitudes (Astoria, Oregon). Note the amount of net radiation that passes to *LE* in the moist coastal locations. By contrast, the arid example shows much of the net radiation's becoming sensible heat. High temperatures will obviously prevail because of this effect.

Aspect and topography The combined influences of the steepness and direction faced by a slope determine its aspect. The importance of aspect is best seen in differences that occur on north-facing and south-facing slopes in the Northern Hemisphere. A north-facing slope may still have snow lying on it while a south-facing slope is quite clear; the north slope gets less intense radiation and, as the sun gets lower in the sky, will be in shadow long before the south-facing slope.

FIGURE 3-6 Average annual variations of the components of the surface energy heat budget at three locations (From W. D. Sellers, *Physical Climatology* [Chicago: The University of Chicago Press, 1965], p. 106. Copyright © 1965 The University of Chicago Press. Reproduced by permission.).

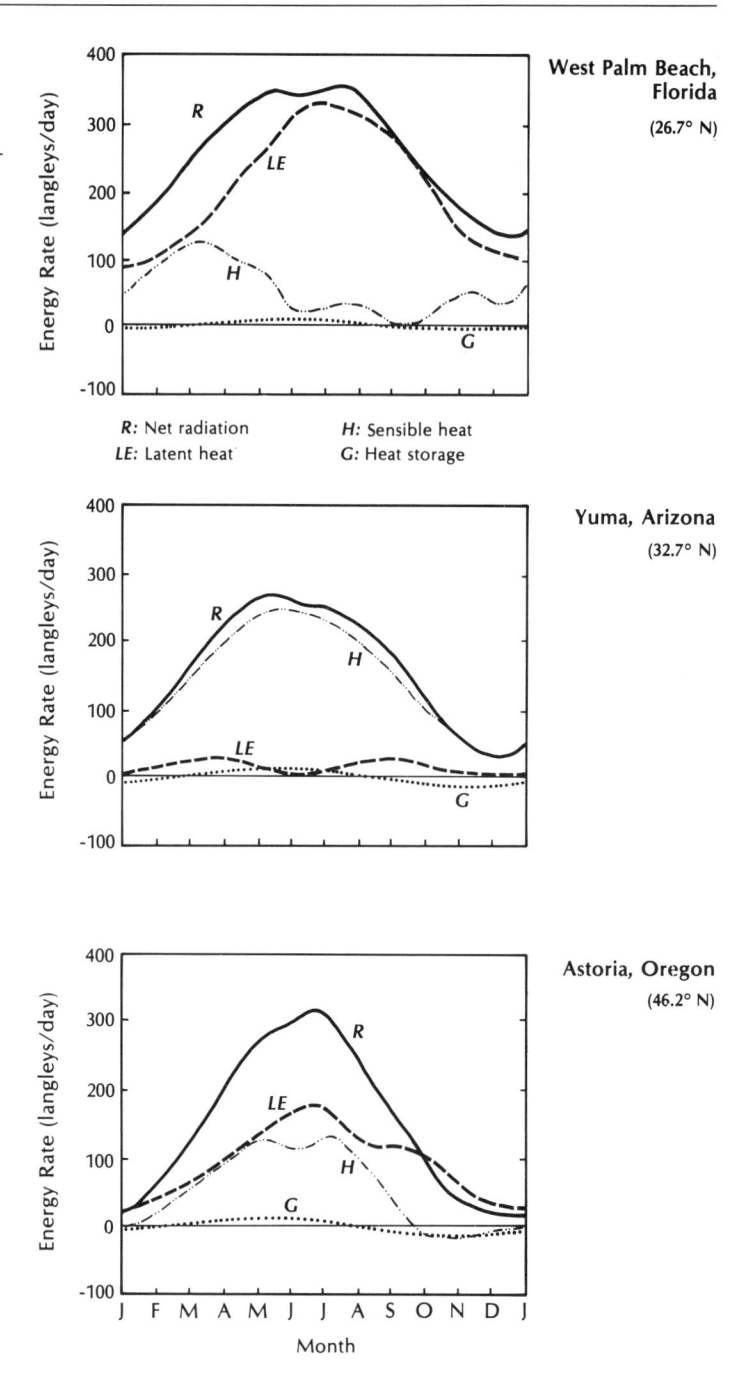

West Palm Beach, Florida (26.7° N)

R: Net radiation *H:* Sensible heat
LE: Latent heat *G:* Heat storage

Yuma, Arizona (32.7° N)

Astoria, Oregon (46.2° N)

The influence of aspect is seen in many ways; for example, the height of the level of permanent snow and ice on mountains will vary from one slope to another, while vegetation levels (e.g., the tree line) will also be affected. Similarly, the depth of snow and frost are found to differ on north- and south-facing slopes.

Topography also plays an important role in the nature of climates of neighboring lowlands. On a continental scale, mountain ranges that run north-south have a very different effect from those that run east-west. Thus, the lack of any extensive east-west barrier in the United States permits polar and tropical air to penetrate great distances into the continent. One result of this unobstructed flow of air is the high incidence of tornadoes in the United States (see chapter 7).

Dynamic Factors

The imbalance of energy between the tropics and poles (see figure 2-10) means that an exchange of air must occur through a dynamic process. The mechanisms for the exchange are those energy balance factors that have already been outlined: the transfers of latent heat *(LE)*, sensible heat *(H)*, and the heat that is stored within the water of the oceans *(S)*. The heat balance for the components for latitudes was presented in figure 2-10, and a similar representation can be completed for temperature. Table 3-4 provides the theoretical planetary temperature for sea level, assuming the atmosphere is at rest and the observed mean annual temperature for every 10° of latitude. The greatest differences are at the equator and for those latitudes above 60°. Tropical latitudes are cooler than the theoretical value, while high latitudes are warmer. The differences between the actual and theoretical values result from the transport of energy over the globe by air masses and ocean currents. Every storm system, circulation pattern, and evaporation/precipitation event contributes toward the redistribution of energy and resulting temperature regimes that prevail over the earth's surface. Such factors are dealt with in more detail in subsequent chapters.

THE DISTRIBUTION OF WORLD SURFACE TEMPERATURES

All of the above factors play a role in determining the distribution of average temperatures over the earth's surface. The resulting distributions for July and January

TABLE 3-4 Theoretical planetary (atmosphere at rest) and actual temperatures by latitude.

Temperature °C(°F)	Equator	10°	20°	30°	40°	50°	60°	70°	80°
Northern Hemisphere									
Planetary temp.	33(91)	32(89)	28(83)	22(72)	14(57)	3(37)	−11(12)	−24(−11)	−32(−26)
Actual temp.	26(79)	26(80)	25(78)	20(69)	14(57)	5(42)	−1(30)	−10(13)	−18(−1)
Difference	−7(−12)	−6(−9)	−3(−5)	−2(−3)	0	+2(+5)	+10(+18)	+18(+24)	+14(+25)
Southern Hemisphere									
Planetary temp.	33(91)	32(89)	28(83)	22(72)	14(57)	3(37)	−11(12)	−24(−11)	−32(−26)
Actual temp.	26(79)	25(78)	22(73)	17(62)	11(53)	5(42)	−3(26)	−13(8)	−27(−17)
Difference	−7(−12)	−7(−11)	−6(−10)	−5(−10)	−3(−4)	+2(+5)	+8(+14)	+11(+19)	+5(+9)

are shown in figures 3-7 and 3-8, respectively; while figure 3-9 shows the annual average range of temperature.

In each case, the general decline in temperatures from tropics to poles is evident, illustrating the basic influences of latitude. However, significant variations from a simple zonal pattern exist, and these result from a combination of other temperature controls. Such controls can be readily identified:

1. The maps show the extremes of temperature that occur over continental land masses. The largest land mass, Asia, experiences average temperatures that range from 4°C (40°F) in January to more than 15°C (60°F) in summer. The effect is clearly seen in figure 3-9, which depicts the average annual range of temperature.

2. The oceans exhibit the results of heat transport by ocean currents. The warming effects of the North Atlantic and North Pacific Drifts are seen in the poleward displacement of isotherms. Cold ocean currents, such as the California and Canaries currents, cause temperatures to be lower.

3. The maps show actual temperatures (many world temperature maps have isotherms reduced to sea level), so that the effect of altitude can be determined. Notice, for example, the isotherms over South America, where the Andes cause a large equatorward displacement of colder temperatures.

4. The closest approximation to zonal temperatures occurs over the southern ocean (and Antarctica). This area is the most extensive location where homogenous surfaces encircle the globe without interruption.

5. By comparing figures 3-7 and 3-8, it can be seen that the hemispheric temperature gradient is greatest in their respective winters. For example, in January the Northern Hemisphere gradient is more than 60°C (110°F); in July it is approximately 10°C (50°F). This pattern has highly significant consequences to atmospheric circulation (chapter 6).

SUMMARY

Temperature is used to measure the amount of heat energy available over the earth's surface and is a function of the net radiation at a location. There is no direct relationship between maximum and minimum inputs of solar radiation and highest and lowest temperatures. A lag that can be identified on both a daily and monthly basis occurs between them.

Care must be taken when using mean temperature data, for means do not provide a full assessment of the temperature regime; some measure of the variation of data about the mean is necessary for complete description. Similarly, the use of the mean for a given time period obscures the fact that temperature changes within the time period.

The vertical distribution of temperature in the troposphere is represented by lapse rate conditions, on average, amounting to a decrease in temperature of 6.5°C for each kilometer. The horizontal distribution of temperature over the earth's

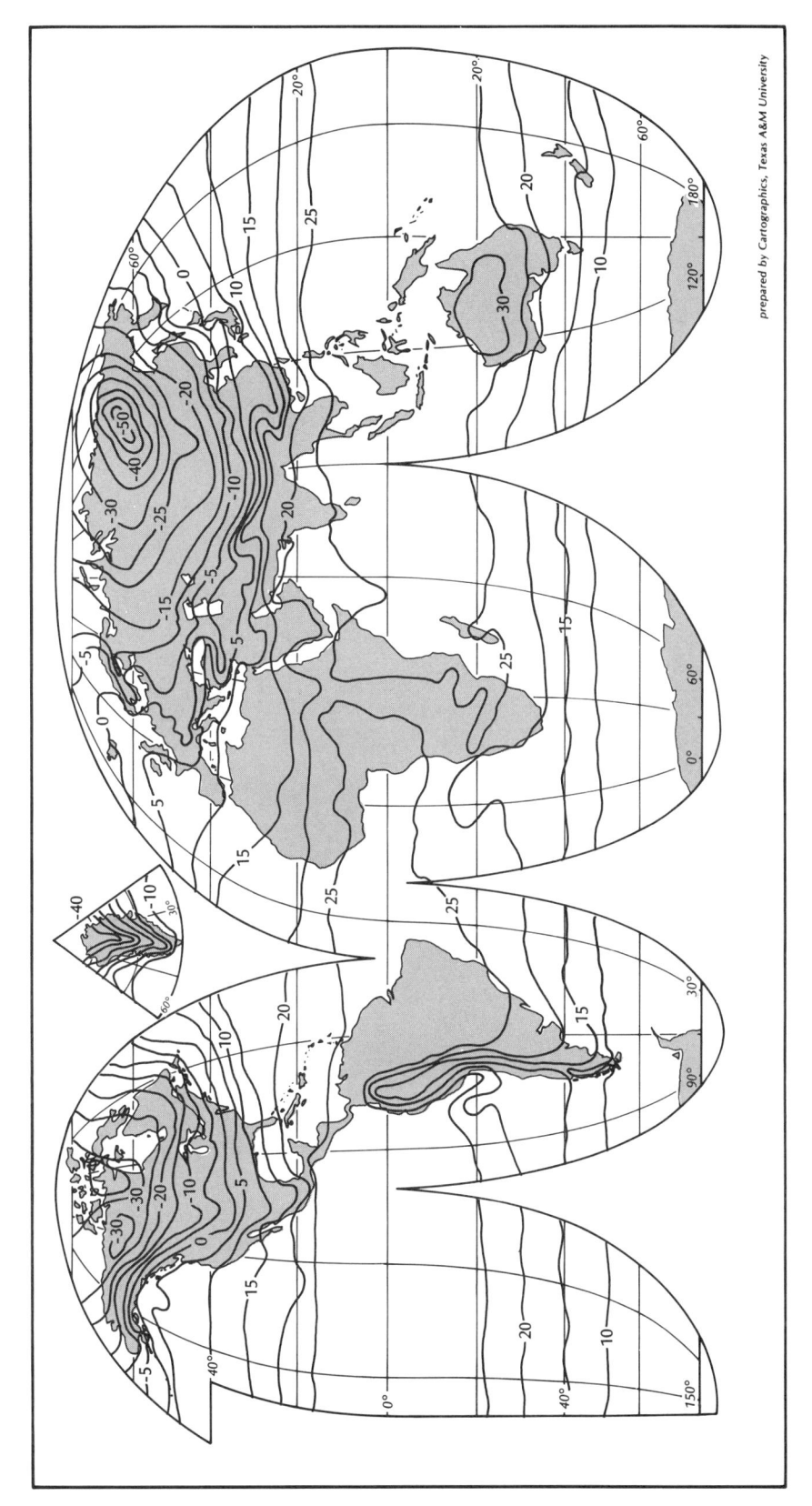

FIGURE 3-7 Distribution of average July temperatures over the world.

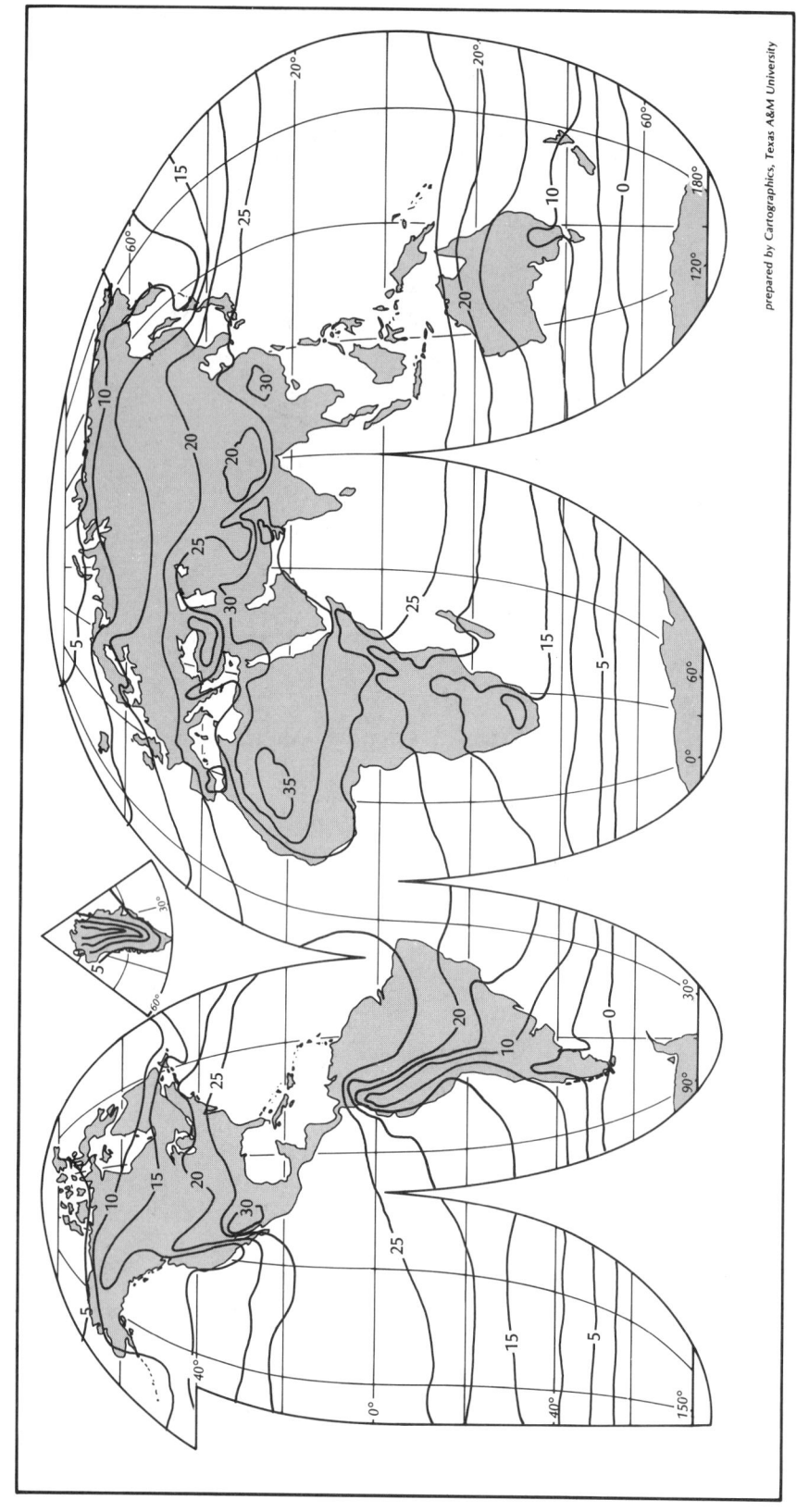

FIGURE 3-8 Distribution of average January temperatures over the world.

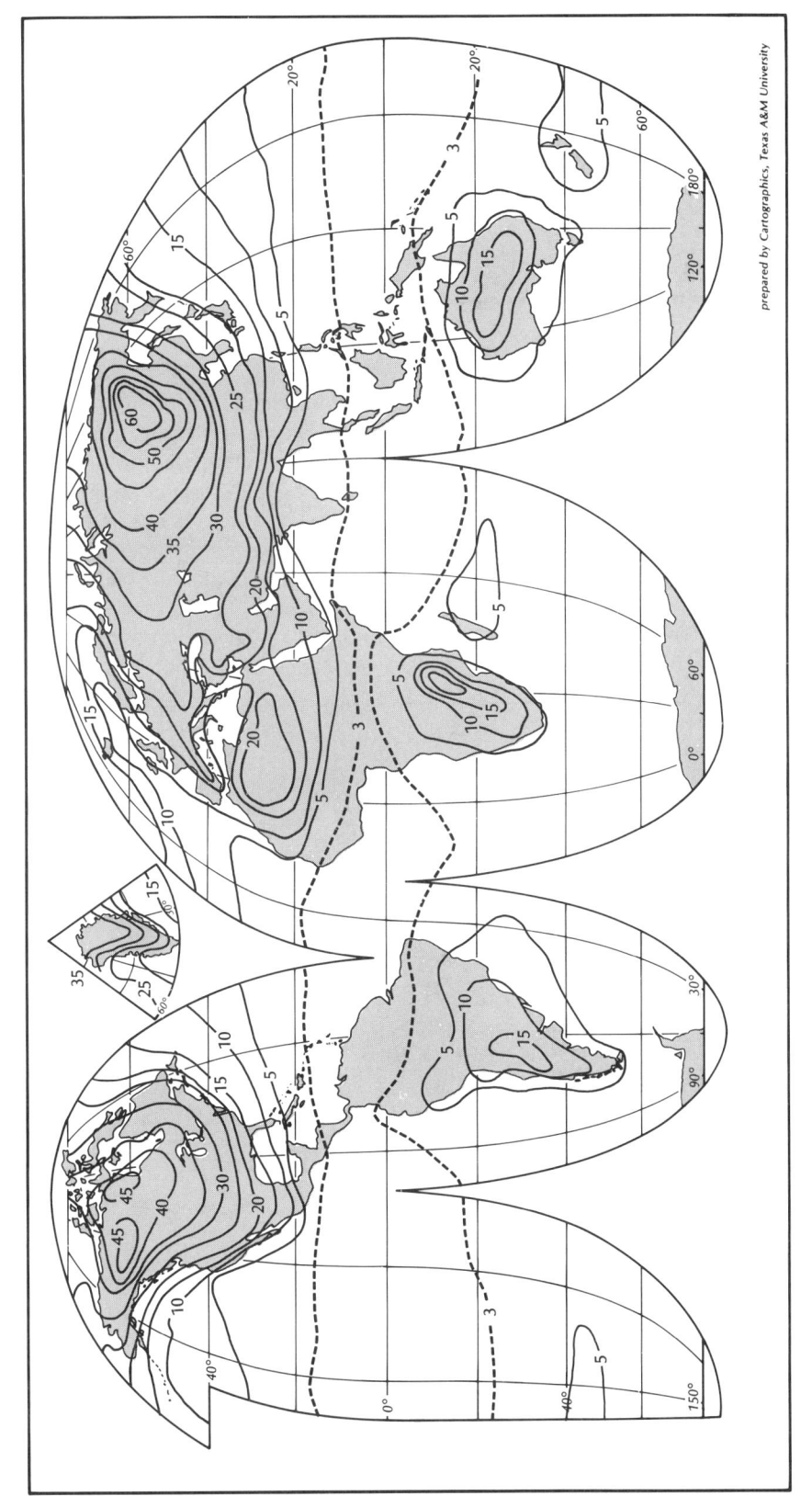

FIGURE 3-9 Distribution of mean annual temperature ranges over the earth.

PHYSICAL AND DYNAMIC CLIMATOLOGY

surface is a function of both locational properties and the dynamic movement of the atmosphere. The former is represented by such factors as latitude, surface properties, and aspect; the latter, by energy transferred not only by the atmosphere but also by the oceans.

APPLIED STUDY 3

Temperature Thresholds

The significance of temperature as an environmental factor is well illustrated by the identification of temperature thresholds. The most obvious threshold is 32°F (0°C), the temperature at which moisture passes from a water to an ice phase. (The Fahrenheit scale is retained in this study because it is in common use in the reporting of temperature threshold data.) This, however, is but one of many critical temperature levels that are used to derive activity guides to prevailing conditions. The following provide examples of the climatological use of derived values.

The frost-free season is the period, usually given in days, between the last frost of spring and the first of fall. Its significance lies in the fact that in order to

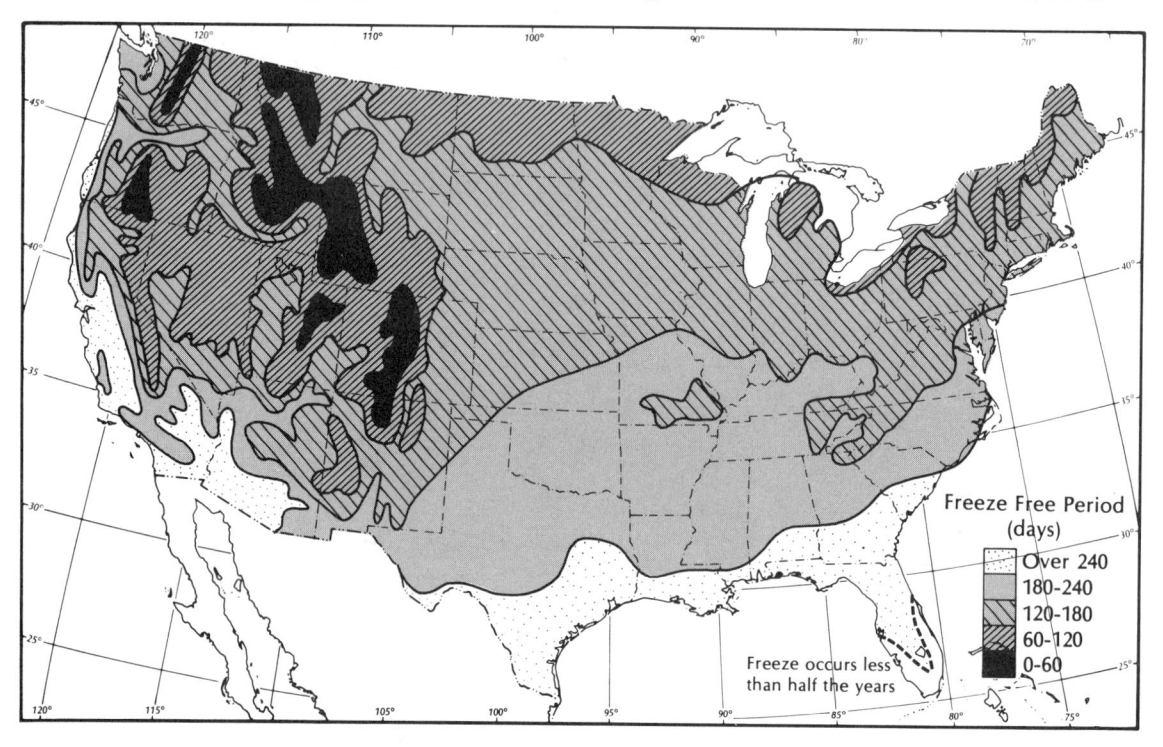

FIGURE 3-10 The length of the growing season in the United States as expressed by the frost-free period (After NOAA, From J. E. Oliver [1981, p. 49]. Used by permission of V. H. Winston and Sons.).

reach maturity, domesticated plants require a given time period without frost. The length of the frost-free season varies appreciably in the United States, as figure 3–10 illustrates. Such distribution places restrictions upon the growth of crops in the United States. Thus, production of cotton, which requires some 200 frost-free days, is restricted to the southern states.

In many ways, the frost-free season is related to the concept of the growing season. Such an analogy assumes that plant growth and development begin when temperatures rise above freezing. This is not completely true, and the actual growing season for individual plants is based upon different thresholds. Certainly, some plants, such as wheat, barley, rye, and oats, begin their growth at temperatures slightly above freezing. But hot-season crops, such as sorghum and melons, require temperatures of about 60°F (15°C) to begin growing. The growing season of a crop must thus be evaluated in terms of the temperature requirements of that particular plant.

Threshold temperatures are also used to derive degree-days. The degree-day is based upon the accumulated temperatures that occur above a given temperature value. Probably the best-known is the heating degree-day. This was initially devised by heating engineers to measure potential heating energy demands. When the temperature falls below a threshold, in this case 65°F, it is

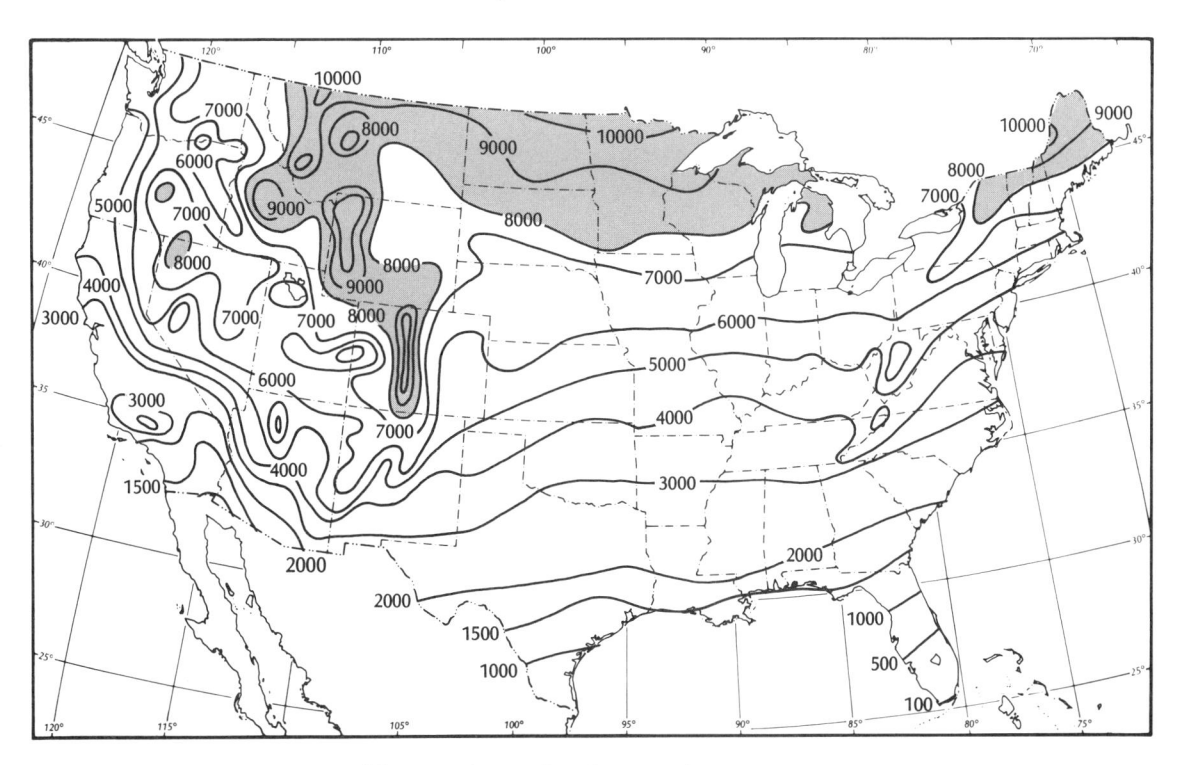

FIGURE 3–11 Average seasonal heating degree-days from 1941 to 1970. The base temperature is 18°C (65°F). (After U.S. National Weather Service, from J. F. Griffiths and D. M. Driscoll, *Survey of Climatology*, Charles E. Merrill, 1982).

assumed that homes and businesses will require some form of heating. The lower the outdoor temperature, the more heat is required. To obtain a guide to relative severity of a cool period, the *degree-day* is derived by subtracting the air temperature from the selected threshold. Thus, if on a given day the average temperature is 60°F, five (65 minus 60) degree-days are accumulated. If the next day's temperature is 54°F, then eleven (65 minus 54) degree-days are recorded. This procedure is carried on throughout the cool period, and all degree-days are added to provide a seasonal total of degree-days. The latitudinal extent of the United States causes the average number to vary appreciably, as figure 3-11 (page 65) shows.

A comparison of the average number of degree-days with those that actually occur provides a method of evaluating the relative severity of a given cold season. In west-central Indiana, for example, the 10-year mean number of degree-days for 1958–1969 was 5538. This is appreciably more than other decades (e.g., 1919–1929 was 5175, 1929–1939 was 5109) and is indicative of the severe winters that occurred during this period.

A similar concept to assess summer cooling requirements is given by the cooling degree-day. This is often based upon a threshold of 72°F, above which many air conditioners will be turned on and a large energy demand created. Obviously, the values for cooling degree-days are spatially opposite those of the heating degree-days, with the highest values in the warmest places.

4 The Physical Basis of the Hydrologic Cycle

PHYSICAL PROPERTIES OF WATER
THE HYDROLOGIC CYCLE
ATMOSPHERIC HUMIDITY
AIR MASSES
EVAPORATION AND TRANSPIRATION
CONDENSATION
Dew and Frost
Fog and Clouds
SUMMARY
APPLIED STUDY 4
The Hydrologic Response to Winter Conditions

PHYSICAL AND DYNAMIC CLIMATOLOGY

*W*ater in the atmosphere is of major importance in several ways. For one, it is a major element in the energy balance of the earth. In the form of clouds, it is a major reflector of solar radiation; as water vapor and as clouds, it is an absorber of earth radiation. Through the processes of evaporation and condensation, it is a major means by which energy is transferred from the surface of the earth to the atmosphere and, subsequently, from place to place over the earth. Water vapor in the atmosphere is essential for, and the basis of, precipitation.

PHYSICAL PROPERTIES OF WATER

Water has a number of unique and unusual attributes that enable it to play such important roles in the climate of the earth.

Specific heat Water has a high specific heat, the highest of any liquid except ammonia. *Specific heat* is the amount of energy needed to change the temperature of the substance. As a result of this high specific heat, a large amount of heat is involved when water changes temperature. Specific heat of water has a tremendous impact in moderating earth temperatures from day to night, through the seasons, and from equator to pole. The ocean currents carry an enormous amount of energy to the polar regions, thus warming the Arctic and cooling tropical regions. It is the ability of water in the atmosphere to absorb and radiate heat that is the basis of the greenhouse effect.

Surface tension Water has the highest surface tension of any liquid except mercury. *Surface tension* is the attraction of molecules for each other; this affinity manifests itself in many ways, including the formation of water droplets, dew, fog, raindrops, and waves. Were it not for this surface tension, large waves would not form in water. It is this surface tension that is responsible for the development of large waves associated with hurricanes, tsunamis, and other marine storms.

Conductivity Water conducts heat more readily than any liquid except mercury. It is through conduction and convection that uniformity of temperature is attained in water bodies.

Density differential Water changes density with temperature, and fresh water has maximum density at around 4°C. It expands at temperatures both above and below this point. When it fuses (freezes), it expands rapidly, adding about 9 percent by volume, a strange but tremendously important attribute. A fresh water body, such as a lake or stream, cools very slowly once the temperature drops below 4°C because the surface water is less dense than the water beneath, and it forms a thermal blanket. The insulating effect is greatly enhanced by the formation of ice.

In sea water the process operates a little differently, and again it is a critical difference for life on earth. The fusing point of sea water occurs near −2°C, and salt water reaches its maximum density just above the same temperature. One result of this is that surface water in the Arctic seas may cool up to 6°C more than would otherwise be the case before ice forms. The colder Arctic water is more dense than fresh water or sea water nearer the equator, so that it sinks to the bottom and settles in the deeper parts of the ocean basins. This brings about colder temperatures than would be expected at great depth in the ocean. This settling cold water helps promote circulation in the ocean, which in turn helps equalize the energy between equator and poles. It also aids in vertical transport of dissolved gases, such as oxygen and carbon dioxide, and carries green matter and plant nutrients to lower depths in the ocean. It is also this latter property that is partially responsible for reducing the accumulation of CO_2 in the atmosphere. Carbon dioxide is fairly soluble in salt water, and solubility is greater in cold water. The colder waters of the ocean dissolve the carbon dioxide and then carry it to the depths of the ocean. Without this process, CO_2 content in the atmosphere would be much higher, and the greenhouse effect of the CO_2 would be greater.

Change in state Water is the only substance that exists on the planet in all three physical states, solid, liquid, and gas; and the earth is the only planet in our solar system on which water is found in all three forms. The range of temperature found on the earth includes the fusing point of water; since temperatures fluctuate through time and space around the freezing point, water is continually freezing and thawing. The range of atmospheric pressure also includes the vapor pressure of water, which allows water to circulate in the gaseous state. Water is thus found in the solid state as snow and ice, in the liquid state as free water, and in the gaseous state as water vapor.

Incorporated in these changes from one state to the other are massive amounts of heat exchange, which play such a major role in the earth's energy balance (figure 4-1). The quantity of heat required to vaporize water is greater than for any other substance. It takes 590 times as much heat energy to change the state of water from liquid to gas as to change the temperature of water 1°C. The heat of fusion of ice is greater than for any substance except ammonia. The importance of this heat exchange in the environment cannot be overstated. It modifies temperatures under virtually all kinds of conditions. The large amounts of heat needed to vaporize water reduce evaporation; otherwise levels of lakes and ponds would fluctuate wildly between rains and between seasons. In desert environments, the formation of dew and fog helps reduce nocturnal cooling by releasing large quantities of heat into the lower atmosphere. In arctic and highland regions, the freezing of water and the formation of frost retard cooling.

THE HYDROLOGIC CYCLE

The distribution of water in the earth environment is far different from that perceived by most individuals. There are good reasons for this misconception. Most of the world's population lives on land and hence thinks of the earth as dominated by land;

PHYSICAL AND DYNAMIC CLIMATOLOGY

Moisture changes of state at 0 °C

Sublimation 680 cal/g
Releases heat to atmosphere*

Solid — Freezing 80 cal/g ← Liquid — Condensation 590 cal/g ← Gas (invisible gas)

Melting → Evaporation →

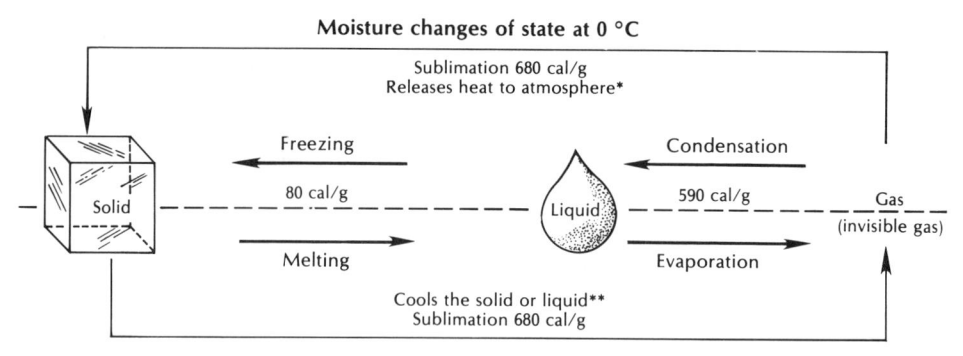

Cools the solid or liquid**
Sublimation 680 cal/g

*Applies to changes shown above broken line
**Applies to changes shown below broken line

Heat transfers associated with phase changes of water

Phase Change	Heat Transfer	Type of Heat
Liquid water to water vapor	540-590 cal absorbed	Latent heat of vaporization
Ice to liquid water	80 cal absorbed	Latent heat of fusion
Ice to water vapor	680 cal absorbed	Latent heat of sublimation
Water vapor to liquid water	540-590 cal released	Latent heat of condensation
Liquid water to ice	80 cal released	Latent heat of fusion
Water vapor to ice	680 cal released	Latent heat of sublimation

FIGURE 4-1 Changes of state of water (From Navear OO-8OU-24. Government Printing Office).

whereas in reality, the majority of the surface of the planet is covered by water. Second, people need and use the water in lakes and streams and assume that an unlimited supply will always be there. The planetary water supply is dominated by the oceans. Ninety-seven percent of all water is in the ocean (figure 4-2). The ocean covers 71 percent of the surface of the earth, more than double the surface covered by land. The climatological significance of this lies in the fact that the major interface between the surface of the earth and the atmosphere is a water surface, not a land surface. The ocean contains a volume of water equivalent to some 3000 years' evaporation, and average residence time of water in the sea may be as long as 5000 years.

While the bulk of all water is in the ocean, water is continually cycling through the environment in ways that involve change of state and change of position (figure 4-3); and, as Applied Study 4 shows, such changes have an important effect on water supply. Evaporation and condensation account for about two-thirds of the heat exchange between the earth's surface and the atmosphere. To maintain the present temperature, about 880 mm of evaporation and condensation must occur each year. It is the constant supply of solar energy reaching the ocean that powers

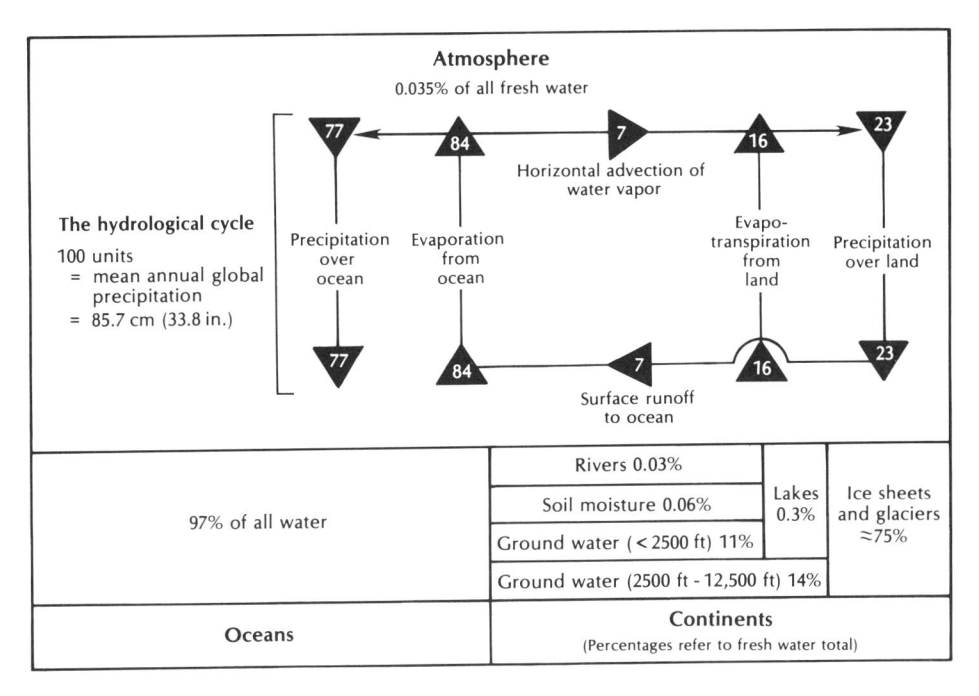

FIGURE 4-2 Estimates of water transfer within the hydrologic cycle. Exchanges and storage units of the cycle assume a mean of 100 units of global precipitation (From More, R., "Hydrologic Models and Geography," in *Models in Geography*, Chorley R. J. and P. Haggett, eds., 1967, p. 146, Methuen. Used by permission.).

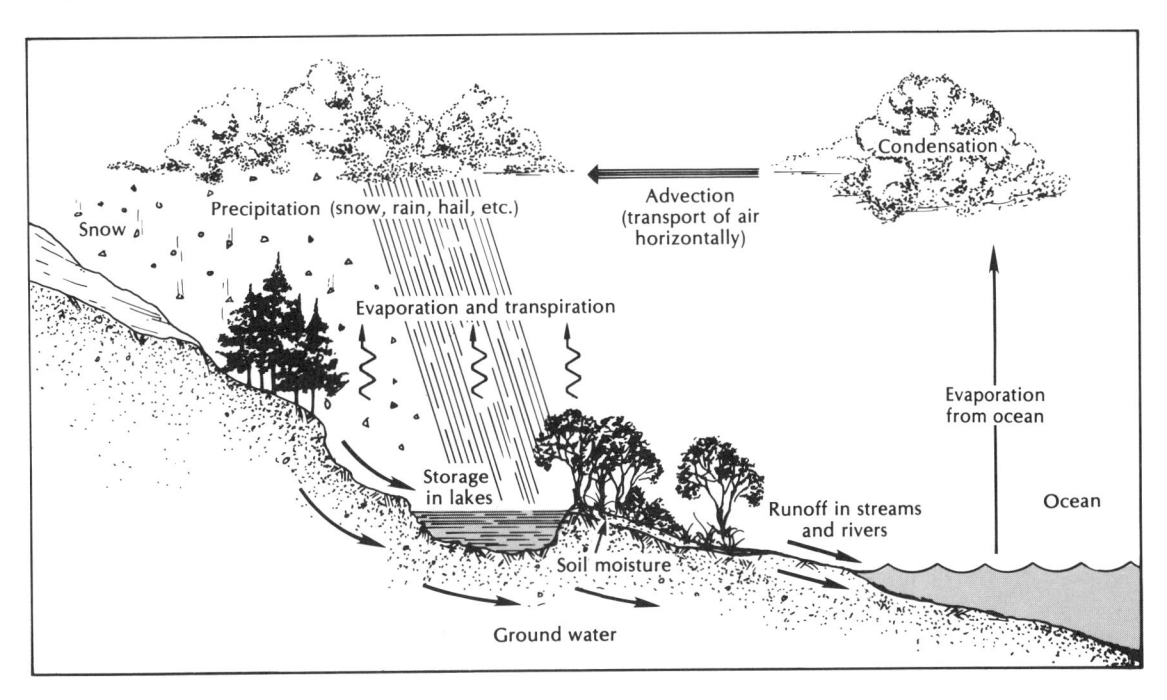

FIGURE 4-3 The hydrologic cycle.

the evaporation process. As moisture is lifted away from the surface, energy is needed to overcome the force of gravity; the source of energy is again solar radiation. The atmosphere does not contain a large amount of water at any time. Less than 0.1 percent of the total planetary supply of water is normally found in the atmosphere. The average *precipitable water* (the total depth of water that would result if all water in a column of air were condensed) is approximately 25 mm. Water in the liquid and gaseous forms is transported by the atmosphere from place to place, which uses more energy. Of the amount of water present in the air at any one time, only a small proportion precipitates. In mid-latitudes the amount is less than 25 percent. The average residence time of water in the atmosphere is only 10 days.

Water on the land has a varying residence time, depending on the route it travels. If it evaporates from the surface, the residence time may be measured in minutes, hours, or days. If the water is taken up by plants, it will be transpired within a few hours or become part of the plant tissue. If the water becomes part of a major stream, it will probably reach the ocean in no more than 3 weeks. Should the water enter the ground water reservoir, residence time could be measured in hours or thousands of years. Eventually water returns to the ocean, and the cycle begins again.

ATMOSPHERIC HUMIDITY

It was stated earlier in this chapter that less than 0.1 percent of the planetary water supply is found in the atmosphere at any one time. Water vapor in the atmosphere becomes part of the gaseous mix, and there is a limit as to how much water vapor the atmosphere can hold. The upper limit is not a constant value but varies with the air temperature. Thus warm air can hold more water vapor than can cold air. Further, the amount of water vapor that the air can hold not only increases as the temperature increases but increases at a geometric rate. Table 4–1 lists the amount of water vapor the air can hold for selected temperatures. For example, it indicates that air can hold 3.8 g/kg at 0°C, or freezing. At 10°C, it can hold twice as much as at 0°C. On a warm summer day when the temperature is 30°C, the air can hold more than ten times as much water vapor as on a winter day when it is −10°C.

There is an adage in northern regions that states, "It's too cold to snow." In theory it is not correct; but in the real world, there is a lot of truth in the statement. As temperatures drop below freezing, the capacity for the atmosphere to hold moisture goes down extremely rapidly. It can be further surmised that most heavy snowfalls occur when air temperatures are near freezing.

A number of different measures of atmospheric humidity are used. Some of these are given in table 4–2. The most commonly used measure is *relative humidity*: the ratio of the amount of water vapor in the air to what the air can hold at a given temperature. This suggests that much of the time the atmosphere is not saturated with water vapor, and such is the case. The actual amount of water vapor present in the air can vary from nearly zero up to the saturation level, and in some cases a little higher than that.

TABLE 4-1 Saturated mixing ratio of water vapor (at 1000 mb pressure).

Temp. (°C)	Vapor (g) per Kg Dry Air	
	Over Water	Over Ice
50	88.12	...
45	66.33	...
40	49.81	...
35	37.25	...
30	27.69	...
25	20.44	...
20	14.95	...
15	10.83	...
10	7.76	...
5	5.50	...
0	3.84	...
−5	2.644	(2.518)
−10	1.794	(1.627)
−15	1.197	(1.034)
−20	0.7847	(0.6456)
−25	0.5048	(0.3955)
−30	0.3182	(0.2375)
−35	0.1963	(0.1396)
−40	0.1183	(0.0803)
−45	0.0695	(0.0450)
−50	0.0398	(0.0246)

AIR MASSES

In chapter 3, the manner in which air temperatures vary from place to place was considered. It was established that the moisture content of the atmosphere also varies substantially from place to place. These differences in temperature and moisture have led to the development of the concept of *air masses:* large bodies of air of considerable depth that are relatively homogeneous in terms of temperature and moisture. They form over large areas of water or land that are fairly uniform in temperature. The overlying air assumes the temperature and moisture characteristics of the surface.

Two basic categories of air masses are recognized on the basis of temperature, and two are based on moisture. Air masses are classified as *tropical* if the source is in low latitudes and *polar* if the source region is in high latitudes. Air masses originating over land masses—and therefore relatively dry—are labelled *continental air masses;* those originating over oceans, being relatively moist, are labelled *maritime air masses*. This categorization identifies four individual kinds of air masses: maritime tropical (mT), maritime polar (mP), continental tropical (cT), and continental polar (cP). Two additional categories are sometimes used in reference to the

TABLE 4-2 Measures of atmospheric moisture content.

Vapor pressure (e)	The partial pressure of water vapor in the atmosphere. $$e = \frac{W}{W + E} \times P$$ W = mixing ratio; E = a constant (0.622); P = air pressure. Units: e.g, millibars
Mixing ratio (W)	Ratio of mass of water vapor to a unit mass of dry air. $$W = \frac{M_v}{M_d}$$ M_v = water vapor mass; M_d = mass of dry air. Units: e.g., g/kg
Saturation values	Saturation vapor pressure (e_s) can be visualized as the equilibrium attained between a water (or ice) surface and the air contained in a closed box. Saturation mixing ratio (W_s) is the ratio of a mass of water vapor in saturated air with respect to the same mass of dry air. $$W_s = \frac{M_{vs}}{M_d}$$ M_{vs} = water vapor mass for saturated air
Specific humidity (q)	The ratio of the mass of water vapor to a given mass of air. $$q = \frac{622e}{P}$$ e = vapor pressure; P = air pressure. Units: e.g., g/kg
Absolute humidity (P_w)	The mass of water vapor per unit volume of air. $$P_w = \frac{P}{RT}$$ P = air pressure; R = gas constant for water vapor; T = temperature. Units: e.g., g/m³
Relative humidity (RH)	The ratio of actual mixing ratio to saturation mixing ratio. $$RH = \frac{e}{e_s} \times 100$$

extremes of the continental polar air masses and the maritime tropical air masses. Continental arctic (cA) indicates exceptionally cold, dry air; equatorial (E) indicates very warm, moist air. More details about air masses are given in part three, where they are used to describe regional climates.

EVAPORATION AND TRANSPIRATION

Evaporation is the process by which water changes from a liquid state to a gaseous one. Water is sufficiently volatile in solid and liquid states to pass directly into the gaseous state at most environmental temperatures. The process of change from ice to water vapor is called *sublimation*. Since the source of water vapor is at the earth's surface, the amount of water vapor present in the atmosphere decreases with height. Most atmospheric moisture is found below 10,000 m.

The amount of water that actually evaporates from a given water surface in a given time depends upon the following factors:

Vapor pressure of the water surface This factor is directly related to water temperature: the higher the water temperature, the greater the surface vapor pressure. When water temperature is greater than the air temperature, evaporation will always take place.

Vapor pressure of the air The greater the vapor pressure of the air, the less evaporation there will be. The rate of evaporation varies directly with the difference between the vapor pressure of the water surface and the vapor pressure of the air.

Air movement Air movement is usually turbulent, with moist air being removed from near the water surface and replaced by dry air from above. Evaporation thus varies directly with the velocity of the wind: the higher the wind velocity, the greater the evaporation.

Salinity of the water Dissolved material reduces the rate of evaporation: the greater the salinity, the lower the rate of evaporation. The rate of evaporation from sea water is about 5 percent lower than from fresh water.

Depth of the water In deep water in mid-latitudes, the temperature tends to stay near 4°C at depths below approximately 6000 m. Because of the high specific heat of water, the temperature changes above 4°C involve huge amounts of heat absorbed by the water during warming or released by cooling. The temperature lag of deep-water lakes and the sea is greater than that of the air, an important factor in determining evaporation rates from free water surfaces. In late fall and early winter, the vapor pressure of the water is greater than the vapor pressure of the air, since the water temperature is likely to be substantially greater than the air temperature. However, in spring and summer, the opposite situation more often exists. During the summer months, the air is warmed more rapidly than water, so that the vapor pressure of the water is less than that of the air, and evaporation is not as great as in winter. Visible evidence of this process is the steam seen rising from lakes, streams, and the ocean on very cool mornings and before the rivers and lakes freeze in the winter.

Rates of evaporation can be measured directly by the use of evaporation pans or other instruments. Evaporation can also be calculated when the relevant variables can be measured. Figure 4-4 shows the distribution of evaporation over the ocean and for some land areas.

The analysis of terrestrial water balances is not quite as simple as that of water bodies. For land areas the net atmospheric transfer of water vapor depends upon the amount of water available and the processes of evaporation and transpiration. The term *evapotranspiration* combines the processes by which water is trans-

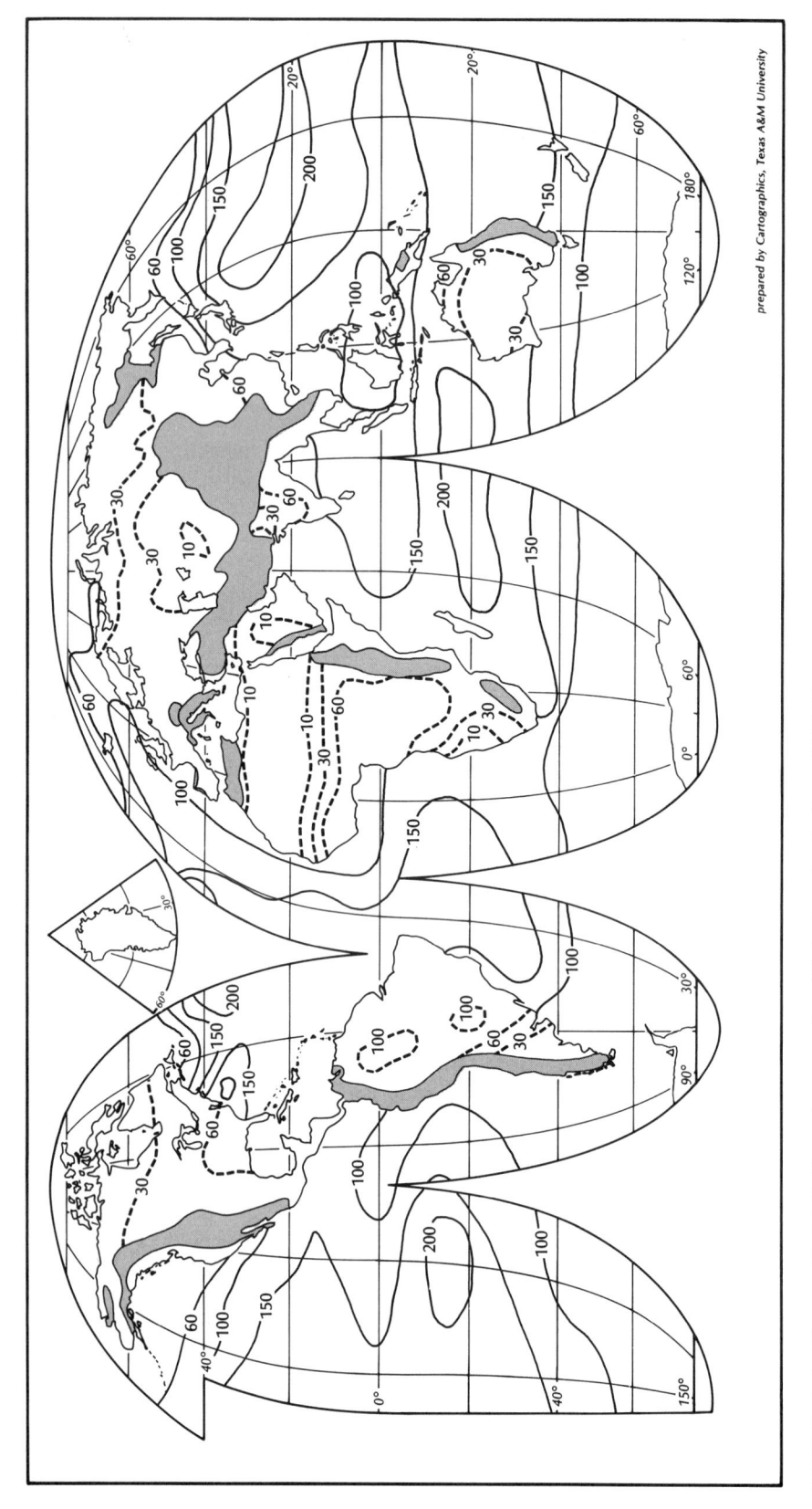

FIGURE 4-4 The world distribution of evaporation (Data from R. G. Barry [1969]).

ferred to the atmosphere from a mix of water, soil, and vegetation. The term is used when evaporation and transpiration cannot be effectively separated. Evapotranspiration is affected by a wide variety of local factors. Among them are the following:

1. Radiation intensity
2. Atmospheric temperature
3. Atmospheric dewpoint
4. Length of day (photoperiod)
5. Wind velocity
6. Type of vegetation
7. Soil moisture conditions
8. Type of precipitation

It is very difficult to measure evapotranspiration because of the many factors related to it. It is also difficult in many cases to isolate the effects of one or two variables to determine their effect on the total of evapotranspiration. Generally methods for estimating evapotranspiration fall into one of three categories: (1) theoretical methods utilizing physical principles of the process, (2) analytic approaches based on the balance of energy amounts, and (3) empirical methods based on both measured amounts and local climatic characteristics.

Of the methods drawing upon physical principles, the best known is that devised by Penman. His equation for the determination of evapotranspiration uses vapor pressure, net radiation, and the drying power of air at a given temperature. To derive such a formula, Penman had first to obtain equations to express both net radiation and drying power. The result is a rather complex set of equations. However, their use is simple; and the method is now used as a basic measure by researchers of the United Nations.

Thornthwaite's method is probably the best known and widely used empiric method, at least in the United States. The system evolved in such a manner that its use is based essentially upon the availability of temperature data, for variables used are mean temperature and daylight hours. Of major significance in the Thornthwaite method is the use of the concept of *potential evapotranspiration* (PE), the maximum amount of water that can be removed from the surface, if an unlimited supply of water is available. This would be equivalent to evaporation from a water surface. If water is limited, actual evapotranspiration (AE) will be less than potential evapotranspiration. The Thornthwaite system, especially as applied to local water budgets, is outlined in Note 6 of Appendix 1.

Measurement of evapotranspiration is essentially confined to agricultural research stations because of the difficulty of instrumentation. Figure 4–5 shows an evapotranspirometer, which measures evapotranspiration, and a lysimeter, which can be used to measure either the potential or actual rate. Drawing upon values derived from such instruments, the relative accuracy of estimating equations can be assessed. In the example shown in table 4–3, it is evident that the Thornthwaite equations, derived for use in a humid middle-latitude climate, underestimate evapotranspiration in a dry climate.

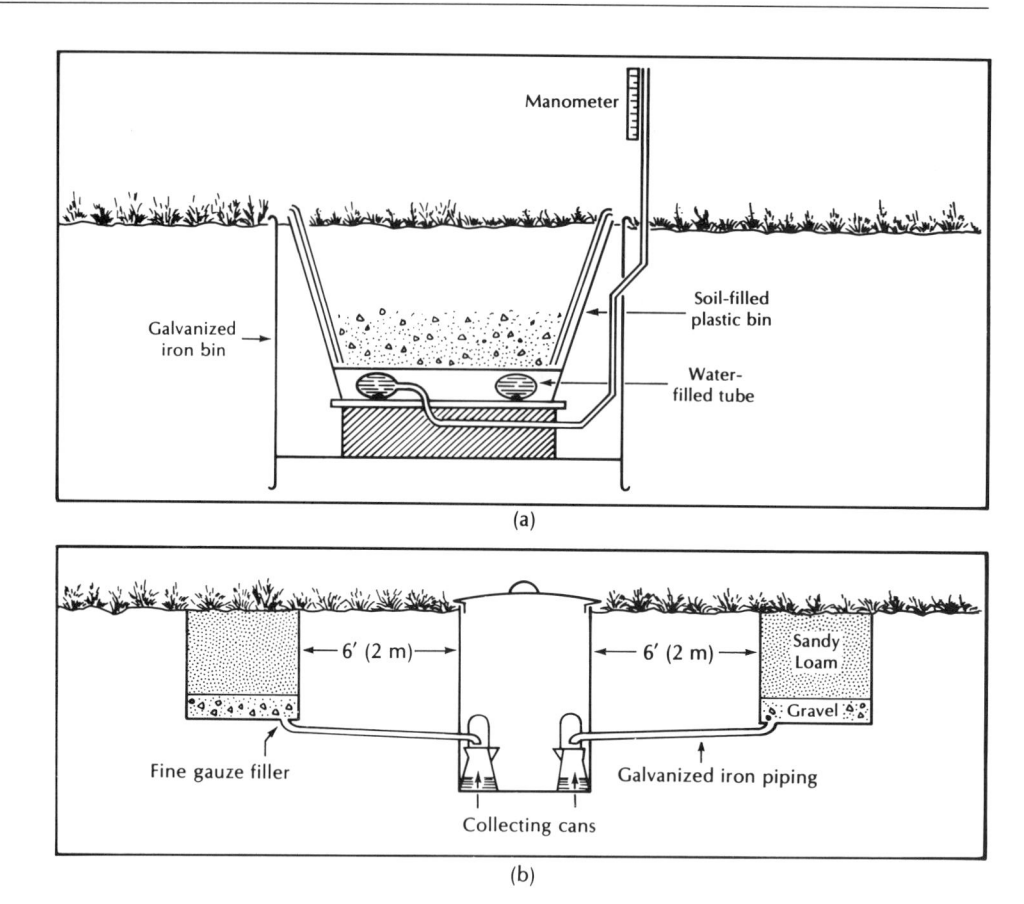

FIGURE 4-5 Sections through (a) a weighting lysimeter and (b) an evapotranspirometer (From *Climate and Man's Environment*, John E. Oliver, Copyright © [1973, John Wiley & Sons]. Reprinted by permission of John Wiley & Sons, Inc.).

Evidence indicates that most evapotranspiration takes place in different air masses than those from which precipitation occurs. It is also fairly well established that moisture added to the atmosphere over a land mass affects only the microclimate of the atmosphere and does not reappear as precipitation in contiguous land areas. Studies of water vapor transport over the eastern United States indicate that in excess of 95 percent of precipitable moisture comes directly from the Gulf of Mexico or from the Atlantic Ocean.

Evapotranspiration from a mixed land-and-water surface is nearly always less than evaporation would be from a similar-sized water surface. This phenomenon is well illustrated on a global scale by the fact that for the continents, evapotranspiration is some 470 mm per year; while from the ocean, it is 1300 mm per year. Average evapotranspiration from the continents varies a great deal through time and space. The major variables are the amounts of water and energy available. In tropical areas where there is ample water and energy, evapotranspiration rates are very high.

TABLE 4-3 Measured and estimated potential evapotranspiration at an Australian station.

Month	(a) T(°C)	(b) Measured Evapotranspiration (mm/day)	(c) Estimated Potential Evapotranspiration (mm/day)
Jan.	23.3	7.76	4.47
Feb.	21.1	5.62	3.51
Mar.	19.6	4.21	2.80
Apr.	17.2	2.94	2.04
May	12.6	1.31	1.09
June	10.9	0.99	0.80
July	10.0	0.93	0.74
Aug.	11.0	1.47	0.89
Sept.	13.0	2.30	1.50
Oct.	15.8	4.07	2.09
Nov.	17.8	5.26	2.73
Dec.	10.1	5.99	3.47
Annual total		1296	793

(a) Temperature
(b) Based upon observed values of seven lysimeters
(c) Calculated potential evapotranspiration, using the Thornthwaite method

In the lower Amazon valley and the central Congo River basin, rates of 1200 mm per year very nearly approach that of evaporation over the open ocean. In parts of the Atlantic and Gulf Coastal Plain of the United States, the amount is almost as high. Evapotranspiration is probably greatest in the Sudd and in the Chad basin in sub-Saharan Africa. Here the rate may reach 2400 mm per year, far in excess of local rainfall. These two extensive areas of swamp and shallow lakes are supplied by rivers. The Sudd is fed by the White Nile, and Lake Chad is fed by a series of rivers from the south. Solar radiation is intense and the air dry, factors that contribute to evapotranspiration. On the other hand, where temperatures are lower (such as in Northern Europe), evapotranspiration rates drop to as little as 200 mm.

CONDENSATION

The reverse process of evaporation is condensation. Condensation will occur when air becomes saturated. There are basically two means by which air can become saturated; one is by cooling the air, and the other is by adding additional water. In the atmosphere, cooling processes are predominant in producing condensation. When air is cooled, the ability to hold moisture decreases; if cooled enough, it will become saturated (table 4-1). The dewpoint is the temperature at which saturation of a given air mass will occur.

Dew and Frost

Dew, a common form of condensation in the environment, forms when the air at the surface cools by conduction until it reaches the dewpoint; the surplus water changes from the water vapor to liquid state and adheres to the cool surface. Frequently the air above ground level will still be above the dewpoint, so condensation does not take place directly in the free air. Dew forms on the ground and on solid objects before condensation occurs in the air (e.g., as fog) because the ground cools more rapidly than the air. By the same token, dew forms on automobiles and other metal objects first because metal cools more rapidly than soil or vegetation.

Dew generally results from nocturnal radiation cooling. Just as the atmosphere is heated from below, it is also cooled from below. Condensation on the ground takes place most readily in still air and in a clear atmosphere. Wind and cloud cover retard the formation of dew. Dew can form at any temperature above the freezing mark; in the tropics, it often forms at temperatures as high as 21°C.

Frost is condensation that forms on solid surfaces when the dew point is below freezing. In the formation of frost, water vapor sublimates as a solid on a solid, passing directly from a gas to a solid. Frost can form at any temperature below freezing. Occasionally, vegetation is damaged when the temperature drops to freezing or below, but there is no visible evidence of frost. These frosts, called *black frosts* or *dry frosts,* are not true frosts, as there is no condensation of moisture. Humans have attempted to alter the atmospheric environment to make it more comfortable and more economically profitable. Frost suppression in areas where commercial fruits and vegetables are grown is economical, as the costs of atmospheric modification are less than the cost of losing the crop to an unseasonable frost.

Frosts occurring from advection tend to be widespread, heavy, and persistent. These frosts are of such a nature that present methods of reduction are not sufficient, and damage still results. Radiation frosts of short duration and local extent can be reduced, and techniques for this purpose are available at the present time.

Fog and Clouds

While condensation occurs without solid material upon which to condense, it requires subdewpoint temperatures. The atmosphere normally contains hygroscopic particles in sufficient amounts to provide nuclei for condensation. When these hygroscopic particles are not present, the relative humidity can be increased several hundred percent before condensation will begin. The hygroscopic nuclei vary a great deal in size. The smaller are ions, and the largest are raindrops.

The *degree of saturation,* or relative humidity, at which condensation will occur varies with the number, size, and kind of nuclei present. Condensation will take place most readily when there are large numbers of particles present, and particularly when the particles are large and hygroscopic in nature. The more favorable the nuclei present, the lower the relative humidity necessary to produce condensation. Most condensation in the atmosphere takes place on the larger nuclei (which are also often the more absorbent), such as smoke particles, dust, and salt. Condensation is a process of continuous growth or accretion of water on such nuclei.

Initial condensation may form particles microscopic in size, though, of course, the large nuclei will form larger particles. The larger nuclei are extremely important because by the time relative humidity reaches 100 percent, the particles are already the size of fine drizzle—large enough to precipitate in fairly calm conditions. This effect is sometimes seen when large fires occur. If fire-fighting equipment pours large quantities of water on a fire, some of the water is vaporized and carried upward by the heated air. The ashes and other particulate matter aloft make ideal nuclei, and condensation may begin at levels of 80–85 percent relative humidity. As the moisture cools, it can readily condense; and a mist can be observed downwind from the fire. This process actually dissipates heat from the area of the fire.

Fog is condensed water droplets in the lower atmosphere. There are basically three kinds of fogs: radiation fogs, advective fogs, and steam fogs. *Radiation fogs* are warm-weather phenomena and result from cooling moist, stagnant air by radiation cooling. When the air is still, or nearly so, fog will be patchy and low. Air movement mixes the air enough to distribute the fog to greater depth and more evenly over the surface. *Advective fogs* are a product of moving moist air over a cool surface. These fogs often occur with January thaws in the northern United States as warm air flows north over snow-covered land. This type of fog is responsible for the fogs common in the Golden Gate of San Francisco. In this case, warm, moist air from the Pacific Ocean flows across the cold Alaska current just offshore, creating the thick fog. The third type of fog is the *steam fog* that occurs over water. The fog in this case is literally steam from a high rate of evaporation. These fogs are common in fall over water in middle and high latitudes.

Fog dispersal is becoming a more and more necessary form of weather modification as time goes by. Fog has a major impact on transportation networks, and delays caused by fog become quite costly in terms of time and money. Techniques for dispersing fogs have been experimental since World War II, and some methods are now economically practical.

Clouds consist of water droplets formed by condensation aloft in the atmosphere. Condensation in the air near the ground surface is usually a product of cooling by radiation and conduction, but the primary cause of condensation aloft is adiabatic cooling. When a parcel or column of air is lifted in the atmosphere, it expands because there is a reduction of pressure on it from the surrounding air. The expansion is accompanied by cooling of the air, and the cooling takes place at a fixed rate. When a mass of unsaturated air is lifted through the lower atmosphere, it cools adiabatically at a rate of 1°C for each 100 m it rises. If unsaturated air is lifted far enough, the air will be cooled to the dewpoint, and condensation in the form of clouds will take place. Saturated air cools adiabatically just as unsaturated air does, but condensation releases latent heat of evaporation; and the net rate of cooling of the rising air is reduced below that of unsaturated air. The average wet adiabatic rate (saturated air) is 0.6°C for each 100 m of rise, though the actual rate depends upon the amount of condensation that takes place.

As explained in Appendix 1, Note 5, air rising through the atmosphere can become either unstable or stable, depending on the temperature and density characteristics of the rising air and the surrounding air. The unstable, or buoyant, condition

PHYSICAL AND DYNAMIC CLIMATOLOGY

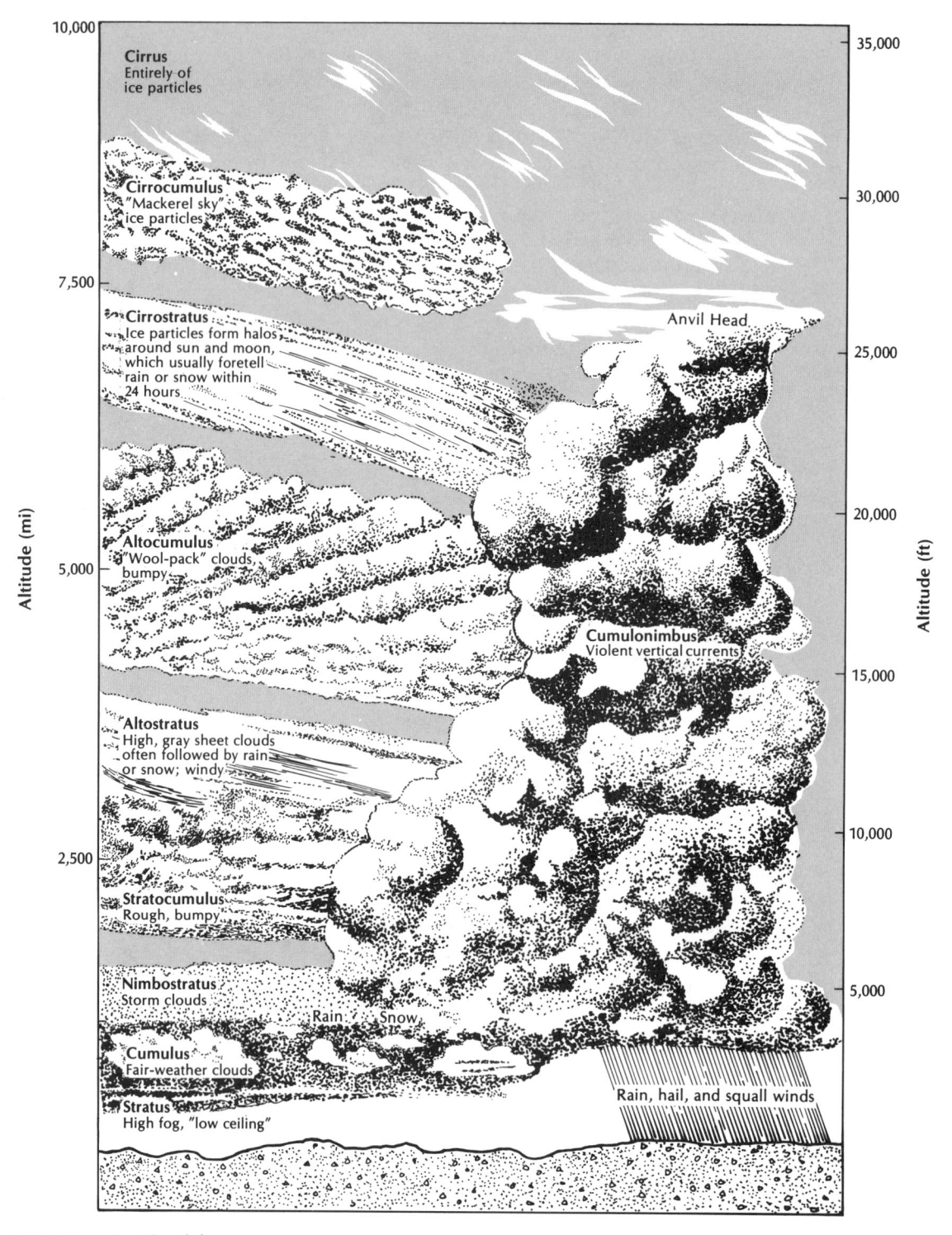

FIGURE 4-6 Cloud forms.

occurs when the density of the rising air is less than that of the surrounding air. When it occurs, air will continue to rise, just as an ordinary cork placed at the bottom of a container of water will rise to the top. If the density of the rising air is the same as that of the surrounding air, then the air will cease to rise and become stable. The notions of stability and instability are indications of the tendency for vertical motion to exist within the atmosphere.

The form and shape of clouds are illustrated in figure 4-6. Essentially, the type of cloud that forms depends upon the type of lifting that gives rise to the cloud. Thus *stratiform* (layered) clouds are indicative of more stable conditions than cumuliform clouds. Droplets are continually forming by condensation at the cloud base, and evaporation at the top and sides of the cloud formation is such that there is a fairly constant cloud layer existing. In a small cumulus cloud, it is likely that a droplet will remain in existence only a matter of minutes. In a column of slowly rising air, such as in stratiform clouds, individual cloud particles may exist for an hour or more.

The surface-atmosphere energy balance is greatly affected by the presence of clouds, as illustrated by figure 4-7. Clouds reduce the amount of direct solar

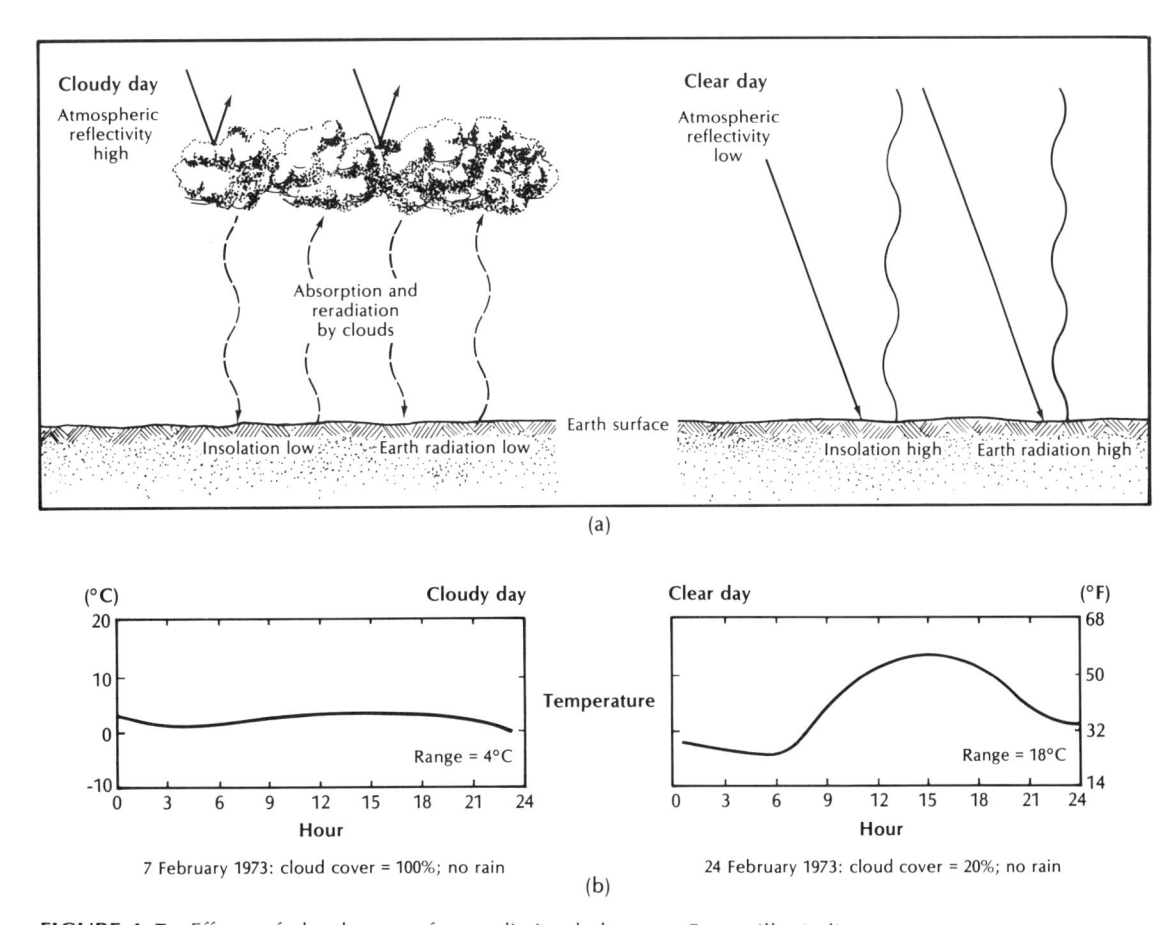

FIGURE 4-7 Effects of clouds on surface radiation balance at Evansville, Indiana.

radiation reaching the surface as shown in figure 4-7(a) but increase counterradiation of long-wave radiation. This process results in a very small change in temperature from day to night. When clouds are absent, as in figure 4-7(b), both solar radiation reaching the ground and earth radiation to space increase, resulting in a greater diurnal range in temperature.

SUMMARY

Water vapor is the most abundant of the variable gases found in the atmosphere. Water in the liquid form is also present in the form of fog, clouds, and precipitation. It has a number of characteristics that make water in the atmosphere extremely important in the energy balance of the earth, both in terms of the earth-atmosphere energy balance and in terms of the movement of heat from equatorial areas toward the poles. The hydrologic cycle is both a means of moving water physically from place to place and a means of moving energy from one part of the environment to another. The evaporation process is a method of cooling the earth surface and placing water vapor into the atmosphere. Condensation releases the latent heat of evaporation to the atmosphere, providing one of the major means of heating the earth's atmosphere. Condensation on the ground in the form of dew and frost is a mechanism that helps retard cooling.

Clouds result from condensation brought about by cooling associated with the lifting of masses of air. Clouds play a major role in both the earth-atmosphere energy balance and the earth-space energy balance. They absorb and reradiate earth radiation, keeping long-wave energy from escaping from the lower atmosphere. Clouds are the most efficient reflector of solar radiation found on the planet. Just as they keep earth radiation in, they keep solar radiation out.

APPLIED STUDY 4

The Hydrologic Response to Winter Conditions

During the winter season in North America, precipitation falls as both rain and snow. Temperatures vary from one winter to the next; as a result, the amount of precipitation that falls as snow varies from winter to winter. In a relatively warm winter, a greater percentage of the precipitation will be in the form of rain than in a cold winter. The hydrologic impact of changes in the amount of snowfall from year to year is potentially very large.

The water content of snow depends on several factors, but on average, 250 mm of snow melts to 25 mm of rain. Implicit in these data is the fact that a small change in the amount of precipitation that falls as snow has a marked effect on the total amount of snow that occurs. For each 25 mm of precipitation that falls as snow instead of rain, snow depth increases by 250 mm. If 100 mm of precipitation is converted to snowfall by cooler temperatures, the snow depth is increased by 1 m.

When winters are below normal in temperature, the amount of precipitation tends to decline as the result of cooler atmospheric temperatures and subsequent lower absolute humidity of the air. This relationship is illustrated in figure 4-8. The temperatures for four successive Januarys over the Kankakee River watershed in Indiana are shown in figure 4-8(a) and the resulting amounts of precipitation in 4-8(b). The pattern of each of the graphs is similar. The temperature declines are accompanied by a drop in precipitation. At the colder times, more of the precipitation fell as snowfall, due to the colder temperatures; and snowfall increased from less than 100 mm to over 700 mm in the same 2-year period (figure 4-8[c]). January, 1978, set a record for snowfall in the area. Temperatures rebounded from the low of 1977 but were still below freezing by more than 5°C. The amount of precipitation increased substantially; and because temperatures were below the freezing mark, most of the precipitation fell as snow. Much of the snow was quite dry, which increased the snow depth. The ratio of snow to water reached almost 20:1.

Analysis of the runoff on the Kankakee River watershed indicates that distinct changes in flow accompany these hydrometeorological changes of rain to snow. The lowest runoff normally occurs during the months of August and September, with October and November having only slightly higher flow in most years. Beginning in December, runoff increases slowly, as soil moisture is replenished and evaporation and transpiration decrease. The peak runoff on the Kankakee usually occurs during the months of March and April, when precipitation is abundant and soil moisture levels are high after the soaking of the winter months. A secondary rise often occurs during the months of January through March because of the melting of snow, if any accumulation has taken place. The change in precipitation affects the flow of the Kankakee River in three apparent ways. One, total runoff follows the pattern of temperature and moisture: the warmer and wetter the winter, the more runoff. Two, the colder it is and the more snow there is, the later that spring high water occurs. This effect is shown

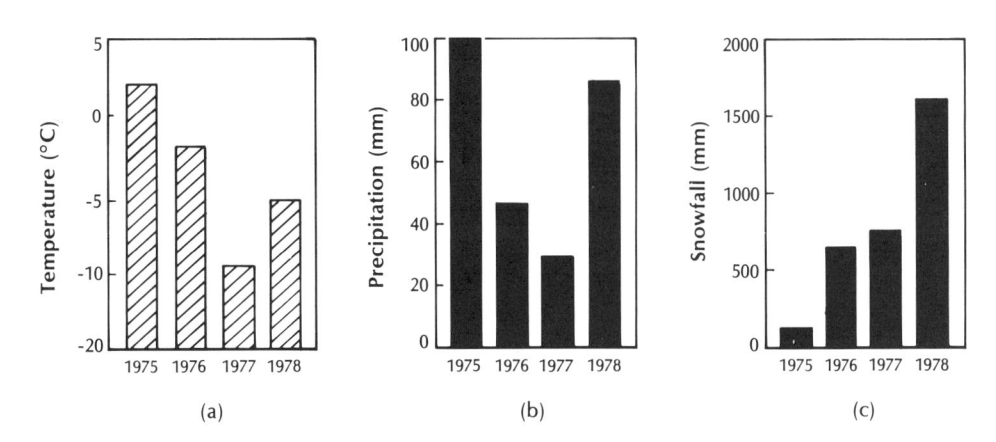

FIGURE 4-8 Temperature, precipitation, and snowfall during the winters of 1975 through 1978.

PHYSICAL AND DYNAMIC CLIMATOLOGY

FIGURE 4-9 Dates of the winter runoff peak on the Kankakee River at Davis.

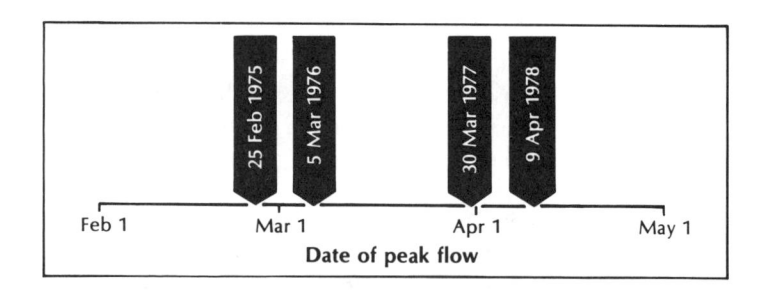

FIGURE 4-10 Yield of the Kankakee River watershed during the winter half-year 1974–1975 through 1977–1978.

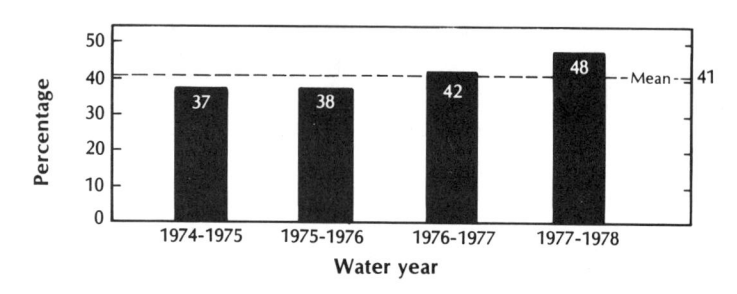

clearly by the sequence in figure 4-9. The third response is an increase in *yield*, the proportion of total precipitation that appears as runoff, depending on temperature and amount of snowfall. Yields increased from 37 percent of precipitation in the winter half-year to 48 percent of precipitation in 1978 (figure 4-10). The underlying explanation for this increase is the reduced evapotranspiration of the colder winters. Later applied studies will deal with the human impact of the cold and snow.

5 Precipitation and the Water Balance

VERTICAL TRANSPORT OF WATER VAPOR
Convective Lifting
Cyclonic Lifting
Orographic Lifting
SPATIAL VARIATION IN PRECIPITATION
TEMPORAL VARIATION IN PRECIPITATION
Floods
Droughts
REGIONAL AND LOCAL WATER BALANCES
SUMMARY
APPLIED STUDY 5
The Impact of Snowfall

PHYSICAL AND DYNAMIC CLIMATOLOGY

Clouds are comprised of water droplets that are microscopic in size. They are so small that the friction between the droplets and the rising air is sufficient to overcome the force of gravity. This is because the *terminal velocity* (the constant velocity reached by a falling object when the pull of gravity is balanced by the drag on the object) of a cloud droplet of 10-μm diameter is 1 cm/s. Assuming it did not evaporate, it would take such a droplet more than 80 h to fall 3 km. The precipitation process thus depends upon the growth of droplets in the cloud. The way in which this occurs depends upon the temperature at which the growth takes place.

In clouds with temperature above freezing, the *collision-coalescence* process, which is basically a joining of water drops when they meet, occurs. In laboratory experiments colliding droplets bounce off each other unless an electrical field is introduced. When a field is present, coalescence of the colliding particles is favored if the droplets have opposite electrical charges (figure 5-1). The earth's electrical field and atmospheric electricity thus play an important role in the collision-coalescence process. The growth of the droplet is enhanced when different sizes are involved. The different terminal velocities cause the larger droplets to overtake those that fall more slowly. Precipitation from warm clouds is most important in low-latitude regions.

In most rain clouds outside of the tropics temperatures are below the freezing level, and a different process operates. Not all water becomes ice at below freezing temperatures. Water that remains liquid at below freezing temperatures is known as *supercooled water*. In a cloud, above the freezing level, a zone occurs in which ice crystals and supercooled water exist side by side. The saturated vapor pressure is less over ice, when both are at the same temperature. For example, in saturated air at a temperature of 10°C, the relative humidity of the air over supercooled water is 100 percent; over ice at the same temperature, it is 110 percent. An inequality such as this cannot persist, and moisture will move from supercooled drops to the ice crystal. In this way the ice crystal will grow, its terminal velocity will increase, and it will begin to fall to the ground. Frequently, in passing through warmer air, the ice will melt and reach the ground as a water droplet. This *Bergeron-Findeisen process* is illustrated in figure 5-1(b).

Precipitation from clouds can take a number of forms. It can fall as rain when it is in the water state or in a variety of forms when it is in the solid state. The latter are differentiated in a number of ways and include the following:

Ice pellets: Sometimes called *sleet*, small ice pellets are produced when falling rain freezes as it passes through a lower level of cold air.

Snow: Snow consists of ice crystals grown directly by condensation of water vapor in clouds that are well below freezing.

Hail: Rounded lumps of ice in which the internal structure, formed in layers like an onion, indicates sequential growth into a nucleus.

PRECIPITATION AND THE WATER BALANCE

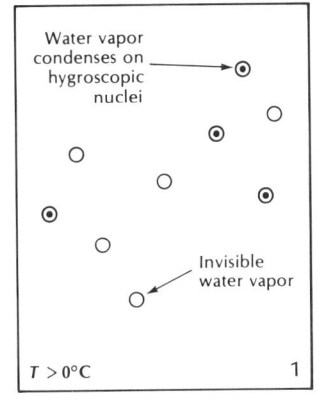

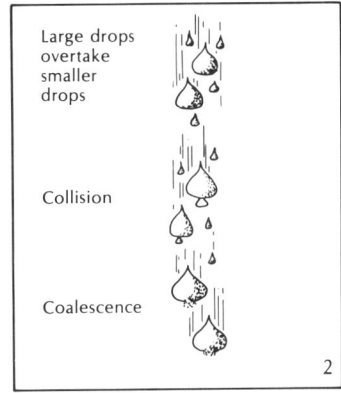

(a) Growth of water droplets by collision and coalescence

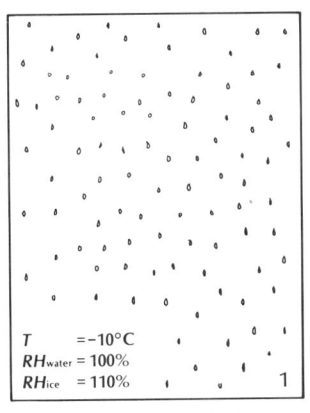

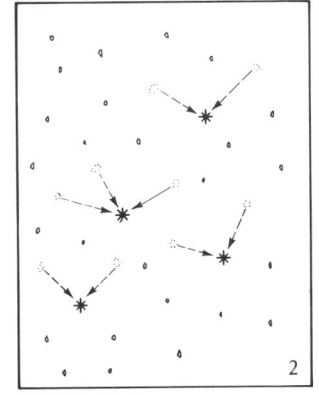

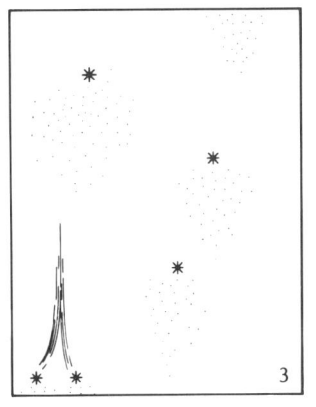

(b) Growth of ice crystals at the expense of water droplets

FIGURE 5-1 Precipitation processes in (a) warm and (b) cold clouds (From R. A. Anthes, John J. Cahir, Alistair B. Fraser, and Hans A. Panofsky. *The Atmosphere*, third edition [1981, p. 154]. Used by permission of Charles E. Merrill Publishing Co.).

VERTICAL TRANSPORT OF WATER VAPOR

For precipitation to occur, there must be some mechanism present to transport large volumes of moist air upward. There are three primary mechanisms for lifting the air that account for most precipitation: *convective*, *frontal*, and *orographic*. Orographic precipitation is primarily cyclonic or convective in nature but is intensified or

PHYSICAL AND DYNAMIC CLIMATOLOGY

increased by topographic lifting along coastlines, hills, or mountain ranges. In some cases orographic precipitation is produced by lifting stable air over high barriers.

Convective Lifting

Convective lifting is typical of warm, moist air and is instigated by heating from the ground surface. When the surface is very warm, the air immediately above is heated. It expands in response to the heating and, with a lower density than the surrounding air, becomes buoyant. If the air becomes buoyant or unstable, the lifting may continue until condensation and precipitation occur. Convective precipitation is highly variable in intensity, producing light showers as well as some of the heaviest downpours that occur in the atmosphere. Thunderstorms are the result of extreme convection; their violent manifestations will be discussed in greater detail in chapter 7. They will be considered here as precipitation producing storms, which develop in warm, moist air over both water and land.

There are several conditions that must be present for thunderstorms to develop. The absolute humidity and relative humidity must both be quite high near the surface, the lapse rate must be such that the air is close to becoming unstable, and some initial lifting mechanism must be present. Intensive heating of the surface provides the initial lifting mechanism in many thunderstorms. As the air is heated from the surface, it expands; large masses of less dense air rise away from the surface to be replaced by cooler air from surrounding areas. Once the rising mass of air becomes buoyant, the development of the convectional cell becomes a continuous process, with several recognizable stages (figure 5-2) through which the storm passes.

First is the *cumulus* stage. During this stage of growth there is a general updraft within the cloud, and condensation of water droplets takes place. Precipitation does not occur because the water particles are extremely small and are held aloft by the updraft. The second, or *mature,* stage develops only if instability continues in the rising air mass. It is at this stage that copious condensation occurs; precipitation begins within the cloud and may subsequently reach the ground. The last major

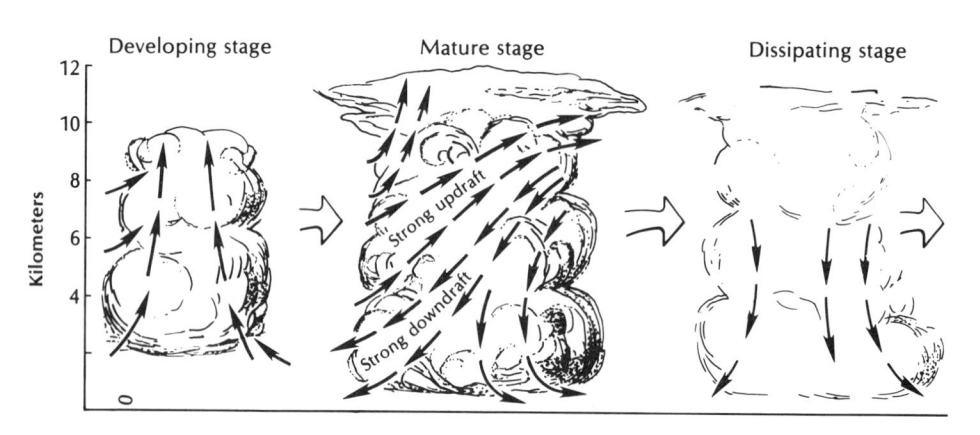

FIGURE 5-2 The three-stage development of a thunderstorm.

PRECIPITATION AND THE WATER BALANCE

stage in the life cycle of the thunderstorm begins when the downdraft spreads over the entire cloud. This process marks the *dissipating* stage because there is no longer moisture being carried aloft. Condensation slows and precipitation stops.

Cyclonic Lifting

Along the boundary between polar air streams and tropical air streams eddies often form and develop into a low pressure storm referred to as a *mid-latitude cyclone*. Figure 5-3 shows a plan view of the development of one of these cyclones. These cyclonic storms are primarily responsible for the day-to-day variation in weather in mid-latitudes. They may produce rapidly changing temperatures, wind direction, and humidity. A well-developed low will move across the United States or Canada in 3-5 days. In figure 5-4 the location of the respective fronts associated with one low pressure system, which crossed the United States 17-20 March 1971 are shown.

These lows are a major means of producing precipitation. The precipitation is brought about by lifting of air along fronts associated with the low pressure systems. The boundary between two unlike air masses is referred to as a *frontal surface*, and the line of intersection between the frontal surface and the ground as a front (figure 5-5). The *leading*, or advancing, edge of a cold air mass is a cold front. As cold air moves into warmer areas it moves in underneath the warm air it displaces because it is more dense than the warmer air. The warm air is forced aloft; if sufficient moisture is present, condensation and precipitation result. When a fast moving cold front displaces unstable moist air, the weather at the surface may

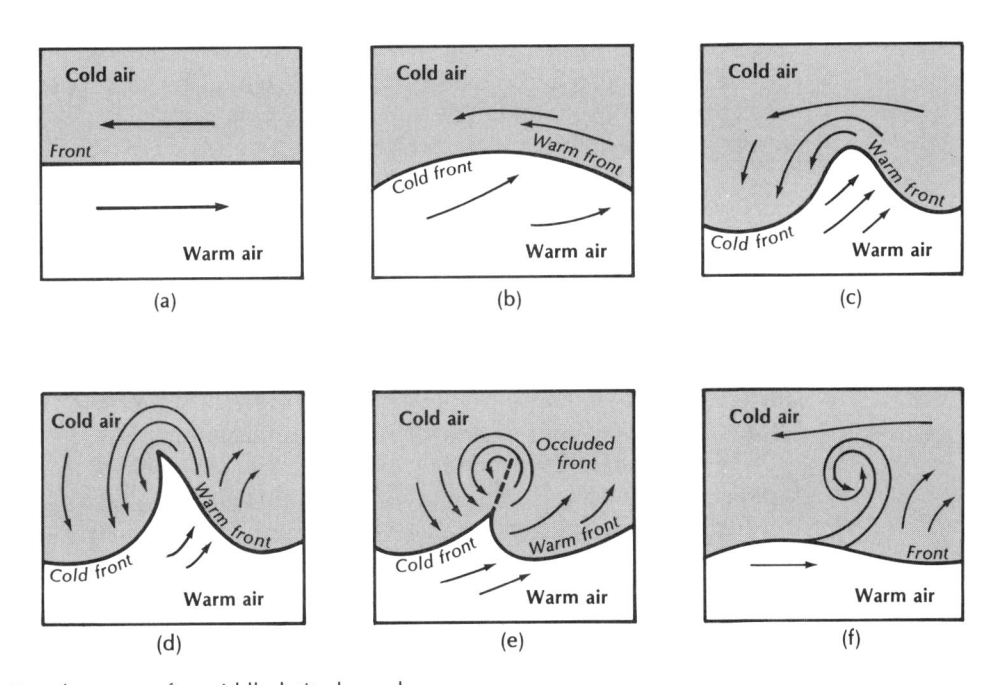

FIGURE 5-3 Development of a middle-latitude cyclone.

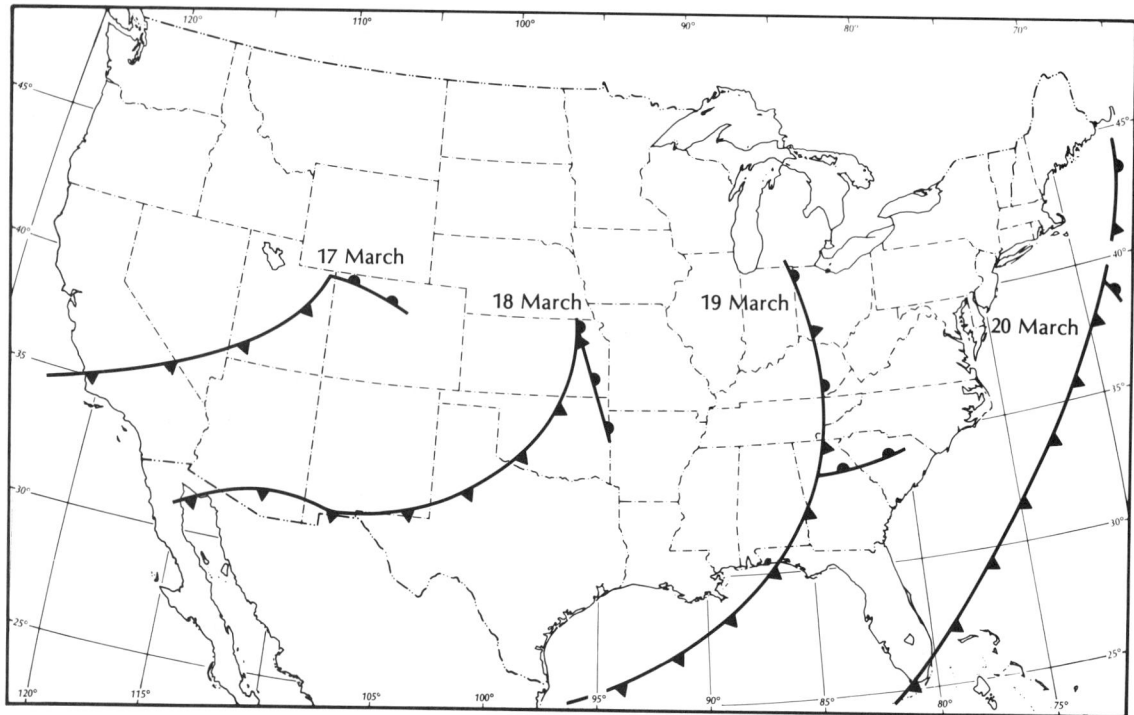

FIGURE 5-4 The movement of a mid-latitude cyclone across the United States: 17-20 March 1971.

become quite turbulent, with gusty winds and intense thunderstorms. The passage of a cold front is often accompanied by a wind shift, rising atmospheric pressure, and decreasing relative humidity. A slow moving front displacing stable warm air may result in a slow clearing of the weather without any precipitation.

A warm front develops when the leading edge of a warm air mass displaces colder air at the surface. Its lower density prevents it from penetrating a cold air mass from underneath. Instead, it overruns the cold air. The weather associated with the warm front is more widespread and persistent than cold front weather. The air is lifted more slowly, and as it cools, clouds, which may exist as much as 1000 km in advance of the surface front, form. The forward movement of a warm front can be observed long in advance of the surface front by the gradually thickening and lowering clouds and the development of rain or fog near the front. As the warm front passes, the temperature rises, wind shifts, precipitation stops, cloudiness decreases.

Two other types of fronts associated with cyclonic lifting are the *occluded front* and the *stationary front*, which usually bring poor weather conditions as the air is lifted continuously over an area and may result in prolonged precipitation.

Although these lows do not follow prescribed routes around the world, they do follow certain paths more frequently than others. In North America more low pressure systems move eastward along the United States and Canadian border than any other route. (Applied Study 5 provides an example of how an unexpected storm passage in winter can have several social and economic influences). In Europe the

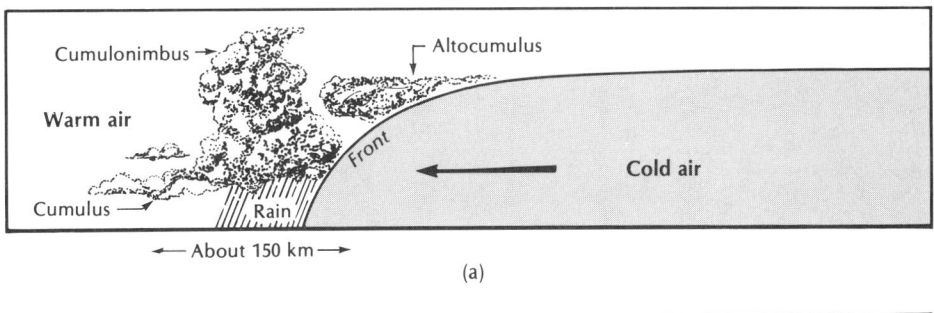

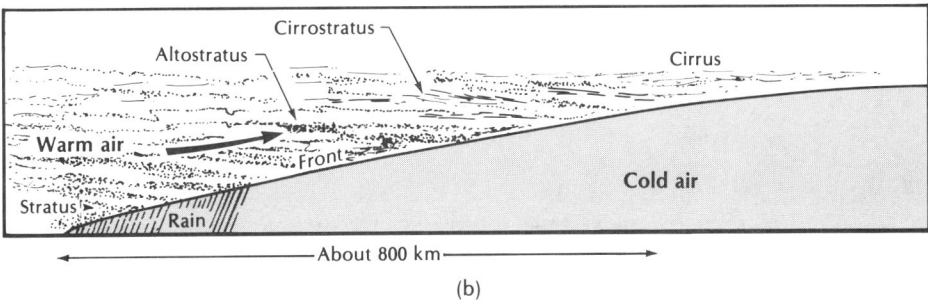

FIGURE 5–5 The characteristics of (a) a cold front and (b) a warm front.

most frequent path is across the British Isles and northern Germany. During the winter season these lows have a higher frequency and intensity, and travel further south. In North America they will travel as far south as the Gulf Coast. In Europe they often traverse the Mediterranean Sea. Some of the larger ones will move across southwest Asia from the Mediterranean Sea, bringing precipitation to the deserts of Saudi Arabia and Iran. Few low pressure systems cross central Asia in winter because of the strength of the high pressure system there at that time of year (see chapter 6). In the late winter or early spring, these storms may cross the center of the Asian land mass, producing blizzard conditions known as the *buran*.

In the summer both the intensity and frequency of the low pressure systems are reduced. As a consequence, summer weather is somewhat less changeable and atmospheric disturbances are less violent. For the most part, the lows in the Southern Hemisphere are stronger than those of the Northern Hemisphere as a result of the air masses' being of greater contrast and similar throughout the year.

Orographic Lifting

Orographic precipitation is convective or cyclonic precipitation that is enhanced by topographic features. This kind of precipitation is often associated with mountain barriers, but the same enhancement occurs in hills and along lakeshores and seashores. Observations have shown that precipitation in mountains appears to increase to a given height and then to decrease again. In Java, for example, the maximum precipitation occurs at an elevation of about 1219 m. Similarly, in the Guatemalan Highlands, the maximum occurs between about 900 m and 1200 m,

PHYSICAL AND DYNAMIC CLIMATOLOGY

TABLE 5-1 Record intense rainstorms.

Time	Amount	Place	Date
1 min	31 mm	Unionville, Md.	6 July 1956
5 min	63 mm	Panama	1911
8 min	126 mm	Bavaria, Germany	1920
15 min	198 mm	Jamaica	1916
30 min	235 mm	Guiana, Va.	24 Aug. 1906
42 min	305 mm	Holt, Mo.	22 June 1947
2 h, 10 min	483 mm	W. Va.	1889
3 h	555 mm	D'Hanis, Tex.	31 May 1935
12 h	813 mm	Thrall, Tex.	9 Sept. 1921
24 h	1.87 m	Cilaos, Reunion Island	15–16 Mar. 1952
5 days	3.81 m	Cherrapunji, India	Aug. 1841
31 days	9.30 m	Cherrapunji, India	July 1861
1 year	26.46 m	Cherrapunji, India	Aug. 1860–July 1861

with lower amounts both above and below this zone. The reason for this difference is probably the different concentrations of moisture at different levels in the middle altitudes and the tropics. Much orographic precipitation in the tropics is derived from cumuliform clouds that have tops below 2700 m. Water droplets in the clouds are probably concentrated near the base of the clouds, so that most of the precipitation takes place at lower levels. In mid-latitudes, much of the precipitation is associated with stratiform clouds. These clouds may extend to considerable heights, and water droplets within them are spread over greater depths.

Cherrapunji, in the Assam Hills of India, is one site where orographic intensification reaches extremes. During the summer when the onshore wind is strongly developed, the flow of moist air over the hills triggers convectional showers. The steady replenishment of water vapor provides the basis for downpours of rain that reach truly torrential proportions. The process led to an unbelievable 26.46 m of precipitation between August 1860 and July 1861. Cherrapunji actually received 3.81 m of rain in a 5-day period in August 1841 (table 5-1).

SPATIAL VARIATION IN PRECIPITATION

The amount of precipitation received at the surface varies because of a variety of factors. The basic element is the amount of water vapor in the air, which varies geographically and seasonally. The mean precipitable water content of the atmosphere at a given moment is 25 mm, with a maximum near the equator of 44 mm and a minimum in the polar regions of 2–8 mm, depending on the season. In latitudes of 40° to 50°, it will range upwards of 20 mm in the summer and drop to around 10 mm in the winter. The presence of water vapor is a necessary but not a sufficient condition for precipitation. There is no direct relationship between the amount of atmospheric water vapor over an area and the resulting precipitation. To illustrate this a comparison can be made between conditions over El Paso, Texas, and St. Paul, Minnesota. The average moisture content above these cities is about the same, and yet the mean annual precipitation is more than three times greater at St. Paul. Other factors must come into play to induce precipitation.

As already noted, if the total amount of precipitation received over the surface of the earth were spread evenly, it would average 880 mm per year. It varies from near zero to almost 12 m. Figure 5-6 shows the mean precipitation over the globe. The total amount of precipitation received depends upon several factors:

1. Whether air *converges* (to give uplift) or *diverges* (spreads out) in the area.
2. Air mass origin, which is an indication of the temperature and moisture conditions of the air.
3. Topographic conditions.
4. Distance from the moisture source: the greater the distance from the source of the moisture the less water vapor will be present in the air because of prior precipitation loss.

Combining the above factors, the areas where precipitation is greatest are mountain areas of the tropics, where there is frequent convergence of air from the ocean. Where such conditions exist, rainfall may reach 12 m per year (table 5-2).

Two general locations where precipitation totals tend to be above average for the earth are found relatively far apart. One is near the equator, where there is a zone experiencing convergence of moist tropical air most of the year. These are the areas underlying the low pressure convergence area caused by the Intertropical Convergence (ITC). The trade winds moving toward the equator pick up moisture over the oceans and, when lifted in the ITC, yield abundant moisture. Precipitation is

TABLE 5-2 World record extremes of precipitation.

Longest period without rain	14 years	Iquique, Chile
Lowest average annual precipitation	0.5 mm	Arica, Chile
Greatest number of days per year with precipitation	325	Bahia Felix, Chile
Greatest number of days per year with thunderstorms	322	Buitenzorg, Java
Highest annual average precipitation	11.98 m	Mt. Waialeale, Kauai, Hawaii
Greatest 24-h snowfall (USA)	1.93 m	Silver Lake, Colo., 14–15 April 1921
Highest one-season snowfall (USA)	28.5 m	Paradise Ranger Station, Mt. Rainier, Wash., 1971–1972

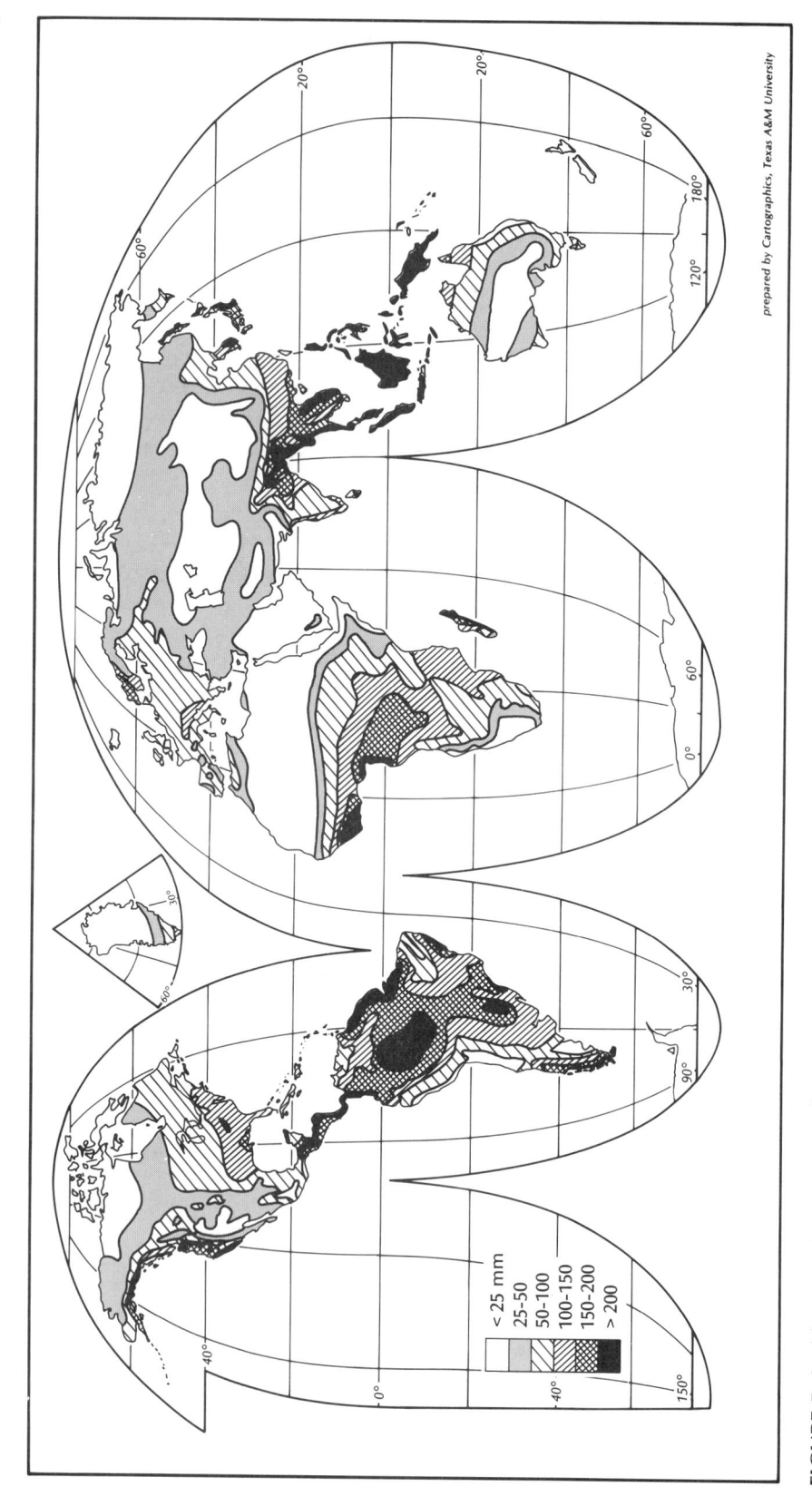

FIGURE 5-6 Mean precipitation over the world.

increased over coastal areas by orographic lifting and increased convection started from surface heating. Average precipitation ranges from 1.5 m to 2.0 m annually but in some cases goes much higher.

The second situation that gives rise to above-average precipitation is found on the west side of the continents in mid-latitudes. Precipitation there is due to convergence of maritime air and orographic intensification. The zone is most well defined in latitudes of 50°–60°. The totals run above 1.5 m along the coasts. The amounts are not as high as in the tropics because the moisture capacity of the air is much lower than that of the maritime tropical (mT) air of the tropics.

The arid regions of the world occur in three great realms:

1. Extending in a discontinuous belt approximately between 20°–30° north and south of the equator. These areas constitute the great tropical deserts, which owe their aridity to large-scale atmospheric subsidence.

2. In the interiors of continents are found the continental deserts, which are arid as a result of their distance from the sea, the major source of the water vapor.

3. The polar deserts of the Arctic and Antarctic constitute the third great area of low precipitation. The low temperatures of these regions, along with subsidence of air from aloft, are probably the most important contributing factor to the low totals.

Mountain ranges play a significant role in the spatial distribution of precipitation. The windward slopes of mountains receive the greatest amount of precipitation. In the lee of the mountain ranges the precipitation decreases markedly to give a rain-shadow effect. The predominant flow of air along the West Coast of North America is from west to east. The mountains produce alternate zones of high precipitation and low precipitation, high on the mountain slopes, and low in the intervening basins and valleys. In California the Coast Ranges are the first barrier to the onshore winds. Precipitation is substantial and forests grow on the windward slopes. Precipitation increases with height to the crest at about 760 m. In the Great Valley precipitation is much lower, and a grassland environment exists. As elevation increases going eastward over the Sierra Nevada Mountains (2600 m), precipitation increases to two or three times that of the Great Valley (figure 5–7). Forests cover the slopes of the mountains near the summit. The air, crossing the crestline, descends slightly and warms adiabatically, and relative humidity drops. Precipitation declines rapidly, and the clouds begin to evaporate. Annual precipitation drops from around 1300 mm at the crest to about 150 mm at Reno, Nevada, giving Reno a desert climate. The arid zone extends from the Mexican border north into Canada between the Sierra Nevada–Cascade Range and Rockies because of the drying of the westerlies as they cross the mountains. The process is repeated as the currents continue to flow eastward over the Wasatch and Rocky Mountains. The Rockies, with a crestline in excess of 3300 m, cause still further drying of the air from the Pacific. Little precipitation falls east of the Rockies from air originating over the Pacific Ocean. The chief moisture sources for the continent east of the Rocky Mountains are the Gulf of Mexico and the Atlantic Ocean.

PHYSICAL AND DYNAMIC CLIMATOLOGY

FIGURE 5-7 Schematic cross-section showing the relationship between elevation and precipitation in western United States.

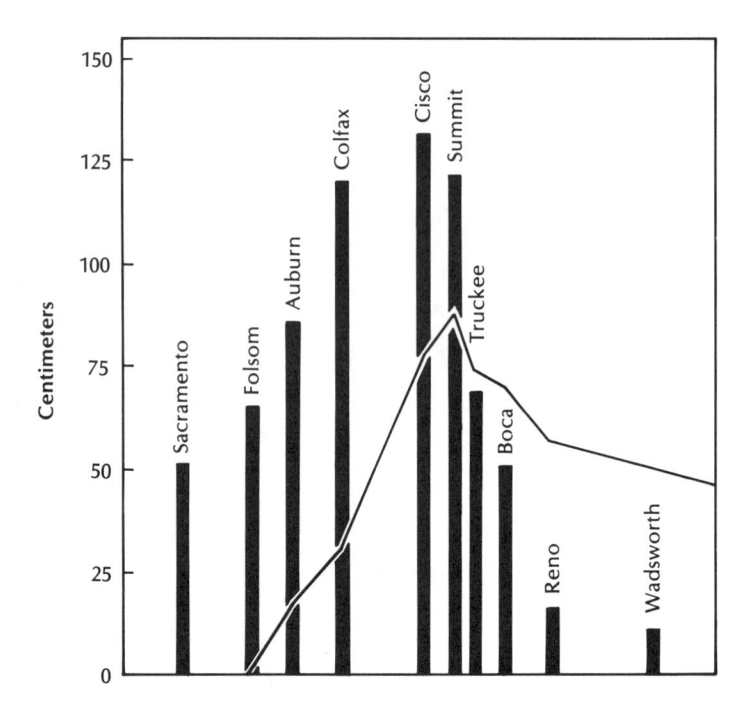

TEMPORAL DISTRIBUTION OF PRECIPITATION

Total annual precipitation alone is an insufficient measure of moisture availability, for it does not take into account the manner in which the precipitation is distributed throughout the year. It is often the temporal distribution that is the dominant factor in determining use of precipitation. There are major discrepancies between the *frequency* of precipitation, measured in terms of the number of days per year on which measurable precipitation falls, and mean annual precipitation. In some parts of the world, precipitation falls nearly every day of the year. Bahia Felix, Chile, averages 325 days per year with measurable precipitation; thus there is an 89 percent chance that it will rain or snow on any given day of the year. Buitenzorg, Java, in the tropics, averages 322 days per year with thunderstorms. At the other end of the scale are those desert areas where rain is an oddity. Arica, Chile, averages only about one rain day annually; at Iquique, Chile, not far from Arica, 14 years passed with no measurable rainfall (1899-1913). It may seem strange that one of the areas with the least frequent rainfall is in the same country as one of the rainiest places in the world. The size and shape of the country, topography of the region, and the general circulation of the atmosphere all play a part in the explanation, which will be discussed further in chapter 10.

Over most of the earth's surface precipitation is of a seasonal nature, with a period of the year when precipitation has a high probability of occurring, and a dry season when the probability is much lower. Surprisingly perhaps, those sites that have the highest annual total rainfall are found in areas with a pronounced dry

PRECIPITATION AND THE WATER BALANCE

season. Cherrapunji, India, is an example. Although an average of more than 10 m of rain falls each year, there are several months when rain is infrequent.

To evaluate the distribution of seasonality of precipitation, a number of index values have been introduced. A *precipitation concentration index,* for example, expresses the monthly concentration of rainfall over a year by using a dimensionless value. Figure 5-8 provides a guide to seasonality on a global scale.

In most parts of the world agricultural activities are adjusted to the rainy season in one way or another. Usually planting takes place near the beginning of the rainy season and harvesting after the rainy season has ended. If the precipitation occurs during the winter season much of it may be in the form of snow. In some locations snow does not become effective moisture until melting takes place in the spring. Much of the precipitation in far western North America occurs in the winter in the form of snowfall. As a result, massive schemes for storing the runoff in the spring and irrigation distribution systems for delivering the water to the fields in the summer have been constructed throughout the west.

Floods

Precipitation is subject to wide fluctuations over time. Greater amounts of precipitation than that which normally occurs in an area will produce floods, while unusually low amounts may bring about drought. Technically, a *flood* may be defined as the condition in any stream or lake when it rises above bank full. Although arbitrary, this is a useful definition of a flood.

All natural-stream floods are due primarily to surface runoff, which may result from heavy rainfall, the melting of snow, or a combination of both. Floods caused by rain can result from either short periods of high intensity rainfall (such as rates of 2.5 cm/h or 25-45 cm/day) or from prolonged periods of steady rains lasting for several days or weeks. Flood runoff from small watersheds usually results from different causes than floods on large drainage basins. Small watersheds are defined somewhat arbitrarily; the average would be 25 km^2 or less. Watersheds of this size can be completely covered by a single convective storm, and most floods on small drainage basins are caused from cloudbursts. The rainfall is so intense that the stream channels cannot carry off the water as fast as it falls. The floods that isolated New Orleans and created havoc in other parts of Louisiana and Mississippi in April 1983 were caused by a series of thunderstorms over a period of several days. The flood that destroyed parts of Rapid City, South Dakota, in 1972 was caused by an extremely intensive storm that covered a very restricted area. The rainstorm lasted less than eight hours and produced flooding on only a few small watersheds. Figure 5-9 shows such a local storm, which struck Kansas City.

On large drainage basins extended precipitation from cyclonic storms or massive snowmelt is required to produce flooding. The Mississippi River floods of the spring of 1973 are a good example. The winter in the upper valley was very wet. The ground was saturated; and when the spring melt occurred, almost all of the water ran off the land surface. To add to the problem, a series of cyclonic storms moved across the drainage basin. The result was one of the highest floods in recent history on the lower Mississippi River.

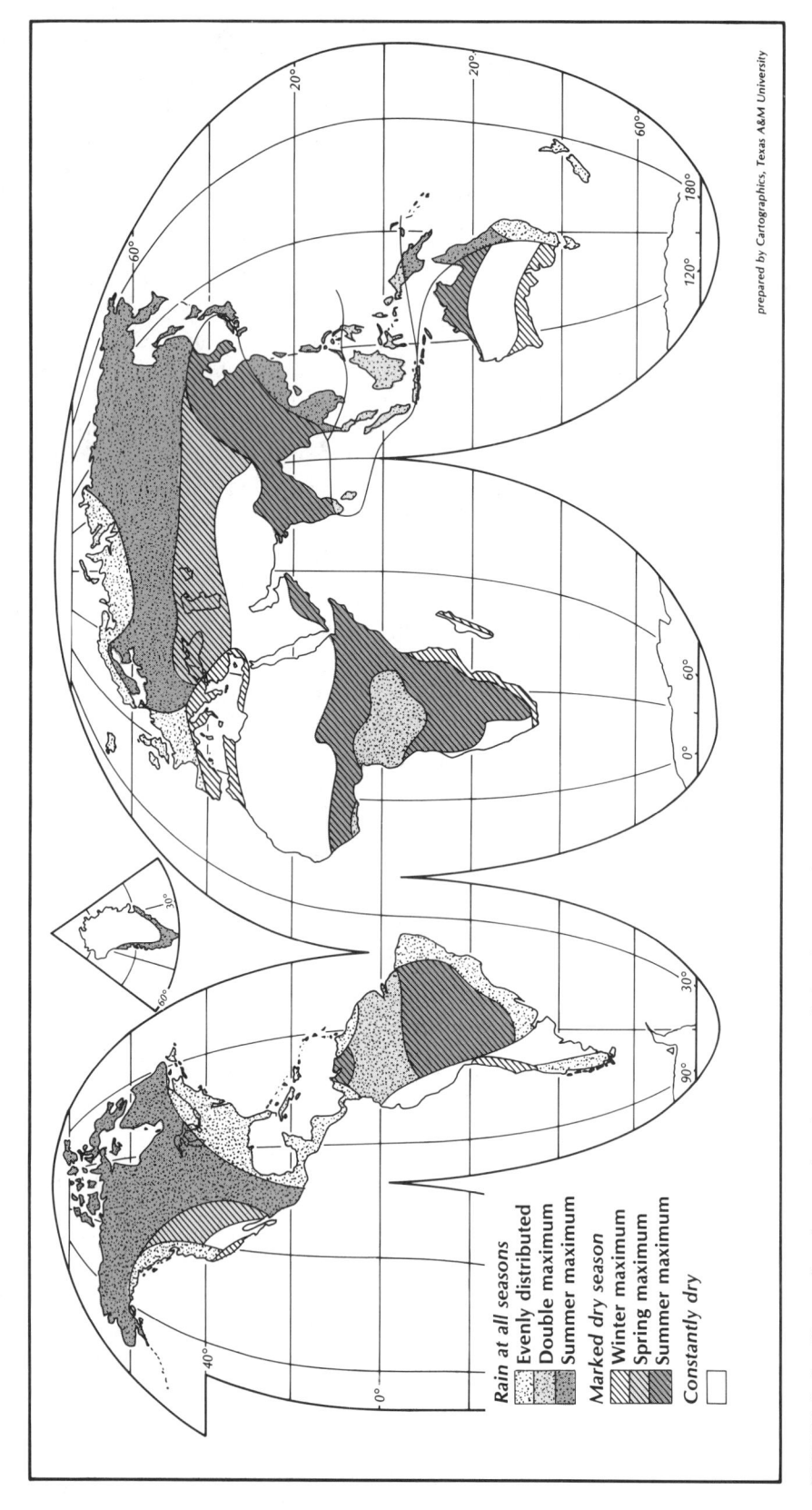

FIGURE 5–8 Generalized map showing seasonal distribution of precipitation (After A. A. Miller, *Climatology*, 1965. Used by permission of Methuen and Co., Ltd.).

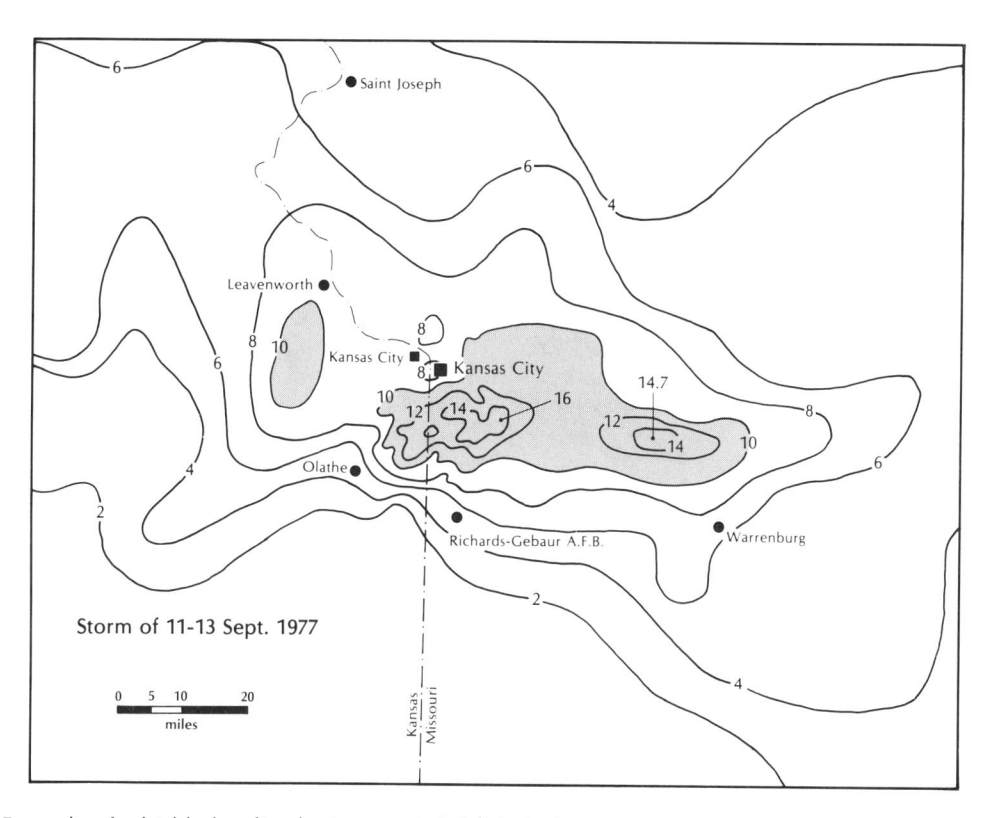

FIGURE 5-9 Example of a highly localized rainstorm (rainfall in inches).

The time of occurrence of floods varies over the earth's surface. All around the Red Sea is a complete network of streambeds that are dry most of the time. During short rainy periods they often flood. The Amazon River basin receives large amounts of rain in all parts some time during the year, and somewhere in the large basin there is usually an area in flood at any given time. The Ganges River of India is frequently flooded during the monsoon season, from April onward, because snowmelt in the Himalaya Mountains and excessive rain combine to produce overflow.

Floods are not a function of modern society; they have always occurred. There are numerous legends of great floods that occurred in prehistoric times, and in North America there is evidence of great floods when the continent was mostly unsettled and forested. There is, however, considerable evidence that flood heights are increasing. If so, either the storms producing the floods are increasing in intensity or the physical characteristics of the drainage basin are changing. There is little evidence of changing storm intensity but a lot of evidence that humans are changing the physical characteristics of drainage basins by changing land use, paving the surface, and expanding cities into the flood plains. Humans have altered the face of the earth primarily by cutting forests and plowing the land for domestic crops.

One result of both the cutting of the world's forests and the cultivation of crops has been increased flood flows. Around the Mediterranean Sea, countless floods in past centuries have been caused by human activities. In 1877, the Hwang

Ho River in China flooded, destroyed 330 villages, took an estimated 7 million lives, and left 2 million survivors homeless. Certainly, there has been no flood so deadly since that time. Among the factors contributing to the flood and its toll in lives were the destruction of vegetation on the drainage basin, the very large population living on the floodplain, and the lack of any kind of warning system. In June 1931, the Yangtze and Yellow rivers flooded in Honan Province, with estimated deaths of over 1 million; some 180 million people were affected by the flood.

Some floods are precipitated directly by structural failures: The Buffalo Creek flood in West Virginia in 1971, the Johnstown flood, and, more recently, the Snake River flood of 1976 are examples. In the Buffalo Creek flood, a dam built across the valley made from the waste from coal mining collapsed during a storm, draining a lake behind it. The Rapid City flood of 1972 was increased by the collapse of a reservoir upstream.

Drought

Drought implies a shortage of water, and yet most areas of the world are accustomed to regular periods of dryness. The three different forms of dryness are based on temporal pattern: perennial, seasonal, and intermittent. *Perennial dryness* characterizes the great deserts of the world, where water is available only incidentally in the form of occasional rain or in the form of an oasis fed by an exotic river or ground water. An *exotic river* is a river, such as the Nile, originating in a humid area and flowing into or through a desert. The oasis of Hofuf in Saudi Arabia is an example of an oasis fed by groundwater. *Seasonal dryness* characterizes areas where most of the yearly precipitation comes during a portion of the year, leaving the remaining weeks or months rainless. Seasonal dryness occurs wherever the climate includes a distinct period of dry weather each year. *Intermittent dryness* occurs whenever the precipitation fails in humid areas or when in areas of seasonal dryness the rainy season does not materialize or is greatly shortened.

Droughts that create a major problem for humans at this stage in our history are not just dry conditions, but abnormally dry conditions; the absence of precipitation when it normally can be expected, and a demand exists for it; or much less precipitation than can be expected at a given time. In some locations where daily rainfall is the usual condition, a week without rain would be considered a drought. In parts of Libya, only a period of two or more years without rain would be considered a drought. Along the floodplain of the Nile River, rainfall is unimportant in determining drought. Prior to the construction of the High Aswan Dam, a drought was any year when the Nile failed to flood. The annual flood provided the soil moisture for agriculture along the river all the way from Khartoum, Sudan, to Alexandria, Egypt. In the monsoon lands of the world, a rainy season producing half the normal precipitation may bring drought; in areas that normally have two rainy seasons, the failure of one would be considered a drought. Since precipitation and water supply vary greatly over the earth, the term *drought* has different meanings in different places. Drought is thus a relative term, and the total rainfall in a place is not a suitable indication of its presence. It is for this same reason that it is so difficult to devise a quantitative index for drought that is widely applicable. Nearly every location has its own criteria for drought conditions.

Drought also has different meanings depending on user demands. There are often distinctions made among meteorological drought, agricultural drought, and hydrologic drought. *Meteorological droughts* are irregular intervals of time, most often of months or years in duration, when the water supply falls unusually far below that expected on the basis of the prevailing climate. *Agricultural drought* exists when soil moisture is so depleted as to affect plant growth. Since agricultural systems vary widely, drought must be related to the water needs of the particular animals or crops in that particular system. There are degrees of agricultural drought that depend on whether only shallow rooted plants are affected or whether deep rooted plants are affected, as well. Both growing season and dry season precipitation can affect crop yields. The Thornthwaite water budget method is often used to assess relative drought. An alternate quantitative index is the Palmer Meteorological Drought Index. The value ranges from $+4$ to -4, as shown in table 5-3. A slightly modified Palmer index is used to develop a generalized map of abnormally dry conditions for crops that is published as part of the *Weekly Weather and Crop Bulletin*.

Droughts are different from some other environmental events such as floods or earthquakes because they are in fact nonevents, since they result from the absence of events that normally occur (Hershfield et al., 1972). Drought often has no distinct onset and takes time to develop. The onset of a drought is often not even recognizable as such. The first clear day following rains may be a grand day: clear, due to the cleansing effects of rain, and with a very fresh and pleasant smell in the air. It is only when a succession of days occurs that concern over water begins to surface. A cry of drought is gradually raised as crops wither, streams dry up, and reservoirs empty.

The temporal and spatial scales of drought vary widely. The duration of droughts varies as much as the timing of their occurrence. How long a drought will last still cannot be predicted, for the same reasons that other irregular oscillations

TABLE 5-3 The Palmer meteorological drought index.

Value	Class (Relative to the Particular Location)
≥ 4.00	Extremely wet
3.00 to 3.99	Very wet
2.00 to 2.99	Moderately wet
1.00 to 1.99	Slightly wet
0.50 to 0.99	Incipient wet spell
0.49 to -0.49	Near normal
-0.50 to -0.99	Incipient drought
-1.00 to -1.99	Mild drought
-2.00 to -2.99	Moderate drought
-3.00 to -3.99	Severe drought
≤ -4.00	Extreme drought

Source: After Palmer (1965).

cannot be predicted with accuracy. Drought simply ends when the rains come and the streams rise. At present we are unable to predict when they will occur or how long they will last or to prevent them from occurring. All that is certain is that they are a part of the natural system and that they will occur again, perhaps with even greater duration and intensity than in the past. They may be very local in extent, covering only a few square miles; or they may be widespread, covering major sections of continents. Even in large-scale droughts, the intensity is likely to vary considerably.

REGIONAL AND LOCAL WATER BALANCES

The basic water balance equation can be stated in terms of the processes at work: evaporation, precipitation, and runoff. For the earth-atmosphere exchange, precipitation equals evaporation. The best estimates place the mean precipitation and evaporation for the earth at about 880 mm per year. It is this transfer of heat in evaporating and condensing this 880 mm of water annually that provides the primary mechanism for heating the atmosphere.

Water is circulating constantly over the earth, though maintaining a basically steady state condition over the entire system. At any one place on the surface, the water balance is a function of the entire global system of circulation of water. The water balance for a particular place can vary a great deal from the average for the planet. The major distinction perhaps is in the difference between the ocean and the continents. The greatest share of the earth's evaporation and precipitation takes place over the ocean. The most important single attribute of the water balance of the oceans is that evaporation exceeds precipitation. Some 1300 mm of evaporation and 1200 mm of precipitation occurs from the sea each year. To balance the system, runoff from the land masses must be added. For the oceans:

$$\text{Evaporation } (E) = \text{Precipitation } (P) + \text{Runoff } (R)$$
$$= 1200 \text{ mm} + 100 \text{ mm}$$
$$= 1300 \text{ mm}$$

For the combined land masses the amount of water exchanged between the earth and atmosphere is substantially less than for the oceans because there is less water available to circulate. The continents must rely upon the transport of water from the sea, namely the excess of evaporation over precipitation that occurs over the oceans.

$$E = P - R$$
$$= 710 - 240 \text{ mm}$$
$$= 470 \text{ mm}$$

where 240 mm depth over the continents equals 100 mm over the oceans because of the difference in areas. The 710 mm of precipitation over land masses is derived from both the water evaporated from the sea and the land. In contrast to the ocean, precipitation is greater than evaporation in the terrestrial environment.

Past earth history reveals that at some stages, when the earth was only a few degrees cooler than at present, glaciers expanded enormously and sea level dropped about 100 m. A substantial cooling would cause the entire circulation system to slow down, with the water taken out of circulation and stored in the form of ice. On the other hand, if temperatures should increase, activity in the cycle would increase. There would be more evaporation, precipitation, runoff, and erosion. Sea level would rise as glaciers melted and added more water to the system. At the present time the system seems to be in relative equilibrium with sea level remaining fairly constant.

There are changes in atmospheric makeup taking place that may alter this balance. The amount of CO_2 is increasing, as is the screen of particulate matter (see chapter 16). A number of scientists have, however, expressed concern that with the changing chemistry of the atmosphere will come a change in the hydrologic cycle. If the nature of the atmosphere is altered enough to change the pattern of radiation reaching the surface of the earth, then it will certainly foreshadow a change in the world water balance. At a more local, immediate level, changes from what is normally expected create a host of problems. Applied Study 5 provides an example of how a heavy snowfall influences everyday life.

SUMMARY

The energy balance of the earth is such that water is present in the atmosphere in all three forms. Of the total amount of liquid and gaseous water, less than 1 percent is found on the continents and in the atmosphere as fresh water and water vapor. The small amount, however, is of far greater importance to life than the proportion would indicate. The world water balance is in relative equilibrium at the present stage of the earth. However, any change in the system might bring about large changes in the atmospheric content of water.

There are fundamental differences from place to place in the water balance. The greatest difference is that between the oceans and the continents. Over the oceans, evaporation exceeds precipitation; and over the combined land masses, precipitation exceeds evaporation. The system is balanced by the movement of atmospheric water over the land masses and the return of an equal amount to the oceans by the many rivers.

At any given place on the land masses of the earth, the water balance may be substantially different from the mean for the continents. The frequency and intensity of precipitation varies greatly from place to place and through time. Where mechanisms that produce precipitation are efficient, rains are torrential. In other areas, precipitation mechanisms fail most of the time and deserts exist. Droughts and floods are evidence of the wide variability in precipitation.

APPLIED STUDY 5

Impacts of Snowfall

Snow can be an economic asset or an economic liability. As examples, one need only consider how the distribution of the skiing industry depends upon the high probability that snow will occur and, on the other hand, how a heavy, unusual snowfall can totally disrupt everyday activities to a surprising extent. It is this latter point that is examined in more detail in this study.

Figure 5-10 shows the average annual snowfall for the eastern United States. For the most part, communities have adjusted their winter activities, transportation requirements, and the like to this average amount of snowfall. Thus it might be anticipated that Minneapolis, which has more than 30 in. of snow a

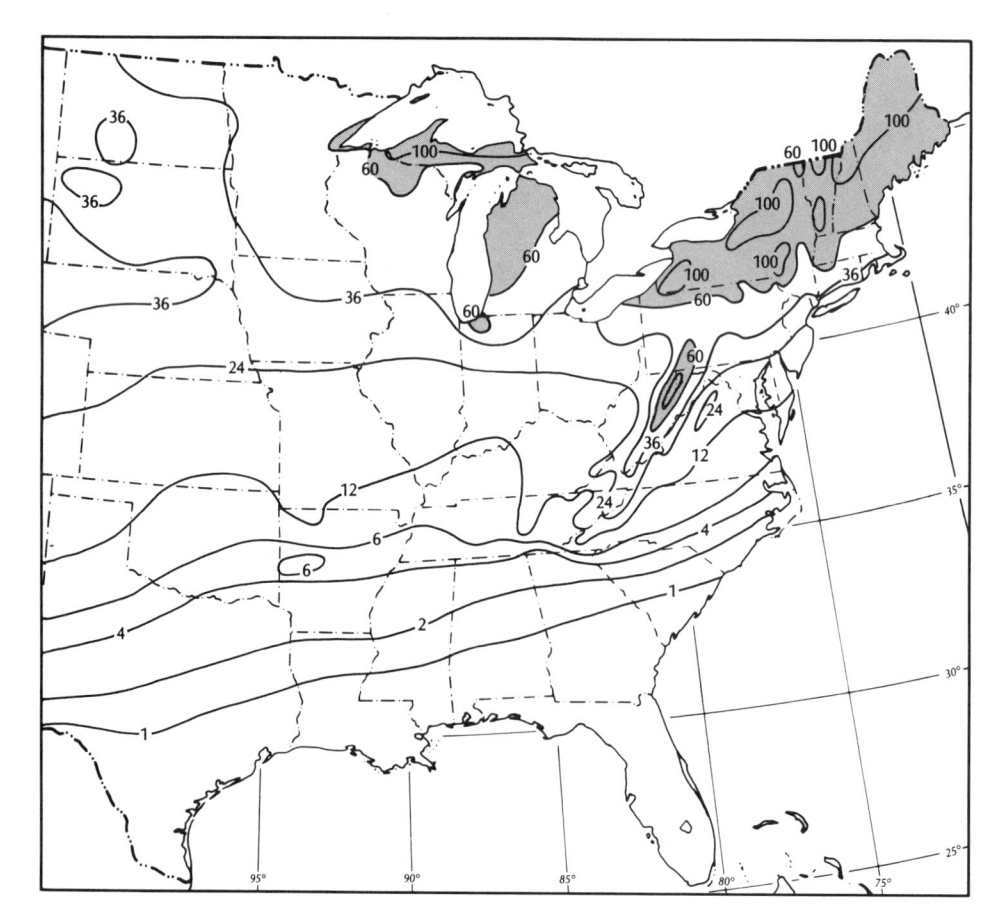

FIGURE 5-10 Average annual snowfall over eastern United States (From *Climatic Atlas of the United States*, Environmental Data Service, Environmental Science Services Administration, U.S. Department of Commerce, 1968).

year, has a system in which snow clearance is a prominent budget item. In Miami, however, the idea of maintaining snow removal equipment is quite unrealistic. Between the two extremes are locations, such as Lexington, Kentucky, where snowfall probabilities are less exact. How to deal with the uncertainty of snowfall is a management problem. It is clear that different locations view snowfall from quite different aspects.

Consider, however, what may occur if there is a large deviation from the normal anticipated snowfall during a single season. Even in areas accustomed to snow, there will be extensive disruption of activities. The extent of this disruption has been cataloged for the state of Illinois. The map on the left in figure 5-11 shows the normal snowfall that occurs in the state. In the winter of 1977-1978, the snowfall was much greater than this average, as shown in the right-hand figure. In places, the normal was exceeded by as much as 50 in. (127 cm), and most places received at least two or three times as much as normal.

FIGURE 5-11 Survey data: (a) mean annual snowfall in Illinois; (b) Illinois snowfall (inches) in the 1977-1978 winter (From S. A. and D. Changnon, *Record Winter Storms in Illinois, 1977-78*. Illinois State Water Survey, 1978.).

TABLE 5-4 Impacts from the severe winter storms of 1977–1978 in Illinois.

1. Transportation systems and vehicles
    Public transportation
        Intercity and in-city buses
            Stopped and delayed, many wrecks and many stuck, greater use of in-city bus service
        Trains
            Delayed, freight trains stuck in snow on branch lines, hauled 50 percent more passengers
        Airlines
            O'Hare Airport at Chicago closed 1 day, central Illinois airports closed 12 days, passengers stranded in terminals
    Other transportation
        Snowmobiles used to rescue motorists
        Helicopters widely used to deliver food and medicine, to take sick people to hospitals, and to rescue trapped motorists
        Automobiles and trucks
            Stranded and stuck by the thousands
            Accidents in the hundreds
    Major roads blocked partially or totally in central Illinois on 12 days
2. Utilities
    Power lines downed, outages, line repairs
    Higher sales of gas and electricity
    Broken water mains
    Unable to get service trucks to line breaks
    Broken telephone lines
    Great use of long-distance services
    Television and radio towers damaged or destroyed
3. Commercial and industrial establishments
    Sales reduced
    Sales of goods delayed
    Stores closed because of power loss
    Purchases of winter gear, clothing, and CB radios increased
    Purchases of space heaters, fireplaces, and firewood increased
    Food stores ran out of supplies
    Service stations sold more gas, did more service work
    Tire stores sold snow tires
    Motel business increased
    Taxi firms bought snow tires
    Roofs collapsed
    Employee absenteeism and layoffs high
    Deliveries of critical materials for manufacturing delayed
    Delays, damaged trucks, perished goods, and business losses hurt shippers
4. Human activities
    Sporting events delayed or canceled
    Stranded travelers by thousands in motels, civic centers, airports, homes
    More sharing with neighbors and others
    Greater use of telephone service and delays in telephone communications
    "Farmer protest" efforts called off and/or delayed

More walking to shops and school
Homes burn—fire protection services unable to reach them
Playing in snow increased
Babies born in homes, restaurants, cars, etc.
Deliveries of goods and services delayed or stopped
    Natural gas deliveries to rural areas delayed, mail between cities stopped, mail to citizens stopped or delayed
    Medical and food supplies delayed, garbage removal halted
    Emergency vehicles delayed or blocked
Many pleasure and business trips delayed or canceled

5. Institutional impacts
Court trials delayed
Illinois license plate deadline delayed
Illinois Emergency Services and Disaster Agency made extensive rescues (80 in one week of January)
Schools closed 7 to 12 days and school extended later in summer
U.S. Postal Service delayed
City, township, county, and state street and highway departments
    Crews overworked and tired, added crew members hired
    Salt supplies exhausted or nearly so, stuck and stranded cars blocked highways and delayed plowing
Lost work cost state in taxable income
Fire departments
    Unable to reach fires, more fires to fight
Police
    Fewer crimes, more accident calls, more minor problems (snowball fights, etc.)

Source: From Stanley A. Changnon and D. Changnon, *Record Winter Storms in Illinois, 1977–1978*. Illinois State Water Survey, 1978.

TABLE 5-5 Injuries and deaths in Illinois due to winter 1977–1978.

Injuries	
Bus accident on January 26	40
Train accident on January 27	317
Car accidents on March 8	47
Car accidents on December 9	24
Car accidents on January 26–28	79
Falls = unknown numbers, estimate	> 2000 injuries
Deaths	
Hit by train (snow blinded)	1
Auto accidents	28
Suffocated in cars	16
Frozen to death	8
Could not reach doctor	3
Fell on ice	1
Shoveling snow	5
Total	62

Source: From Stanley A. Changnon and D. Changnon, *Record Winter Storms in Illinois, 1977–78*. Illinois State Water Survey, 1978.

The resulting impacts were analyzed by Illinois state climatologist Stanley Changnon. His findings are listed in table 5-4 (page 108). This list, though impressive, is probably incomplete. In viewing the list it will be noted that some sectors of the economy actually benefited from the paralyzing snowfall. Service stations reported large profits and motels received increased business from stranded motorists. Some speciality shops, like those selling CB radios and tires, did additional trade. In some cases, the impact on a single industry was mixed. Utility companies, for example, had higher income from sales of electricity but greater costs from damaged lines. Information on injury and death is incomplete, but as table 5-5 (page 109) shows, there was a toll of human life.

The impacts on Illinois were replicated throughout most other midwestern states during the same year. The results clearly indicate problems that arise when weather and climate deviate from the expected norm.

6 Atmospheric Circulation

FACTORS AFFECTING AIR MOVEMENT
Pressure Gradient
Rotation of the Earth
Friction

GLOBAL PATTERNS OF WIND
Tropical Circulation
Mid-latitude Circulation
Polar Circulation
Surface Winds of the World
Seasonal Changes in the Global Pattern

DIURNAL WIND SYSTEMS
Land and Sea Breeze
Mountain and Valley Breeze

APPLIED STUDY 6
Energy From the Wind

In chapter 2 it was pointed out that solar radiation is not distributed equally over the surface of the earth. Between latitudes of about 35° N and 35° S, incoming solar radiation exceeds earth radiation to space; and in the latitudes poleward from these two parallels there is a net loss of energy to space. For incoming solar radiation to exceed outgoing radiation in equatorial areas over a prolonged period, energy must be removed in some other manner. The same is true for polar areas: for outgoing radiant energy to exceed incoming radiation there must be energy transported into the polar regions. The mechanism for transporting this energy is the general circulation of the atmosphere and oceans, and it is a mechanism that tends to equalize the distribution of energy over the surface of the earth.

About 80 percent of the poleward transfer of energy is accomplished by the atmosphere, and the rest by the ocean. This complex circulation system consists of a number of semipermanent areas of convergence and divergence and the airflow in and between them. The energy is transported primarily as sensible heat and latent heat of water vapor. The kinetic energy of the circulation wind systems is eventually converted to sensible heat either by internal friction or by friction with the ground surface, representing an internal energy balance: The rate of kinetic energy generation within the atmosphere is balanced by the energy lost from the system by friction.

That the wind is part of the earth-atmosphere energy system is aptly demonstrated by the fact the wind does represent an energy source for human activity. Applied Study 6 examines some aspects of wind as an energy source.

FACTORS AFFECTING AIR MOVEMENT

Pressure Gradient

Differences in heating and internal motion in the atmosphere produce differences in atmospheric pressure. When such a difference occurs, a pressure gradient is established, and air will move down this gradient from high to lower pressure. Figure 6-1(a) shows a surface pressure situation in which high and low pressures are distinguished. Air will initially move in the direction indicated by the arrows. The rate at which the air moves is determined by the steepness of the gradient. When small differences in pressure occur over large areas, a weak gradient gives rise to weak winds; a steep gradient causes rapid motion, and high winds prevail.

The moving air associated with the pressure gradients will either converge or diverge as it moves from one area to another. Since the air is moving from a high to a low pressure, it follows that areas with low pressure experience convergence. This mergence brings airstreams from different parts of the atmosphere together. As they converge at the surface the only outlet for them is aloft and the air must consequently rise. Conversely, high pressure regions tend to be regions of divergence and, to replace the diverging air, subsidence from aloft must take place. Figure 6-1(b) illustrates these concepts.

FIGURE 6-1 Horizontal and vertical pressure patterns and wind direction (a) Pressure gradient and wind direction. Air flows from high to low pressure, or down the pressure gradient. Data are in millibars (mb). Wind speed is determined by steepness of gradient. Wind speed at (i) will be greater than that at (ii); (b) basic features of convergence and divergence.

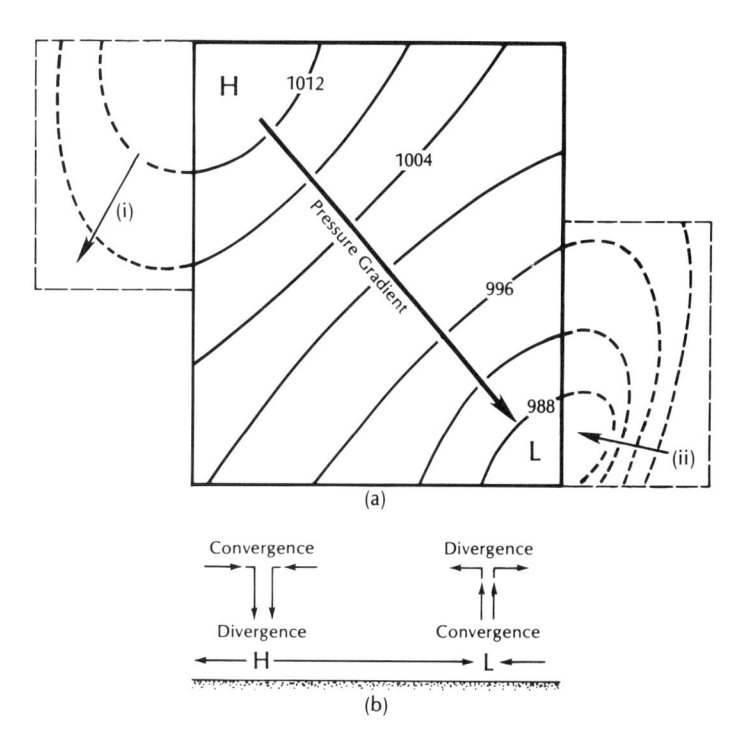

Rotation of the Earth

Air motion begins when a pressure gradient exists. Once the air is in motion, other forces come into play to influence its path. Were the earth not rotating, the air would move directly down a pressure gradient. But the earth is rotating, and this rotation plays a very important part in determining the circulation as it actually exists. Rotating bodies are subject to the law of conservation of momentum, which states that if no external force acts on a system, the total momentum of the system is unchanged. Angular momentum is a function of the mass of the rotating body, the angular velocity (degrees per unit time), and the radius of curvature. The product of these elements tends to stay constant when there is no external interference. A tetherball serves as a good illustration. As the ball goes around the pole and the radius shortens, the angular velocity increases to keep the momentum the same as it was initially (See Appendix 1, Note 6). Since air has mass, it also has momentum. As it moves poleward or equatorward, the rotational velocity changes as its distance to the earth's axis changes (table 6-1). In actual practice, the increase in angular velocity is much less than it might be because the energy is dissipated by friction and by diffusion in the large mid-latitude eddies or cyclones.

The rotation of the earth causes winds in the Northern Hemisphere to appear to be turned in a clockwise direction or to the right, and in the Southern Hemisphere to the left (figure 6-2). The explanation of this process, which had long been observed and recorded by mariners, was first explained by the French mathematician, G. G. Coriolis (1792-1843). Today the rotational component of winds is still

113

PHYSICAL AND DYNAMIC CLIMATOLOGY

TABLE 6-1 Earth rotation.

Time of rotation	24 h
Angular velocity	15°/h
Surface velocity	
Equator	1670 km/h
45°	1180 km/h
60°	835 km/h
90°	0 km/h

referred to as the *Coriolis effect*. The rate of curvature imparted to the moving air is directly related to the velocity of the wind and the sine of the latitude.

$$C = 2V\Omega \sin \phi$$

where C = Coriolis effect

V = wind velocity

Ω = angular velocity

ϕ = latitude

Given the sine ratio in the formula, it is evident that at the equator the Coriolis effect is 0 (sine of latitude $0° = 0$). Although the Coriolis effect is relatively small in itself, it is significant because general atmospheric pressure gradients are small and air streams travel great distances over the earth.

Friction

Winds blowing over the surface are subject to three variable forces: the pressure gradient, the Coriolis effect, and friction. Friction works in opposition to the pressure gradient to reduce wind velocity and subsequently the Coriolis effect (figure 6-3(a)). With the rotating effect reduced, the pressure gradient predominates, and winds blow from high to low pressure at an angle across the isobars. Over the oceans, where friction is less than over land, the winds cross the isobars at angles of 20°–40° and at a velocity of about 65 percent of gradient velocity (figure 6-3(b)).

FIGURE 6-2 The Coriolis effect acts on the wind to deflect it to the right in the Northern Hemisphere and to the left in the Southern Hemisphere. Deflection is in the same direction, regardless of wind direction.

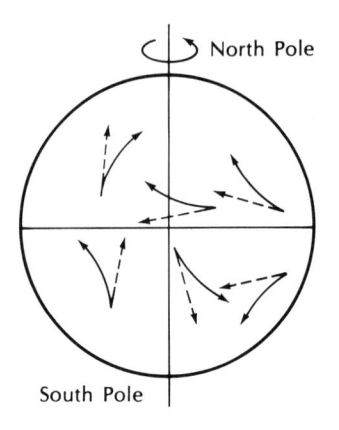

FIGURE 6-3 The forces affecting wind direction and mean direction over land and sea (a) the direction of the three forces that affect surface wind direction and velocity; (b) the mean angle of wind direction relative to the pressure gradient over land and water.

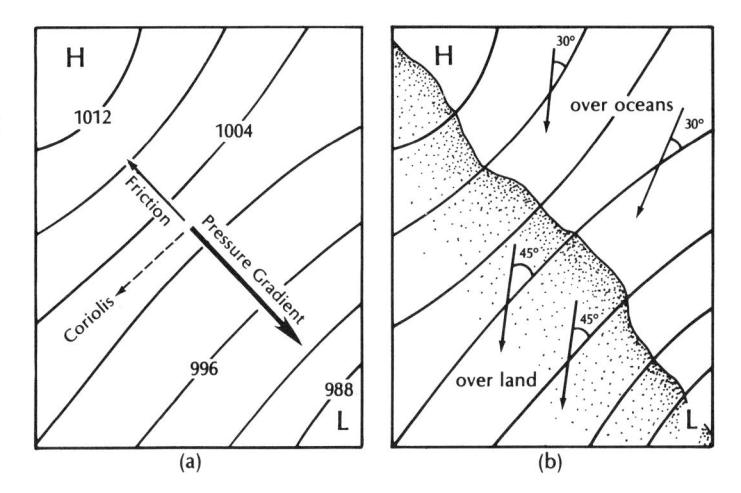

Over the land masses, where the surface is rougher and friction greater, the wind vectors cross the isobars at an average angle of 45° with a velocity of about 40 percent of gradient velocity.

GLOBAL PATTERNS OF WIND

The principles of air motion outlined above apply to the entire circulation of the globe. They are directly related to its entire circulation and are used to establish global models.

In 1735 George Hadley proposed a cellular model to explain the primary circulation of the atmosphere (figure 6-4). It was based upon the fundamental

FIGURE 6-4 The fundamental driving mechanisms of the general circulation (From *Meteorology for Naval Aviators*, 1958, p. 2-6, Navaer OO-8OU-24, Washington, D.C.: Government Printing Office).

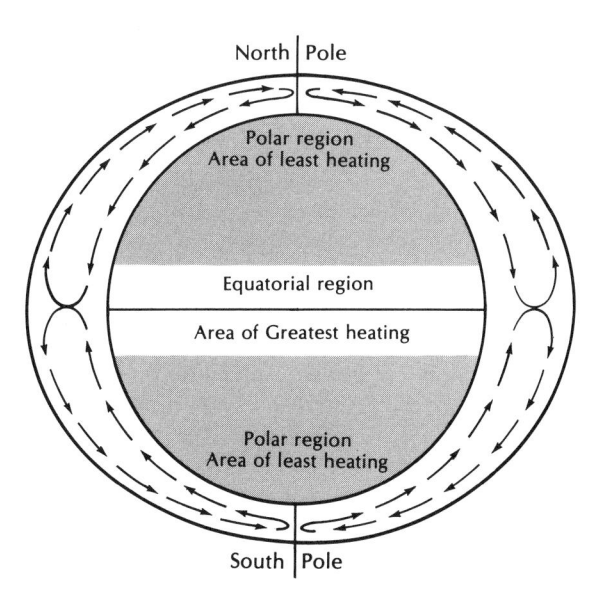

pressure differences brought about by uneven heating of the earth's surface. Hadley postulated that cold air descended at the poles and flowed along the earth's surface towards the warmer equator. This flow would be countered by rising warm air at the equator and a poleward flow aloft, as shown in figure 6-4. This model is very simplified, however, and corresponds only in part to the actual general circulation, as we shall explain subsequently.

The rotation of the earth prevents the development of a single convective cell, and Hadley's model was gradually altered to include a three cell structure between the poles and the equator, still with subsiding cold air at the poles and rising warm air at the equator. This three cell structure shown in figure 6-5(a) is still being refined, but in a crude way it illustrates the essentials of the primary circulation.

The more detailed circulation pattern shown in figure 6-5(b) actually represents a long-term average of the *meridional* (north-south) movement all over the globe. In reality, and as indicated later in this chapter, the circulation is much more complex than this model suggests. While the low-latitude cell exists, the circulation of middle latitudes consists essentially of a west-to-east flow of upper air currents that ultimately determine the position and location of moving high and low pressure systems at the surface. Embedded in these upper air westerlies are corridors of rapidly moving air, *jet streams*, whose intensity and location help define the extent of movement in the upper air currents. Figure 6-6 provides a schematic view of the relationships that exist. The Hadley cells (a) are seen to correspond to the pattern shown in the three cell model, while an outflow of air is identified at the pole. In middle latitudes (b) the circulation is dominated by an upper level air flow making up the westerlies. This flow is not constantly zonal for periodically large waves form in the flow pattern, with a jet stream associated with the leading part of the waves. These are Rossby waves, so named for the meteorologist Carl-Gustav Rossby, whose theoretical work provided the understanding of the mechanisms for their existence. To differentiate the dominance of the circulation patterns, they are called the Hadley and Rossby regimes. The final diagram (c) in figure 6-6 combines the two regimes to provide a depiction of general circulation.

Tropical Circulation

The most equatorward of the three cells closely resembles that proposed in the three cell model and also resembles the original circulation cell proposed by Hadley. The cell extends from around 25°–30° to near the equator. Like Hadley's cell, it features surface flow toward the equator and counterflow aloft, with rising air at the equator and subsiding air near the tropics of Cancer and Capricorn (figure 6-7). This section of the primary circulation is still referred to as the Hadley cell since it operates effectively as Hadley outlined it in 1735, though in a more complex manner. Associated with this tropical cell are two semipermanent belts of surface divergence and above-average sea level pressure located in the vicinity of the tropics of Cancer and Capricorn. These semipermanent high pressure zones on either side of the equator act as a source region for air flowing both poleward and equatorward at the surface. As the air subsides, it is heated adiabatically and is quite warm; since it is subsiding, the relative humidity is quite low. These zones are characterized by high

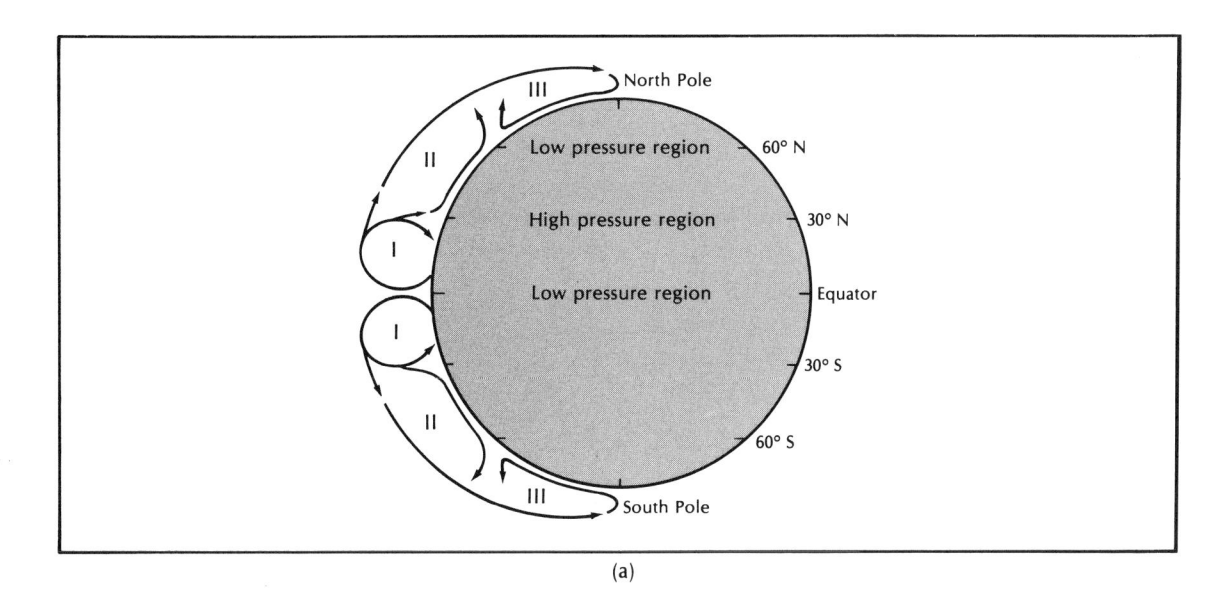

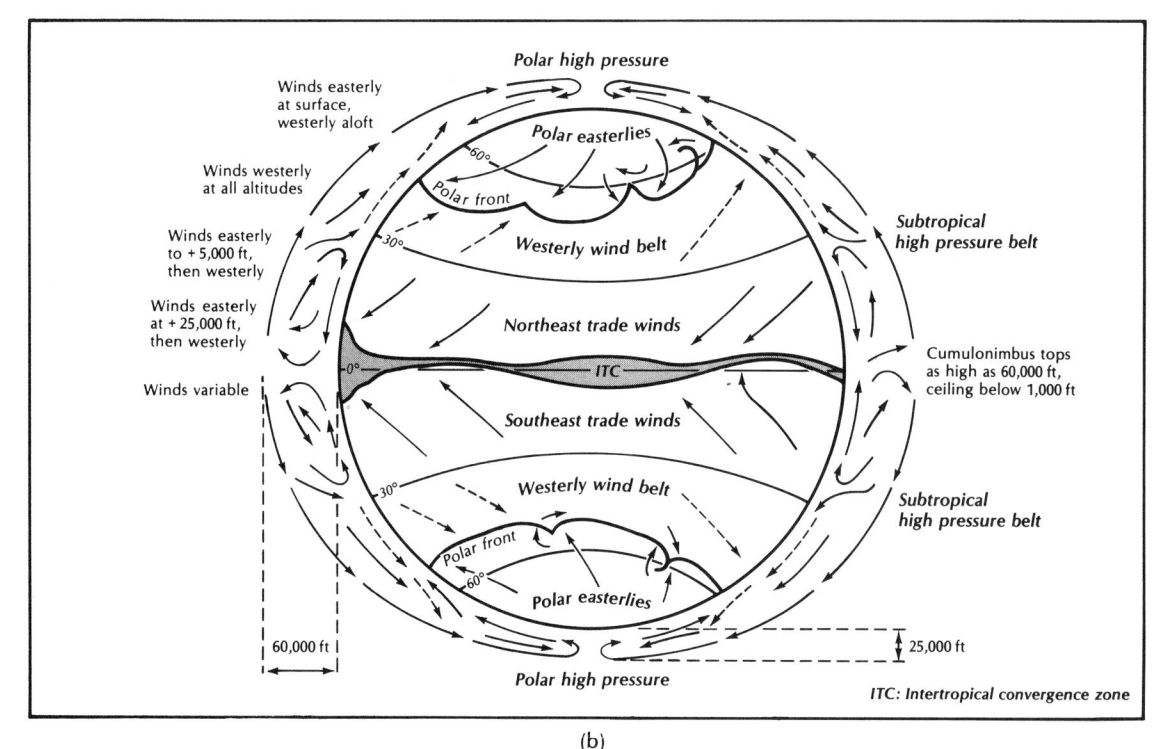

FIGURE 6-5 The three cell model that results from differential heating, the earth's rotation, and the fluid dynamics of the atmosphere (a) the vertical cross-section of the atmosphere and the high and low pressure regions; (b) vertical and horizontal winds in the idealized general circulation (From *Meteorology for Naval Aviators*, 1958, p. 2-9, Navaer OO-8OU-24, Washington, D.C.: Government Printing Office).

PHYSICAL AND DYNAMIC CLIMATOLOGY

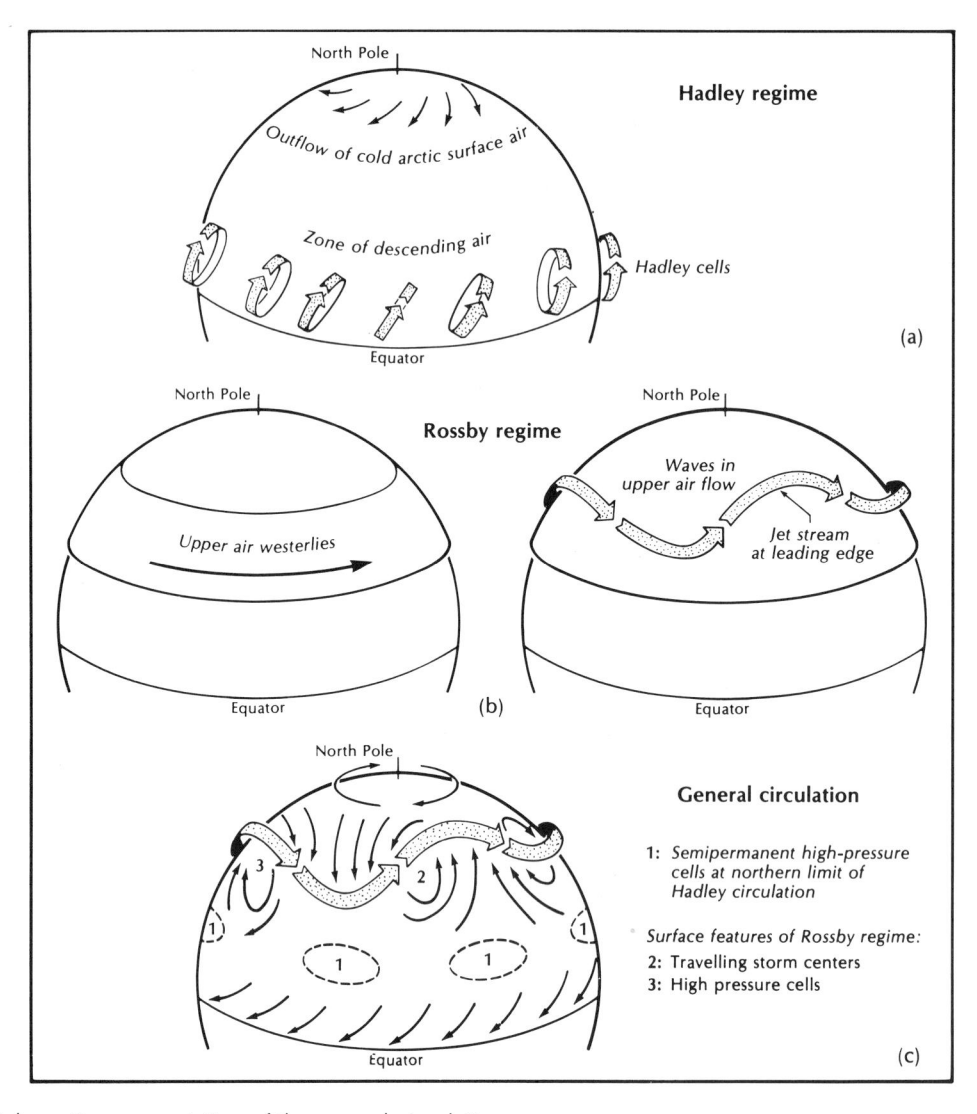

FIGURE 6-6 Schematic representation of the general circulation showing Hadley and Rossby regimes (After N. Calder [1974], from J. E. Oliver [1977, p. 71]. Used by permission of Wadsworth Publishing Co.).

amounts of sunshine, clear skies, and low frequency of precipitation. There are, in fact, major deserts associated with these areas of subsidence: the Sahara and Sonora deserts of the Northern Hemisphere and the Great Australian, Atacama, and Kalahari deserts of the Southern Hemisphere. If the air that flows out of these areas of subsidence has a trajectory over a land mass, it remains warm and dry and is categorized as a continental tropical air mass. The high pressure that prevails is due

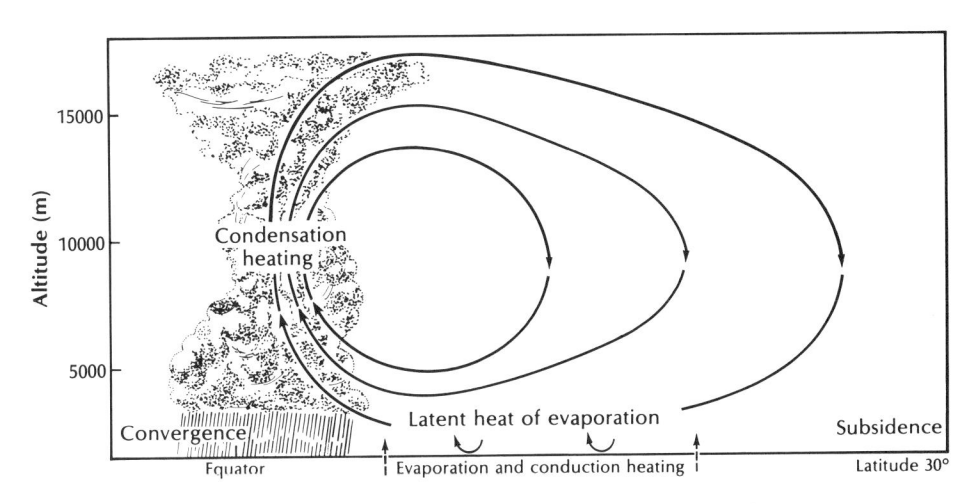

FIGURE 6-7 An idealized cross-section of the Hadley cell with the vertical scale greatly exaggerated (From *Introduction to the Atmosphere* by Herbert Riehl, © 1972, McGraw-Hill Book Company. Used by permission.).

to the subsidence, which necessarily gives rise to divergence at the surface. These zones of subsidence and divergence result more from the motion of the atmosphere and the Coriolis effect than from thermal factors. Over the oceans, the surface winds are light and variable; and in the days of the wind-driven sailing vessels, these zones were known as the *horse latitudes* because of the belief that horses were thrown overboard to lighten ships when they were becalmed.

In addition to serving as a mover of heat, the primary circulation is a mover of moisture. As mentioned in chapter 5, moisture is transferred from low to high latitudes and from the oceans to the land masses by the general circulation. The ocean yields much more water to the atmosphere through evaporation than other areas. Evaporation from the ocean is greatest where there are the greatest differences between the vapor pressure of the air and of the water. Ocean temperatures are the highest at latitudes approximately 30° either side of the equator as a result of the high-intensity radiation and clear atmosphere associated with the divergence there. The high temperature of the water and the low moisture content of the air create a great difference between the vapor pressure of the air and the water in these latitudes. Thus, the subtropical oceans in the vicinity of the divergence zones are the major source regions for moisture, which falls as precipitation on land areas of low and mid-latitudes.

Lying between the subtropical belts of subsidence and the equatorial convergence zone is a region of surface winds flowing equatorward known as the *trades*. These wind systems are best developed in belts 10° in width and centered on 15° either side of the equator and represent the most regular wind systems found at the surface of the earth. The winds, which average 4–7 km per hour, are best developed toward the eastern half of the oceans and are more dependable there than elsewhere.

119

PHYSICAL AND DYNAMIC CLIMATOLOGY

Both poleward and equatorward from 15° these winds are less distinct. Their origin is in the subsidence of the subtropical high pressure zones, and they are eventually lost in the convergence of the equatorial low pressure belt. The Coriolis effect gives the winds a westward turn, so that they are generally northeasterly winds in the Northern Hemisphere and southeasterly in the Southern Hemisphere. The trade winds are often layered, with a surface layer that becomes increasingly moist and unstable as the air moves toward the equator and an overlying layer that is fairly dry and stable.

As soon as an air mass begins to move from the source region, modification of the temperature and moisture characteristics begins. The air masses are modified by heating or cooling from the surface, either by conduction or radiation. Moisture content is altered by addition or removal of water by evaporation, condensation, and precipitation. At any given time, an air mass is a result of the nature of the source region and the changes that the air mass has undergone while in motion over the surface.

As the air from the subtropical high moves equatorward, it not only increases its moisture and instability but adds latent heat at the same time. The following excerpts from *Mutiny on Board the HMS Bounty* illustrate the effects of the sea surface on the overlying air. The entries in the log were made as the *Bounty* sailed south in the North Atlantic toward the equator while in the trades.

> The thermometer was at 82°F in the shade and 81½° at the surface of the sea, so that the air and the water were within a half a degree of the same temperature.
> Monday the 4th. Had very heavy rain; during which we nearly filled our empty water casks. So much wet weather, with the closeness of the air, covering everything with mildew. The ship was aired below with fires, and frequently sprinkled with vinegar; and every little interval of dry weather was taken advantage of to open all the hatchways, and clean the ship, and to have all the peoples wet things washed and dried.
> Monday the 18th. The weather, after crossing the line, had been fine and clear, but the air so sultry as to occasion great faintness, the quicksilver in the thermometer, in the daytime, standing at between 81° and 83°F, and one time at 85°F. In our passage through the northern tropic, the air was temperate, the sun having then high south declination and the weather being generally fine till we lost the Northeast trade wind; but . . . such a thick haze surrounded the horizon, that no object could be seen, except at a very small distance (pp. 30-31).

While the trades are best known for the route across the Atlantic followed by the sailing ships of the age of exploration, these same winds play a major role in the Pacific Ocean as well. It was these same trade winds that carried Thor Heyerdahl and his crew across the Pacific Ocean on the first modern crossing by raft.

> The wind did not become absolutely still—we never experienced that throughout the voyage—and when it was feeble we hoisted every rag we had to collect what little there was. There was not one day on which we moved backward toward America, and our smallest distance in twenty-four hours was 9 sea miles, while our average run for the voyage

as a whole was 42½ sea miles in twenty-four hours (Heyerdahl, 1950, p. 41).

Near the equator the trade winds from both hemispheres converge to form a low pressure trough with a gentle upward drift of air (figure 6-7). This convergence zone is often referred to as the intertropical convergence zone, as it is the area in which the trade winds from the north and south of the equator merge. The area of convergence is quite broad along the equator because of the extensive heating of the surface. Although this trough of low pressure is not very deep, it is quite consistent over the oceans and is sufficient to produce convergence most of the time. Where the converging trades have a trajectory over the ocean the air contains large amounts of moisture, cloud cover is extensive, and precipitation is frequent. Evaporation from the ocean is actually reduced along the equator because there is more water present in the equatorial air and the vapor pressure at the water surface is lower. Sea temperatures are not as high near the equator as they are farther out near the tropics of Cancer and Capricorn because of the addition of fresh water from precipitation and from streams draining adjacent land masses. Insolation is also reduced as a result of the more extensive cloud cover, which further reduces water temperatures.

In the center of the ITCZ over much of the Atlantic and Pacific are found the *tropical easterlies,* which are fairly stable winds of low velocity moving from east to west. These are not to be confused with the trade winds. These easterly winds are quite regular in direction, if not in velocity. Occasionally, a change in circulation will bring quite unstable weather in the form of squalls and general rainstorms. The converging winds on occasion will rise some distance away from the center of the convergence zone. When this happens, quite stagnant conditions may result near the surface. These stagnant conditions of very low velocity winds with ill-defined direction were named *doldrums* by sailors of the sixteenth century. While traveling to the Western Hemisphere from Europe, sailing vessels were occasionally becalmed in this area, just as they were further poleward in the horse latitudes.

A large storm system, if not a particularly violent one, that is characteristic of the tropics is the *easterly wave.* The prolonged periods of drizzle and steady rain in the tropics are largely a result of these weak waves in the atmosphere. They are associated with the trades, but sometimes they are found in latitudes between 20° and 30° on either side of the equator. The tropical wave lacks the sharp contrast in air masses associated with mid-latitude storms, and the Coriolis effect responsible for producing curvature of airflow is not very great. Divergence and dissipating cumulus cloud forms mark the forward edge of the wave (see figure 6-8). Convergence exists behind the wave, with associated high cumulus clouds, and the thunderstorms are often arranged as *cloud streets,* long parallel lines of clouds that form with light winds of constant direction. The direction of air flow and associated cloud streets changes about 90° across the crest of the wave. The wave itself has only a slight pressure drop, something in the vicinity of 2 or 3 mb. These waves travel rather slowly, usually less than 25 kph. There is an extensive area of discontinuous precipitation east of the wave crest. These troughs passing across Africa are responsible for a large portion of the total rainfall of the areas on the equatorward margins of the deserts.

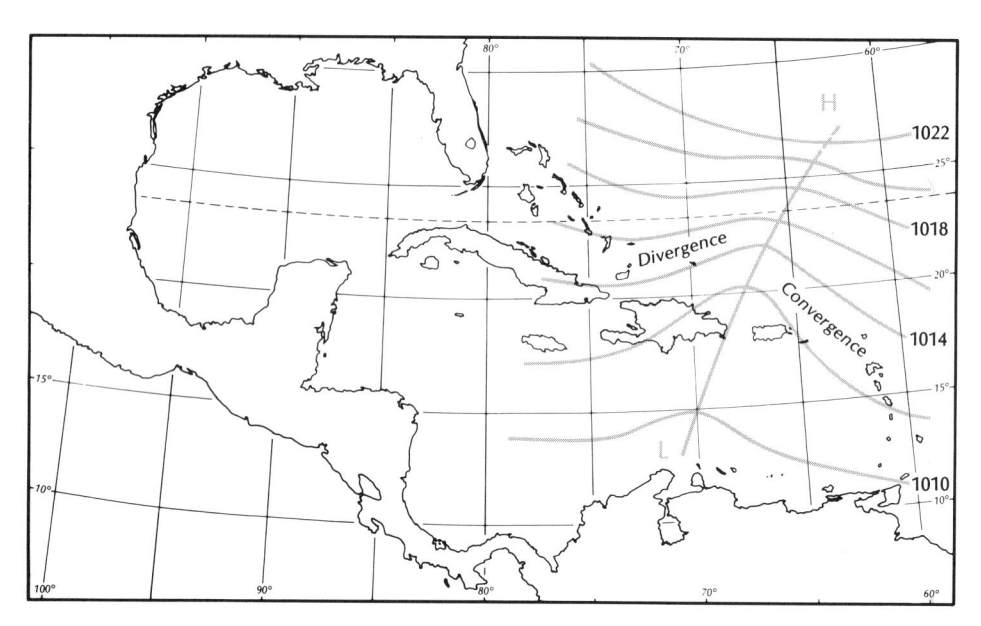

FIGURE 6-8 Model of a wave in the tropical easterlies (From *Climate and Weather* by Hermann Flohn, trans. by B. V. de G. Walden, 1969, p. 138. Used by permission of Weidenfeld Publishers Ltd.).

Mid-latitude Circulation

The three cell model of the atmosphere has inadequacies, and the real world differs most from the model in the mid-latitudes. The model calls for a mean poleward flow of air between the semipermanent high pressure cells at 30° and the semipermanent low pressure trough near 60°. In reality the mean meridional flow in these latitudes is very weak. Surface winds are quite variable in both direction and velocity, but the greatest frequency and highest average velocity indicate a west-to-east flow. Maximum westerly velocities are reached at about 35° N, and velocities at this latitude become more pronounced with height up to the tropopause. Thus, the primary flow in mid-latitudes is zonal (west to east); it is in the latitudes of 35° to 40° north and south that the maximum poleward transfer of energy takes place. With a relatively weak meridional flow, there must be some other means of transporting the energy through the zone. In this region exists the transition zone between the warm tropical air and the cold polar air known as the *planetary frontal zone*. It is here that the mid-latitude westerlies are strongest and the flow increases with height up to about 12 km. In the upper part of the troposphere and the lower stratosphere, long waves develop in the planetary frontal zone that meander around the earth (figure 6-9(a). These waves have an amplitude and wavelength that vary up to about 6000 km. They are always in a state of flux, sometimes being standing waves and sometimes moving with the westerlies.

Near the surface large eddies or vortices develop that are 1000–2000 km across and have lifetimes of several days. These eddies are the cyclones and anticyclones of the mid-latitudes. They are rotational in character and thus carry cold

FIGURE 6-9 Model of the development of waves in the westerlies. When the westerlies are zonal (flowing from west to east), as in (a), there is minimal mixing of tropical and polar air. When waves from air flow become more meridional, there is a great deal of cold and warm air exchanged.

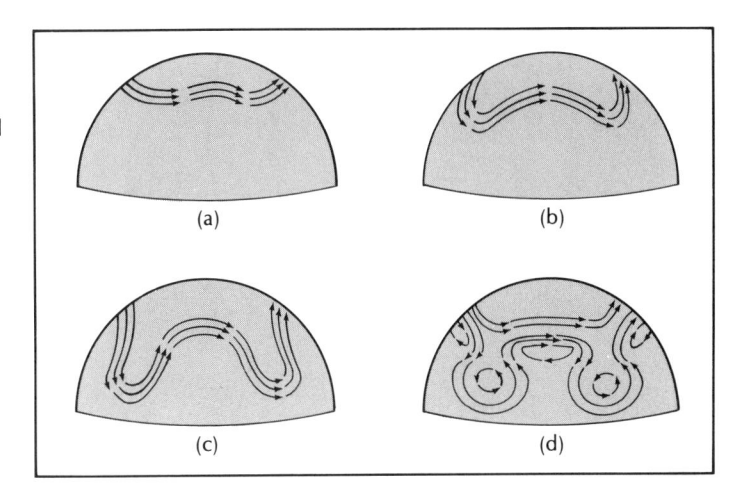

air from the poles equatorward and warm air from the subtropics poleward. These eddies carry energy poleward, but the net flow of air northward is very small. It is basically an exchange process of warm for cold air. Thus there is a high energy transfer with a low net poleward flow of air. It is also the high frequency of these cyclonic cells that forms the subpolar low pressure zone. This convergence zone differs markedly from the equatorial convergence zone in terms of the different temperature and moisture conditions flowing into it.

Polar Circulation

Over the north and south polar regions thermally induced high pressure systems exist, although knowledge of them is rather limited. The high pressure cells are usually displaced from the geographic pole toward the *continental cold pole* (the point of coldest temperature). This displacement in the Antarctic is not large, but in the Northern Hemisphere it is significant. In the Northern Hemisphere winter there is always a heat flow through the pack ice and from the open water areas so that the cold pole is shunted off onto the land mass. Since there are two major land masses in the Northern Hemisphere, there are often two cold poles, one over North America and one over Eurasia. Upper air divergence and temperature inversions are characteristic of the cold poles. There is a predominance of high pressure systems and divergent air flow over polar areas most of the year, but the size of the area affected changes, expanding during the winter months. Outward from the polar highs are frequently found belts of weak easterly winds that tend to be more pronounced in the summer in the Northern Hemisphere. They are strong gravity winds blowing down the ice slope and out to sea. These winds can attain fairly high velocities and be quite persistent, but they are usually shallow, extending upward to 300–400 m.

Surface Winds of the World

The prevailing winds that occur at the earth's surface are illustrated in figure 6-10. The semipermanent highs and lows can be identified, as can the prevailing

(a)

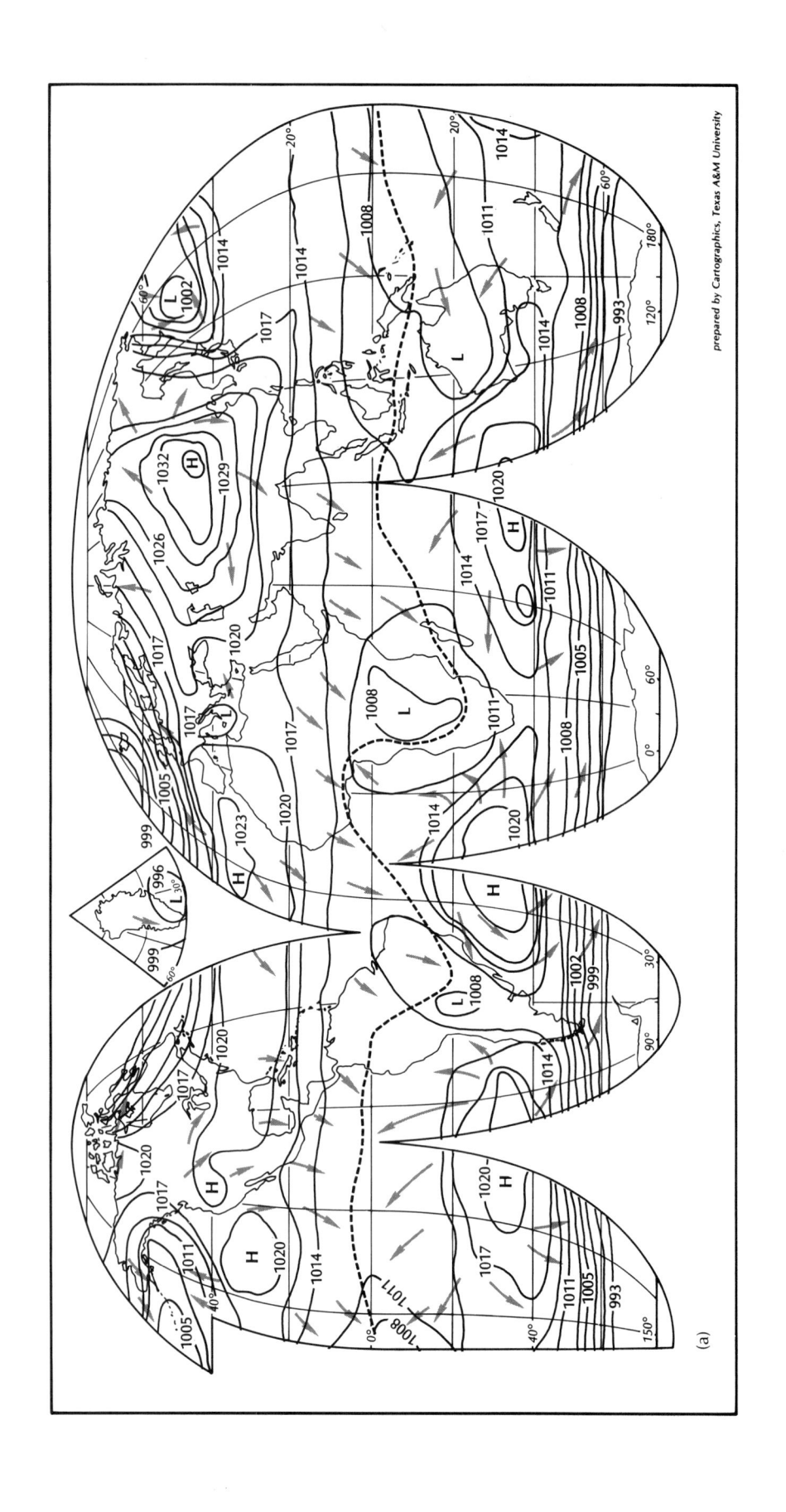

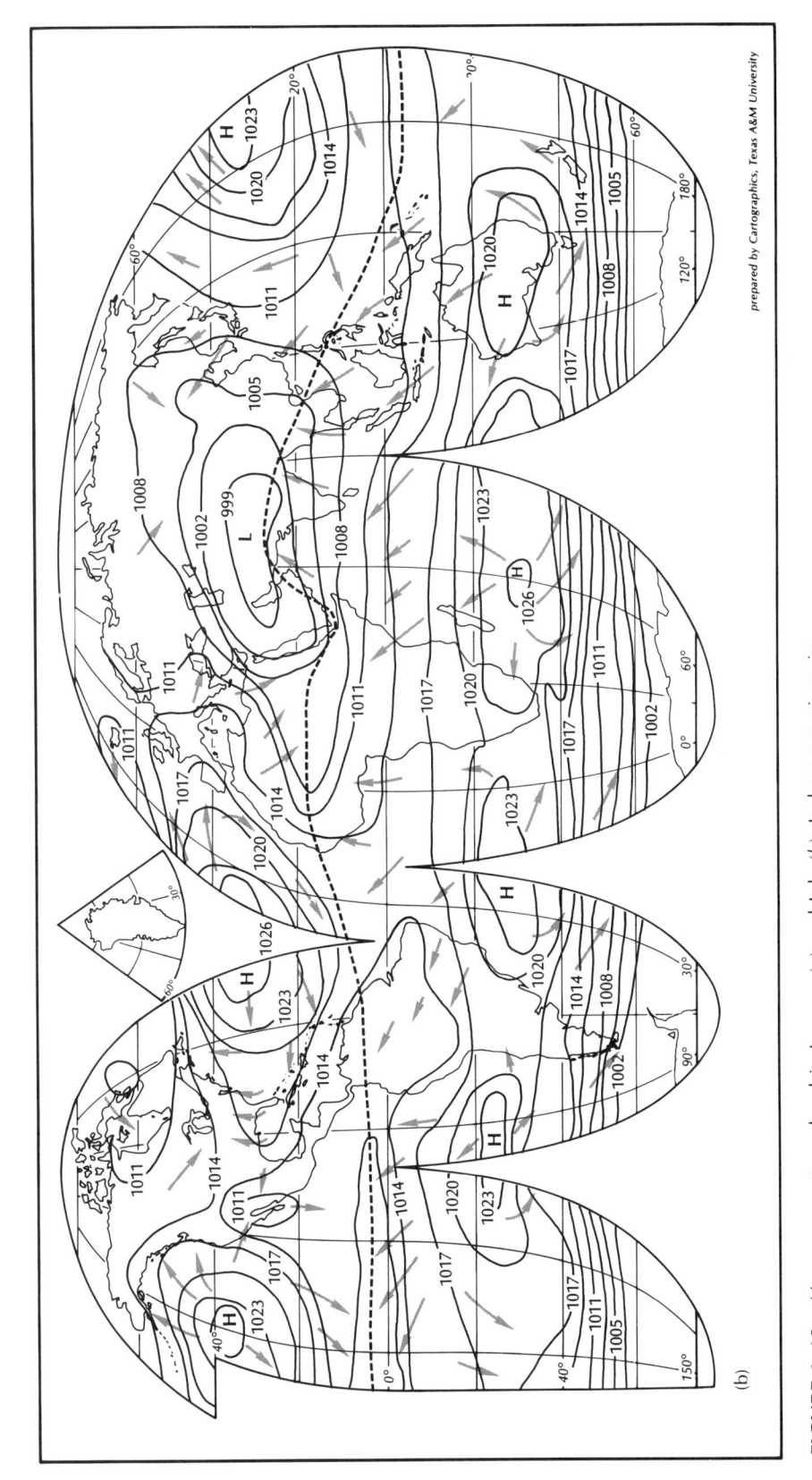

FIGURE 6–10 Air pressure at sea level in January (a) and July (b). Isobars are given in millibars (mb); arrows indicate general air flow.

winds. It can also be noted, however, that the actual system is not the same as the ideal shown in figure 6-6(c) because the earth is not a smooth surface and the distribution of the land masses in the world ocean modifies the actual winds.

Seasonal Changes in the Global Pattern

The major semipermanent pressure zones tend to shift north and south with the seasons. Consider, for example, the location of the ITCZ in figure 6-10. This convergence zone is located mainly south of the equator in January, but in July is located north of the equator. The subtropical highs and subpolar lows also shift location. This seasonal migration of the pressure belts and associated winds is a direct result of the changing distribution of temperature that occurs with the seasons.

The differences in distribution of solar energy in land and water produce major shifts in relative position of the subtropical highs and subpolar lows. These are particularly pronounced in the Northern Hemisphere, where the greatest amount of land is found. Seasonal reversals occur over the land masses in terms of pressure and resulting convergence or divergence. During the winter half-year, the subtropical high and polar high expand and often meet over the land masses that have undergone substantial cooling. In midwinter, high pressure ridges or cells are found over the Northern Hemisphere land masses. During the summer months, the subpolar low expands over the continental areas; and sometimes tropical lows and the subpolar lows merge to form a single trough of low pressure extending from the tropics to 60° to 70° of latitude.

The change in pressure systems brings about a marked reversal in wind and moisture conditions with the season. During the summer months, convergence predominates and high humidity and precipitation are general over the continents. In the winter months, when high pressure systems predominate, divergence over the land is more frequent than convergence. As a result of the latter phenomenon, humidity and precipitation are both lower in winter than in summer. This seasonal shift of pressure, with all its ramifications, is a monsoon (figure 6-11). The seasonal shift in pressure, wind, and humidity is greatest over Asia. During the winter, subsiding dry air covers much of the continent, resulting in a dry season that is comparatively much drier than the North American winters. The subsidence produces offshore winds along much of coastal Asia. As these winds move out over the oceans, they evaporate large amounts of water from the sea surface and are sources of moisture for the offshore islands.

During the summer months, the low pressure system and convergence are strong, and the onshore flow of moisture is great (figure 6-12). Moisture-laden winds blow north from the Indian Ocean and the Arabian Sea. They converge over the northern plains and the slopes of the Himalayas. One of the rainiest regions of the world is found here. Cherrapunji, India, is often used as an extreme example of seasonal rainfall. Here the annual precipitation has reached 25 m (82.5 ft) and most comes in 4 months. Resulting precipitation on the coastal areas and oceanic sides of offshore islands is heavy. In the spring and fall there are intervening periods between the rainy season and the dry season when the weather is unsteady but not too unpleasant. The highest temperatures of the year often occur late in the spring or in

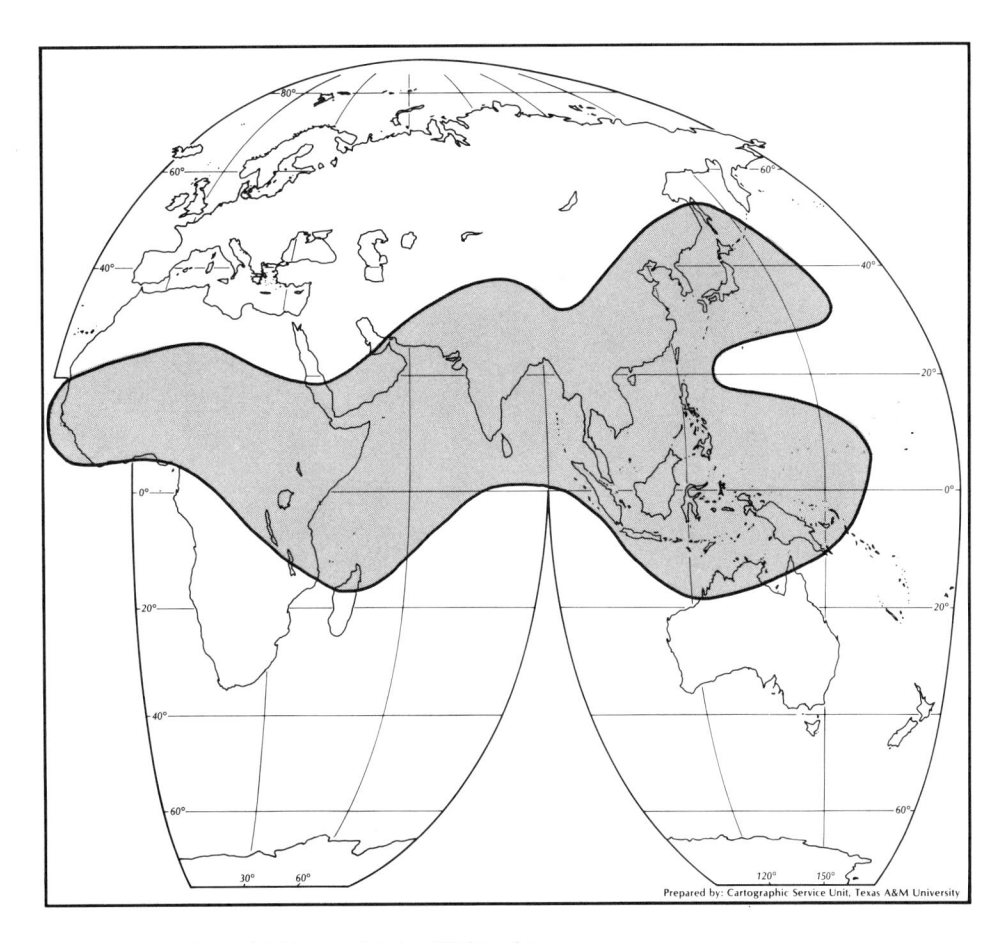

FIGURE 6-11 The monsoon region of Africa and Asia. Within this extensive area, seasonal reversals of wind direction are common and pronounced.

early summer before the high rates of moisture influx begin. The clear skies prior to the development of the onshore winds allow more insolation than does the maritime air that follows later. The rains and humidity, which reduce absolute temperatures, actually increase sensible temperatures.

The monsoon is probably the product of a variety of factors working together: the ITCZ migrates to between 25° and 30°; the land mass interior records a large change in pressure that enforces the migration of the ITCZ; and it is likely that the Himalaya Mountains are a very significant topographic barrier and tend to split the circulation over Asia. In the winter, the jet stream is most frequently located over the mountains. At this time the polar high is expanded and covers much of north-central Asia. Divergence from this strong system north of the Himalayas produces extremely strong offshore surface winds. South of the Himalayas and the jet stream, the anticyclonic flow is much weaker, though offshore flow prevails. In the summer as the temperature differential over the continent diminishes, the jet stream breaks

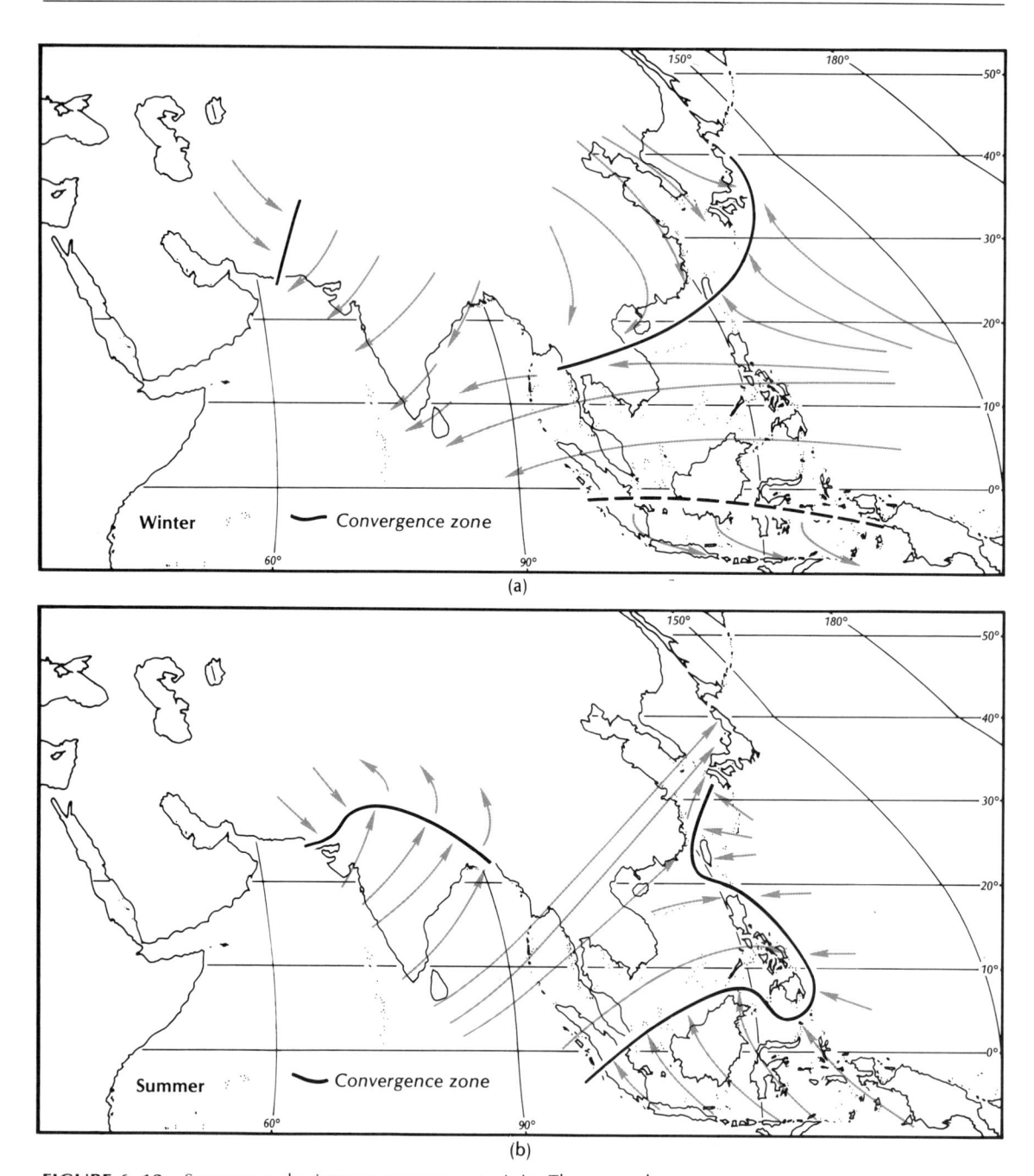

FIGURE 6-12 Summer and winter monsoons over Asia. The general location of the convergence zone shifts substantially, so that most of the area has distinct rainy and dry seasons.

down. The subtropical high and ITCZ shift rapidly northward and bring the sudden shift of winds onshore over the Indian subcontinent. At the same time, north of the Himalayas a weaker convergent system develops with attendant onshore flow of air. The precipitation is primarily convectional rainfall, which is randomly distributed with the flowing moist air. Some traveling cyclones associated with the ITCZ add further rainfall. The peak season of rainfall moves northward over the subcontinent as the zone of convergence moves northward toward the slopes of the Himalayas.

The largest share of the rainfall in northern India occurs during the summer onshore monsoon. The Asian monsoon is quite variable from year to year in terms of the time of year when the onshore and offshore flow begins and in terms of the duration and intensity of the rainy season. The total amount of rainfall varies as a consequence, and the agricultural economies of northern India and Pakistan are cruelly subject to the whims of the system.

The monsoon applies to North America as it does to Asia, although it is more commonly associated with the latter. In summer the subtropical high shifts north over the Pacific coast, and virtual desert conditions prevail over much of southern California. The summer season is the dry season for all of the United States west of the front ranges of the Rockies. On the eastern side of the continent the low pressure trough is more persistent, and convergence is frequent over this part of the continent. Onshore winds from the Atlantic high pressure system centered over the Azores Islands carry considerable moisture. The convergence and subsequent lifting over the land areas produce an appreciable amount of rainfall and a summer maximum over much of the eastern part of the country. In the winter the subtropical high over the Pacific and the subpolar low move equatorward as the solar equator moves southward. As the subpolar zone of convergence moves southward, precipitation increases along the West Coast. To the east of the mountains the winter high pressure systems are more frequent, and the Great Plains experience a season of less precipitation, although not absolute drought.

DIURNAL WIND SYSTEMS

Diurnal variations in atmospheric pressure develop in many parts of the world and lead to distinctive wind patterns. Diurnal variations in wind direction and velocity are generally associated with changing ground temperatures through the day.

Land and Sea Breeze

One of the diurnal winds associated with temperature differences of the ground surface is the land and sea breeze occurring along coastal areas and shorelines of large lakes. The land and sea breeze is a function of the change in temperature and pressure of the air over land in contrast to that over water. In the hours after sunrise, land temperatures rise rapidly while water temperatures rise slowly. When the overlying air above the shoreline is heated, it expands and density decreases. The air rises over the land surface and cold air flows in from the water to replace it. This flow of air landward is the *sea breeze*, which usually begins several hours after sunrise and peaks in the afternoon when temperatures are highest. This air will penetrate inland up to 30 km or more along an ocean coast and to a lesser extent around lakes. The cell is quite a shallow one, ranging up to several kilometers deep under optimum conditions. The sea breeze is generally gusty and brings a decrease in temperature and an increase in humidity. The complementary *land breeze* is less well developed than the sea breeze; it occurs when the land surface cools at night and the water surface remains warm. The temperature and pressure relationships are reversed, but the contrast between water and land is not as great. The water surface does not actually gain heat at night but retains the heat acquired during the daylight period and thus remains relatively warm with respect to the land. There is usually a lag in the development of the breeze that is due to the general inertia of the system.

Richard Henry Dana, Jr., described the effects of the land and sea breeze on commercial shipping in 1840 in the days of sail in the following passage from *Two Years Before the Mast:*

> The brig Catalina came in from San Diego, and being bound up to windward, we both got under weigh at the same time, for a trial of speed up to Santa Barbara, a distance of about eighty miles. We hove up and got under sail about eleven o'clock at night, with a light land breeze, which died away toward morning, leaving us becalmed only a few miles from our anchoring-place. The Catalina, being a small vessel, of less than half our size, put out sweeps and got a boat ahead, and pulled out to sea, during the night, so that she had the sea-breeze earlier and stronger than we did, and we had the mortification of seeing her standing up the coast, with a fine breeze, the sea all ruffled around her, while we were becalmed, inshore. (p. 187)

Figure 6-13 shows the effect of a lake breeze on the temperature at Chicago, Illinois, compared to that of Joliet, Illinois, on the same day. Up until about 3 P.M., temperatures were increasing in a similar manner at both cities. The lake breeze set in at Chicago at approximately 3 P.M., and there was a marked drop in temperature of some 7°F in a little more than an hour. The temperature at Joliet, 56 km southwest of Chicago, did not experience the decline because the breeze was not strong enough to reach as far inland as Joliet. The wind shift producing the drop in temperature was nearly 180°, from west to east. Later in the evening when the lake breeze died there was a rather sharp jump in the temperature in Chicago.

Mountain and Valley Breeze

Similar in formation to the land and sea breeze is the mountain and valley breeze characteristic of some highland areas (figure 6-14). In the daytime, when the valley and slopes of mountains are heated by the sun, the air expands and rises up the sides of the mountains. This breeze, called the *valley breeze* for the place of origin, is a warm wind and a daytime or late afternoon phenomenon. The clouds often seen forming over the hills in the afternoon are the result of condensation taking place as the air rises to cooler heights over the mountains. At night the valley

FIGURE 6-13 The effect of a lake breeze is shown by the difference in temperature over a 24-hour period between Chicago, on the lake, and Joliet, an inland location.

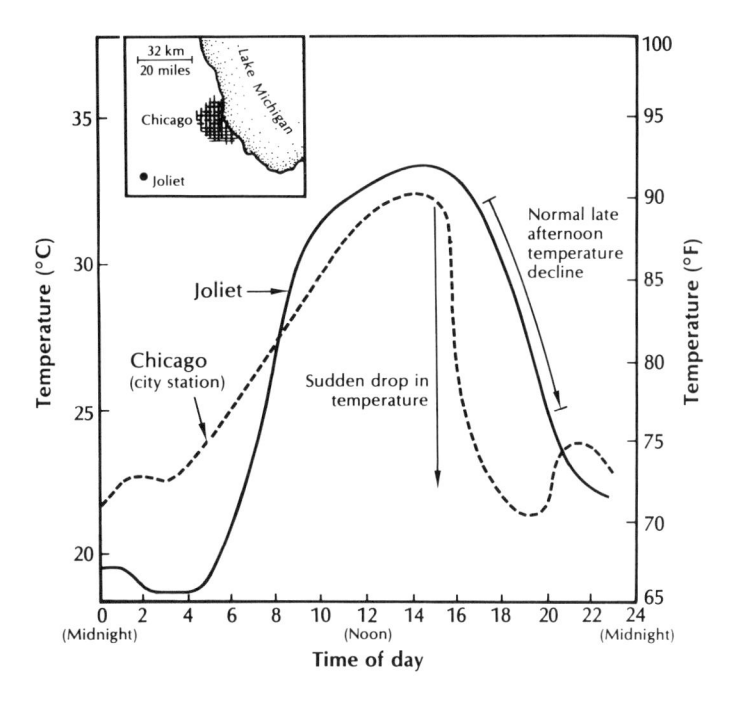

walls cool; and as the air at the surface cools, it flows downslope due to greater density. This is the night breeze. In some mountain regions subject to frost, the valley slopes are the preferred places for orchards. The air moving up and down the valley slopes reduces the probability of the stagnant conditions conducive to frost formation.

SUMMARY

Air motion results when pressure differences occur in the atmosphere. Once air is moving, it is acted upon by the Coriolis effect, which deflects it to the right or left, depending upon hemisphere. On a global scale, the basic patterns are set by imbalance of insolation over the globe and the necessity for interchange of air of

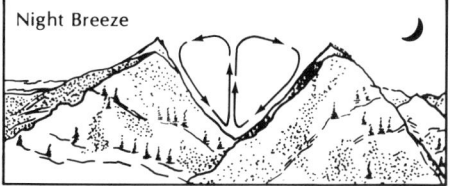

FIGURE 6-14 The mountain and valley breeze. The mountain breeze is a nighttime breeze; the valley breeze is a daytime phenomenon.

different temperatures. The primary circulation is a complex one that is represented by various models that are still being developed.

The general circulation is of primary importance in transporting heat energy to reduce the energy imbalance between tropical and polar areas. The primary circulation also transports water from ocean to land. The seasonal imbalance in energy between the northern and southern hemispheres causes the planetary circulation system to shift north and south. As the solar equator moves north toward the Tropic of Cancer, the general circulation moves northward. Because of the extensive amount of land in the Northern Hemisphere, there is a land-to-sea shift in relative position of the subtropical high and subpolar low. This produces a marked seasonal reversal of wind direction, from onshore in summer to offshore in winter.

Diurnal changes in insolation also produce changes in pressure in winds. The land and sea circulation system and mountain and valley circulation systems are examples.

APPLIED STUDY 6

Energy From the Wind

Until fossil fuels became the main source of energy for industrial societies, wind was widely used as an energy source. The famed windmills of Holland provide perhaps the best-known example of the use of the wind. As part of the search for alternate energy sources, the use of wind is again being considered, and much research is currently in progress. The successful use of wind depends upon the wind characteristics that occur at a given location.

The amount of power generated by a uniform flow of air is given by the equation

$$P = \frac{1}{2}D \times V^3$$

where P is the available power, D is the density, and V is the wind speed. Of importance in this formula is the use of the cube of the wind speed. It indicates that as wind speed increases, the amount of energy generated is geometrically increased and that high-velocity winds are the key to the power generated. For example, a uniform wind moving at 2 m/s (4.5 mph) provides 5.16 Watts (W) on each square meter. At a velocity of 4 m/s (9 mph), the amount is 41.28 Watts/square meter (W/m^2). At double the wind speed, the energy generated is eight times as great.

This relationship means that to evaluate the available wind power at a location, it is important to know the number of times (or percentage of occurence) that the wind falls within a given range. The average wind speed cannot provide a meaningful assessment, a fact that can be easily demonstrated. Consider two locations where the mean wind velocity is given at 7.6 m/s (15 knots). At one site, the wind blows fairly consistently at this speed and, by substituting $V = 7.6$ m/s in the formula, the energy (P) is 282 W/m^2. At the

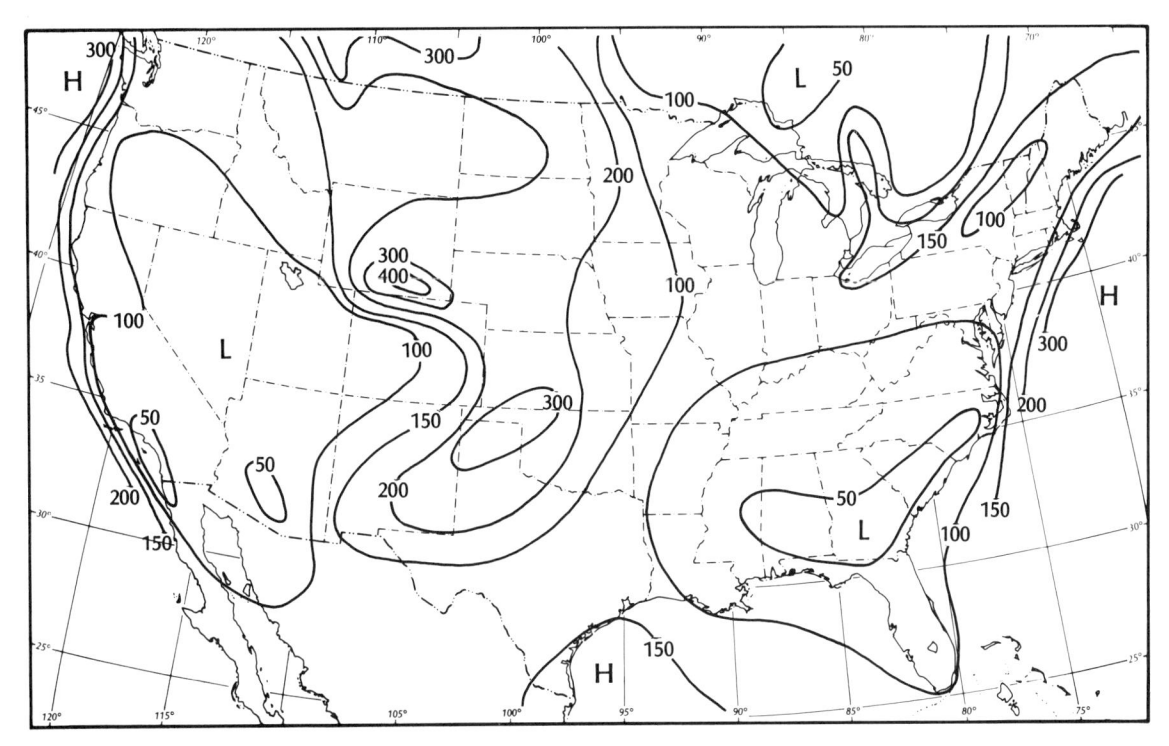

FIGURE 6-15 Average annual available wind power (watts/m²) in the United States (Data from Reed [1974]).

second site the mean value of 7.6 m/s is made up of wind moving at 5.1 m/s (10 knots) for half of the time and 10.1 m/s (20 knots) for the other half. With V at 5.1 m/s, $P = 85$ W/m²; with V at 10.2 m/s the value of P is 648 W/m². Averaging these two gives a value of 384 W/m², which is appreciably higher than the power generated by the 7.6 m/s wind and shows just how significant is the distribution of wind speeds. The geographic distribution of potential wind power in the United States has been determined and is shown in figure 6-15.

A number of high value areas can be identified on the map. The main areas of high potential are in a ridge extending north-south through the western High Plains with peak values in southern Wyoming and the Oklahoma Panhandle. Secondary areas of high potential include the New England coast and the northwestern states. Lowest values are in the southern states and mountainous west.

The distribution suggests that the potential for wind power in populous areas such as Phoenix, Los Angeles, and Atlanta is poor. It is perhaps unfortunate that the areas with highest potential are somewhat remote from major consumer markets.

7 The Climatology of Severe Storms

THUNDERSTORMS
Thunderstorm Formation
Associated Hazards
Hail Regions

TORNADOES
Formation and Characteristics
Distribution and Severity

TROPICAL CYCLONES

SUMMARY

APPLIED STUDY 7
Storms and Planning

Violent storms represent an atmospheric response to the unequal distribution of energy over the earth's surface. They are thus an integral part of dynamic energy exchange rather than simple isolated events. The energy involved in such storms is prodigious (figure 7-1) and the concentration of this energy in a limited area is responsible for the loss of life and devastation associated with severe storms. The fact that storms are mechanisms for global energy exchanges makes them an important climatological element. It might initially be assumed that their study is essentially a meteorological undertaking: The immediacy of the event and the need for forecasting for preventive activities do certainly point to the short-term emphasis. But the long-term analysis of storm activity provides equally significant information, particularly in terms of distribution, probability of occurrence, and potential for destruction.

THUNDERSTORMS

Observers for the National Weather Service consider a thunderstorm to begin when (1) thunder is heard or (2) overhead lightning or hail is observed. The storm is considered to have ended 15 minutes after the last thunder is heard. Note that in this definition there is no mention of rainfall; in fact, in dry climates thunderstorms often occur without measurable precipitation.

Thunderstorm Formation

Irrespective of whether or not precipitation occurs, thunderstorm formation results from an initial uplift of moist, unstable air and the release of sufficient latent heat to cause continued uplift. (Aspects of air stability and thunderstorm formation are dealt with in chapter 4.) The keys to storm formation are thus a source of moist air and a mechanism to produce the required uplift. The significance of the availability of warm, moist air in storm formation is seen in the world distribution of thunderstorms (figure 7-2). The greatest number occur in the moist tropical realms, with Africa having the highest incidence.

The second requirement, a mechanism to initiate the development of cumulonimbus clouds and hence thunderstorms, is more varied and permits the storms to be classified according to their mode of formation. Three main types can be identified: air mass thunderstorms, storms associated with fronts, and those that form ahead of fronts along squall lines.

Isolated thunderstorms that occur in summer in middle latitudes typify the *air mass thunderstorm*. They occur in a somewhat disorganized manner, often consisting of a single cell or several distinct cells less than 10 km wide. It was the study of this type of storm that permitted the life cycle of thunderstorms to be identified; the life cycle from cumulus through mature to dissipating stage has already been described.

While air mass thunderstorms sometimes occur because of difference in surface heating, there is often another trigger effect to cause their development.

Relative energy units*

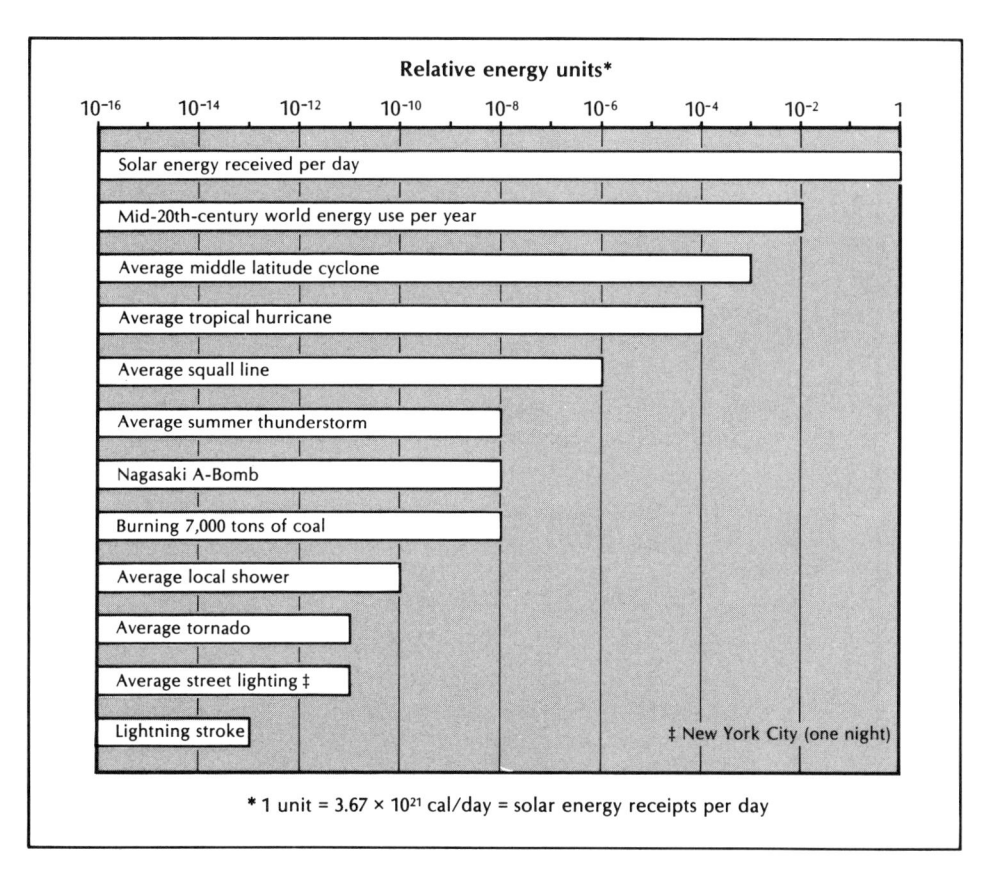

Phenomenon	Approximate relative energy
Solar energy received per day	1
Mid-20th-century world energy use per year	~10^{-2}
Average middle latitude cyclone	~10^{-2}
Average tropical hurricane	~10^{-3}
Average squall line	~10^{-4}
Average summer thunderstorm	~10^{-6}
Nagasaki A-Bomb	~10^{-7}
Burning 7,000 tons of coal	~10^{-7}
Average local shower	~10^{-8}
Average tornado	~10^{-9}
Average street lighting ‡	~10^{-10}
Lightning stroke	~10^{-11}

‡ New York City (one night)

*1 unit = 3.67×10^{21} cal/day = solar energy receipts per day

FIGURE 7-1 Energy equivalents of various phenomena expressed relative to the amount of solar energy received by the earth in one day (From W. D. Sellers, *Physical Climatology* [Chicago: The University of Chicago Press, 1965], p. 106. Copyright © 1965. The University of Chicago Press. Reproduced by permission.).

Alternate mechanisms responsible for growth of air mass storms include converging winds and topography. An example of the former occurs in Florida, where convergence of the moist ocean air along both coasts results in frequent thunderstorms over the peninsula. The role of topography is exhibited on the slope of the Rockies. Air near the slope is heated more intensely than air at similar levels over flat land, resulting in a distinct upslope movement and the potential for growth of cumulonimbus clouds.

Air mass thunderstorms are generally much less violent than those associated with the forced upward motion of air that occurs along cold fronts and squall lines. The nature of air movement along cold fronts (discussed in chapter 4) need not be reiterated here. Instead, emphasis is on the related squall line.

Some of the most severe thunderstorms are associated with *squall lines*. These are lines of moving thunderstorms that occur mostly ahead of cold fronts and,

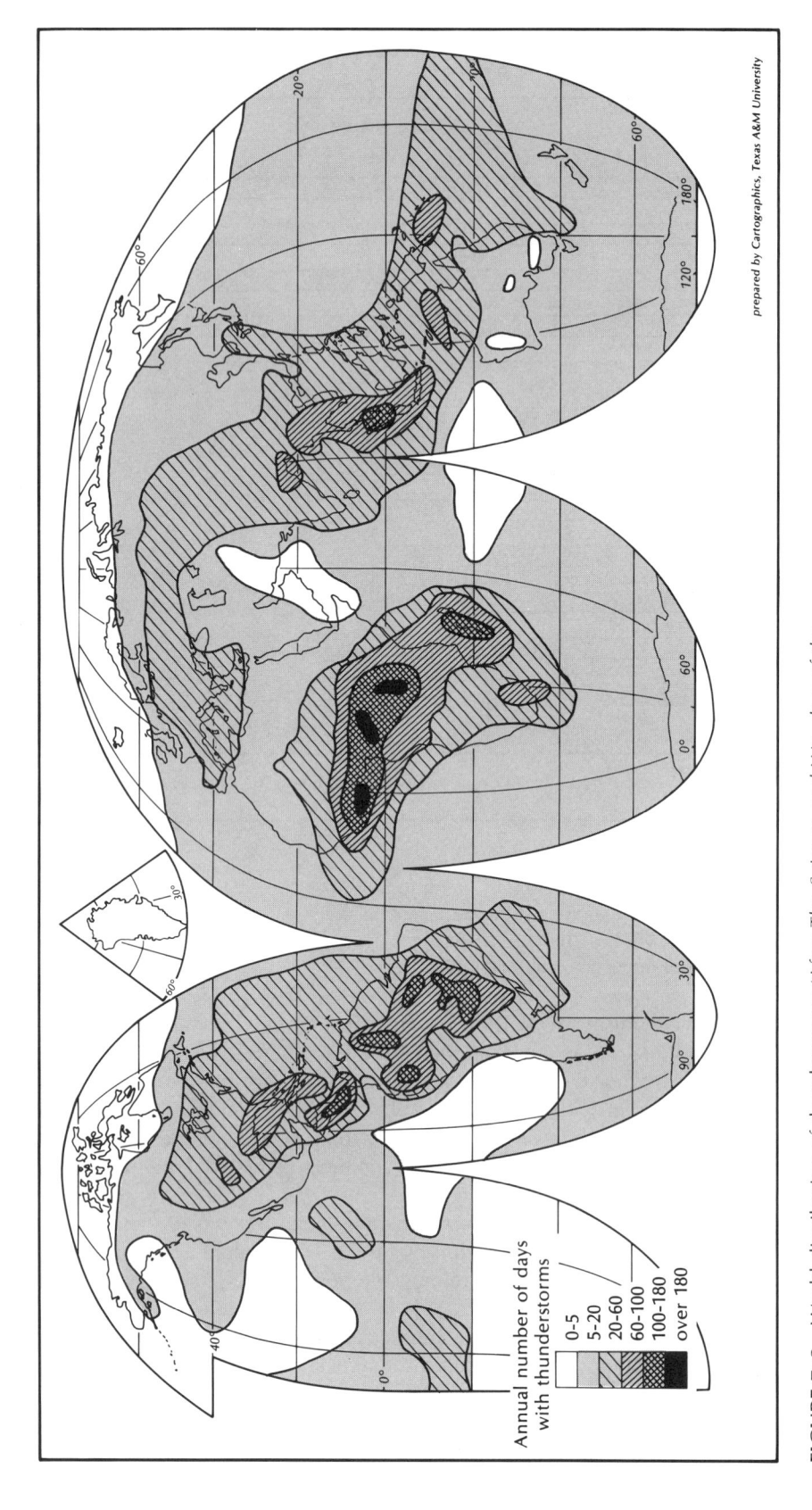

FIGURE 7-2 World distribution of thunderstorms (After *The Science and Wonders of the Atmosphere*, S. D. Gedzelman, Copyright © [1980, John Wiley & Sons]. Reprinted by permission of John Wiley & Sons, Inc.).

unlike the isolated air mass type, are an integral part of large-scale circulation patterns. The schematic sequence given in figure 7–3 provides the essential details of the storm. Warm, moist air at surface levels lies ahead of an advancing cold front. At the 850 mb level (about 1500 m) the flow of this air is from a warm southerly source. In contrast, at 500 mb (about 5500 m), a westerly stream of cool, dry air—with divergence—flows across the surface systems. The combination of the unstable surface air and the divergence aloft leads to extensive vertical development of clouds and a line of thunderstorms. The differential flow of air at varying altitudes adds to the severity of the storms. The westerly high-level flow tilts the top of the storm clouds so that falling precipitation does not cause a slowing of the updrafts (as in the case of the mature stage of air mass thunderstorms); the life of the storm is thus extended and enhances the potential for the growth of hail. At times the size and severity of the storms allow them to be classed as *supercells*.

Note that the squall line may be located hundreds of kilometers ahead of the cold front. In fact, squall lines in tropical climates do not require the presence of a front for their formation. There is no doubt, however, that the association of fronts and squall lines enhances the severity of thunderstorms.

Associated Hazards

The definition of a thunderstorm uses lightning as the key to identify storms; such must be the case because thunder is the result of lightning, a sound resulting

FIGURE 7–3 Schematic depiction of air flows that give rise to severe weather.

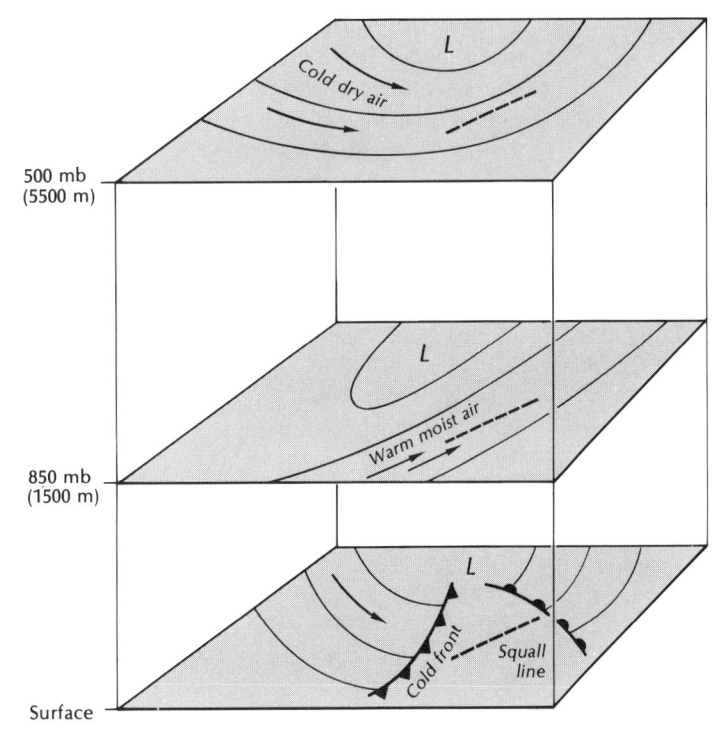

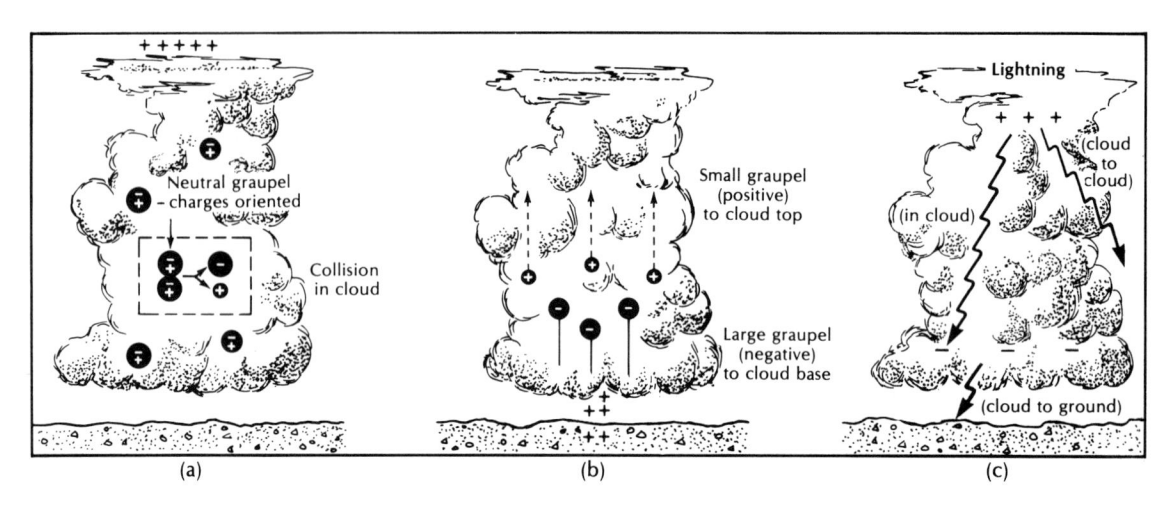

FIGURE 7-4 Lightning formation requires separation of electrical charges in the atmosphere. One potential cause is shown here.

from violent expansion of air close to the lightning stroke. Lightning is thus an integral part of severe storms and is, in itself, a distinctive hazard.

Lightning is an electrical discharge resulting from the separation of positive and negative charges in clouds between clouds and the ground. When the difference in charge is great enough to overcome the insulating effect of air, a lightning flash results. While the reason for the separation of charges is still imperfectly understood, a number of plausible theories exist. One of these, illustrated in figure 7-4, suggests that *graupel* (ice pellets) and hail in the cloud become polarized, with negative charges at the top and positive charges at the bottom of the particle. At this stage, the graupel has no net charge, but, as shown in figure 7-4(a), the larger falling pellets acquire a negative charge. The smaller positively charged graupel is carried aloft while the larger particles accumulate in the lower part of the cloud (7-4[b]). The separation of charges leads to lightning inside the cloud and between the cloud and the positively charged earth below. (7-4[c]).

FIGURE 7-5 Mean number of days with thunderstorms in the United States (From A. Miller, J. C. Thompson, R. E. Peterson, and D. R. Haragan, *Elements of Meteorology*, Fourth Edition, [1983, p. 166]. Used by permission of Charles E. Merrill Publishing Co.).

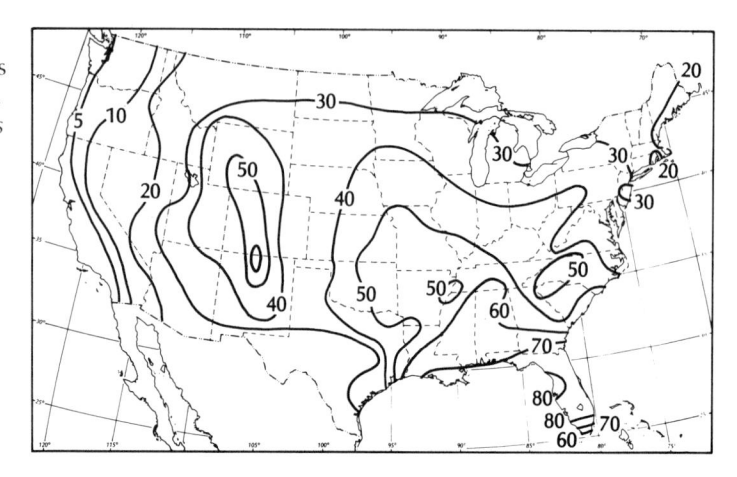

THE CLIMATOLOGY OF SEVERE STORMS

The distribution of lightning is obviously associated with the distribution of thunderstorms (figure 7-5). However, deaths resulting from lightning in the United States do not correlate with this distribution, there being no distinctive area of concentration of lightning deaths. This is probably due to the time of thunderstorm occurrence. A study of injuries and fatalities from lightning shows that 70 percent occur in the afternoon, and only 1 percent occur between midnight and 6 A.M. Thus where there is a high proportion of night storms there are fewer injuries. In regions such as the Great Plains, where night thunderstorms occur, fewer people would be in exposed, open areas during that period.

Hail also results from thunderstorm activity. The mode of its formation is illustrated in figure 7-6. Essentially, the growth of a hailstone is a result of additions of a supercooled water to an initial nucleus, the eventual size of the stone depending upon the length of time spent and its passage through the cloud.

Hail exists in three forms. Graupel, or *soft hail*, is usually less than 5 mm in diameter and has a crisp texture that causes it to be crushed easily when it strikes the ground. Graupel may serve as the nucleus for small hail, a type that is often mixed with rain. Because of the thin layer of exterior ice, a small hail stone remains intact on reaching the ground. Neither graupel nor small hail, both of which are about the same size, is large enough to cause much destruction. Destructive hail is classed as

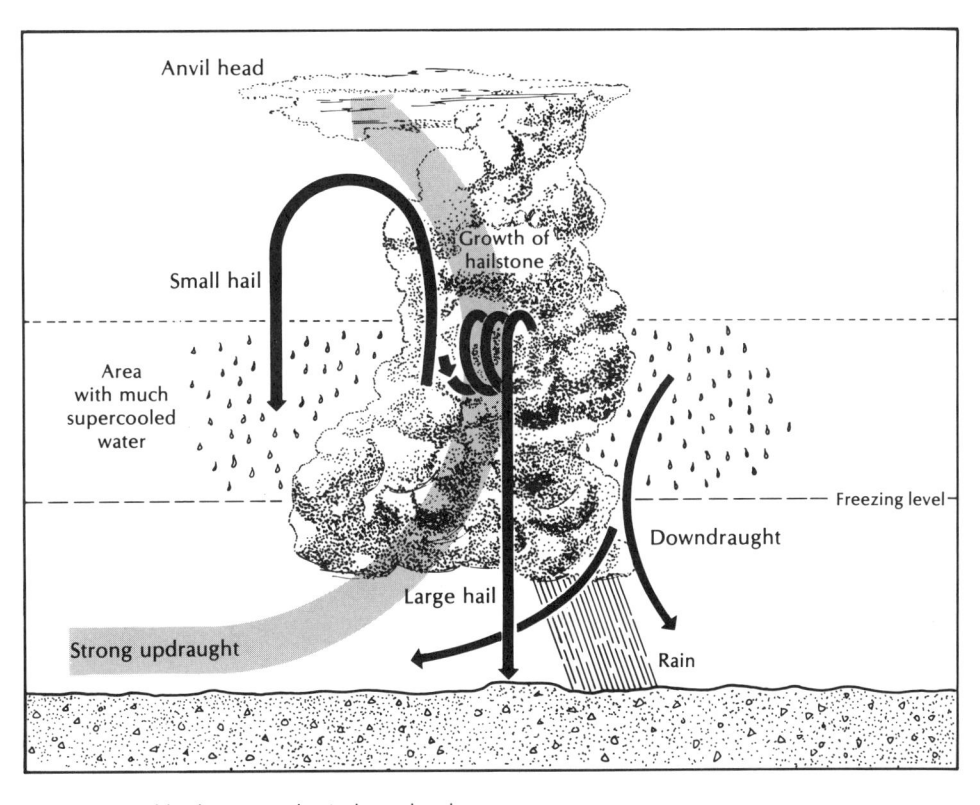

FIGURE 7-6 The formation of hail in cumulonimbus clouds.

true or *severe hail*; it can attain large sizes, as illustrated in the remarkable Coffeyville, Kansas, hailstone that weighed 3.7 kg.

Neither the size nor the number of stones associated with severe hail has been systematically measured for long periods of time throughout the United States. The most general map of hail distribution is that of the average number of days with hail. Such a map is shown in figure 7–7 and is based upon all occurrences of damaging hail, whether large or small. On a statewide basis, Texas, Oklahoma, Kansas, Nebraska, and Missouri (ranked respectively) experienced over one-half of the total severe hail occurrences in the 48 continental states for the period 1955–1967.

Studies have shown that there is a concentrated maximum of the diurnal distribution of hailstorms in the three-hour period from 1500 to 1800 LST. Forty percent of the severe hail that fell in the continental United States occurred within these three hours.

Hail Regions

On the basis of both meteorological frequency and economic factors the United States can be divided into hail regions. These hail regions are delineated on the basis of (1) average frequency, (2) peak hail season, (3) primary cause of hail, and (4) regional hail intensity. Based on these four criteria, 13 hail regions can be identified in the United States. These regions are defined and portrayed in figure 7–8.

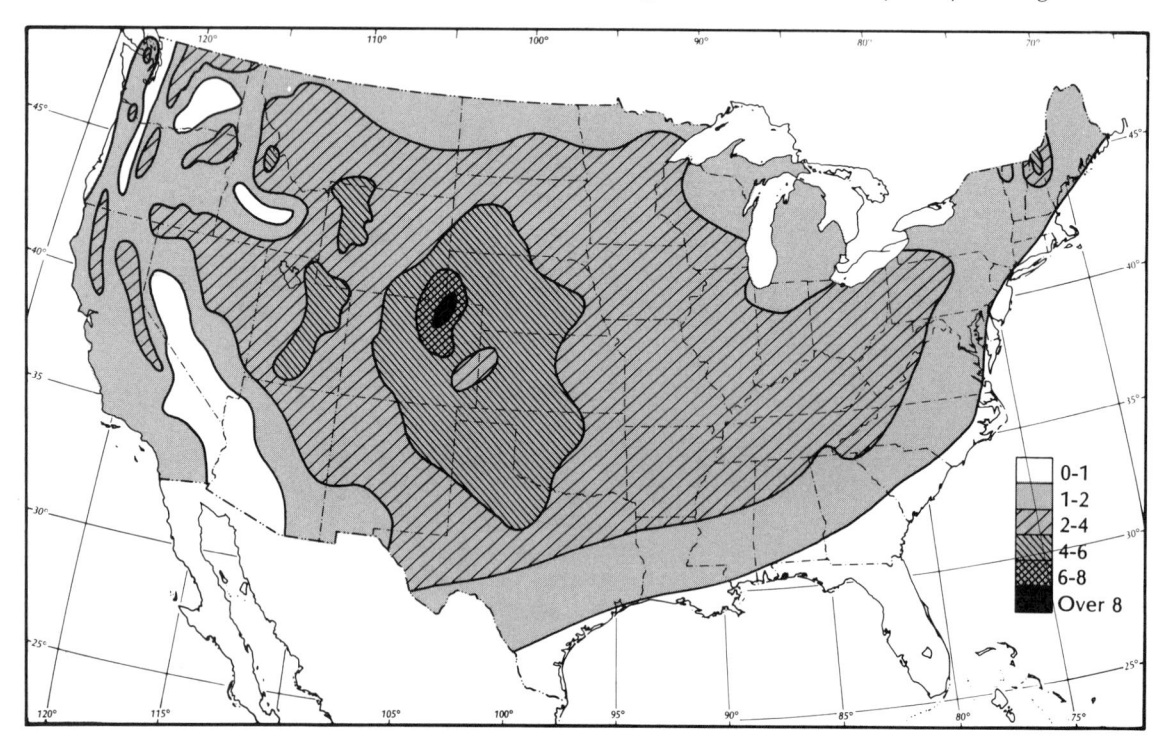

FIGURE 7–7 Average number of days with hail in the United States (From EDS [NOAA]).

THE CLIMATOLOGY OF SEVERE STORMS

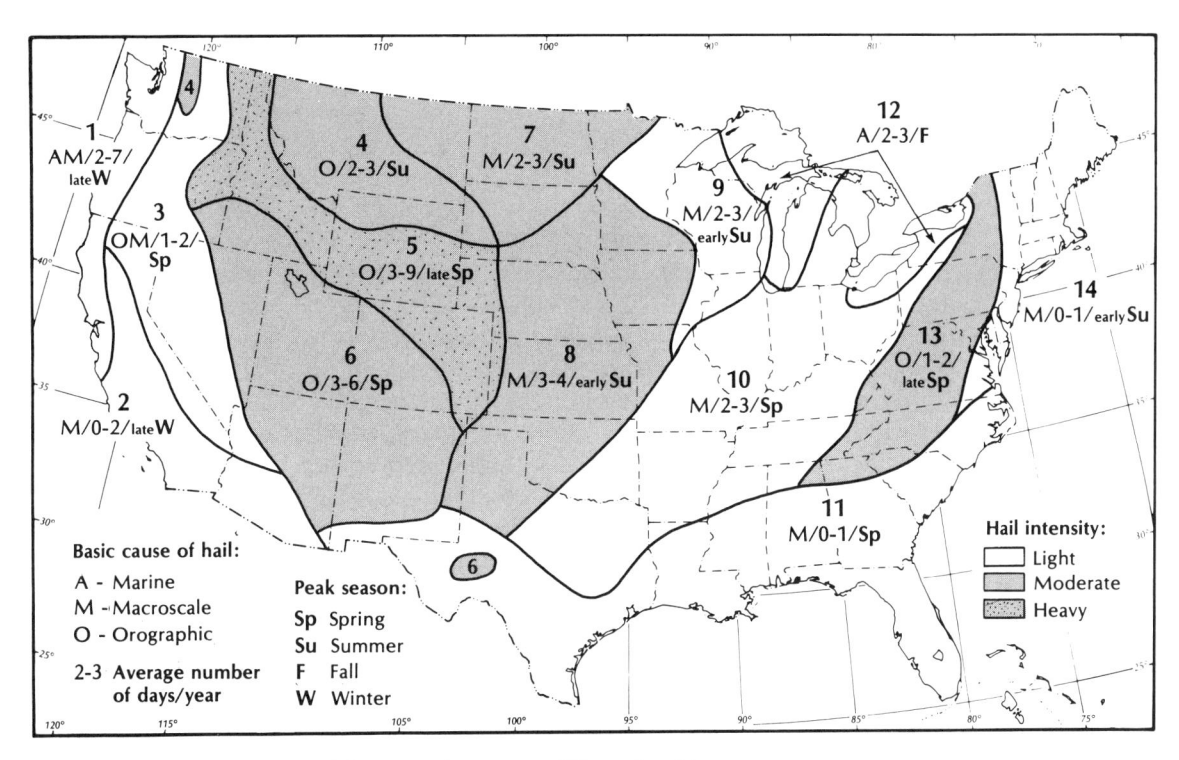

FIGURE 7–8 Hail regions of the United States as differentiated by four hail characteristics: frequency, cause, season, and intensity (From Stanley A. Changnon, *Journal of Applied Meteorology*, 16 [1977, p. 630]. Used by permission of American Meteorological Society.).

TORNADOES

Although tornadoes are associated with cumulonimbus clouds, they are of such violent intensity that they merit separate discussion. A *tornado* is a vortex of rapidly moving air, which is visible when condensation occurs or when debris picked up from the surface is caught up in the spiralling air to form the well-known funnel shape. The funnel appears to hang from a cloud base, at times reaching the ground, but frequently touching down only for a short while and then rising again. Where the funnel does touch down, terrible destruction can occur in a few seconds. Fortunately, the destructive path is limited, usually 0.5 km wide and seldom more than 25 km long. Records show however, that these average conditions can vary appreciably. In May 1977, a tornado traveled 570 km across Illinois and Indiana and existed for a total of 7h 20 min.

The forward motion of a tornado is equally variable. The average rate of motion of a tornado in the midwest is 65 km/h, but some have been observed to move 110 km/h, and others have remained stationary. Within the funnel, revolving winds may exceed 600 km/h. This estimate is based upon the effects of the winds, for direct measurement using conventional methods is impossible because instruments in the path of the tornado are usually destroyed. However, new methods of measuring and monitoring tornadoes are being introduced. One of the most promising is the use of Doppler radar. This not only has the normal capabilities of radar but adds the ability to detect motion directly.

The damage associated with tornadoes is not only the result of wind. In the center of the vortex, air pressure is very low, often on the order of 800 mb. The average sea level pressure is about 1014 mb, and 800 mb represents a substantial change in pressure. When a tornado passes over a building, the very rapid drop in pressure has the same result as suddenly removing a weight from the building, and the roof and walls collapse outward. Although this effect may increase tornado damage, recent research suggests that pressure decrease may not be as important as formerly thought. It has been suggested that the swath of devastation that occurs from a tornado, as opposed to relatively minor damage on either side of actual path, is caused because within a tornado varying degrees of severity occur. An area can be crossed by a tornado and remain relatively unscathed while an adjacent area is utterly destroyed; the latter area happens to be in the part of most violent section of the tornado.

Formation and Characteristics

The events leading to the formation of a tornado are those that have already been described in the discussion of severe thunderstorms. The basic requirements are an active cold front moving into moist, unstable air at the surface with an upper level stream of divergent air (see figure 7-3). The thunderstorm may thus be considered the "parent" of the tornado; as the storm moves, funnels may form at intervals along its path.

Generally, atmospheric scientists are of the opinion that the energy needed to produce a tornado is a combination of both thermal and mechanical forces, but why they occur at specific locations is still not certain. The explanation must, however, explain why the rapid rotation of air occurs, and this analysis requires a knowledge of vorticity and angular momentum. Note 6 of Appendix 1 provides an overview of these aspects.

Since 1950 the National Severe Storm Forecast Center has kept tornado data and compiled attributes of United States storms. Prior to that date, tornadoes were not monitored or recorded in detail, although, of course, many individual storms were described. As a result, an accurate statistical record is available only for approximately the last 30 years. Nonetheless, these data do permit description of storm characteristics and assessment of their relative distribution, times of occurrence, relative severity, and resulting death and destruction.

Distribution and Severity

The two maps in figure 7-9 provide the basic patterns of tornado distribution in the United States. Based upon maps prepared by the National Severe Storm Center, the data represent records of all cited tornadoes for the period 1950-1978. Figure 7-9(a) shows the frequency of tornadoes expressed in terms of the average

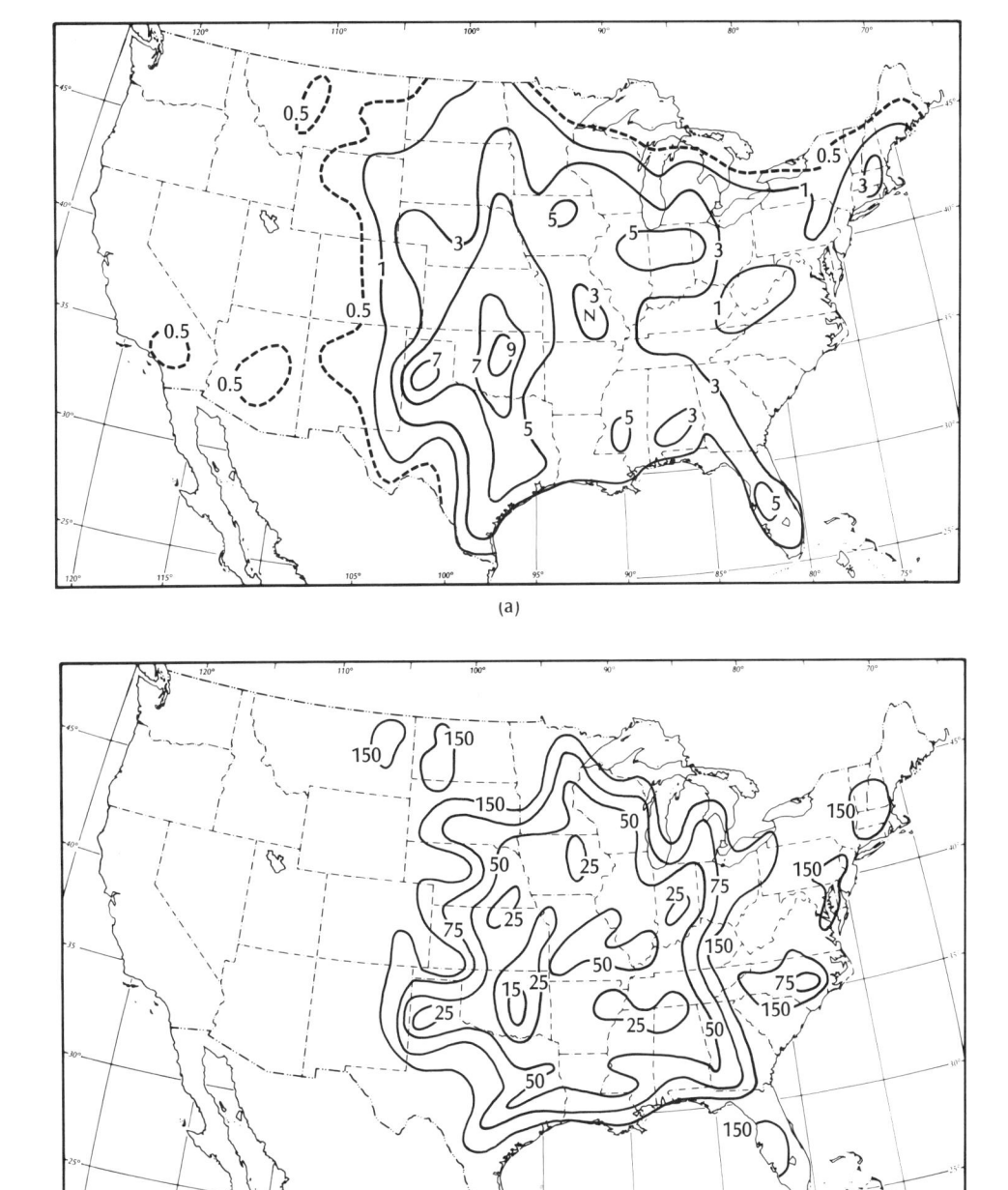

FIGURE 7-9 Some spatial patterns of tornado activity in the United States (a) Frequency of tornadoes within a 90 km (56 mi) radius circle per year. (b) Tornado hazard potential across the United States. Values are the number of units of 2.58 km² (1 mi²) that will contain one 2.58 km² (1 mi²) of tornado damage in 100 years. (Both figures from *Weatherwise*, 3(2), April 1980, pp. 54 and 56, a publication of the Helen Dwight Reid Educational Foundation.)

PHYSICAL AND DYNAMIC CLIMATOLOGY

number of tornadoes reported within approximately 91 km (56.6 mi) of any point in the country. The maximum value occurs in Oklahoma and extends roughly north-south in an area that has been termed "tornado alley." Lowest values are seen in the west and in the Appalachians.

Tornado hazard potential is shown in figure 7-9(b). Data for this figure are derived by dividing the country into sections and computing for each section the annual average area that experiences tornadoes. Results of this computation can be expressed in a number of ways. Figure 7-9(b) shows the number of square miles that will contain 1 square mile (1 mi^2) of tornado damage in 100 years. The lower the number, the greater the tornado hazard. Lowest values of this expression of tornado, less than 50, are located in a zone that corresponds to the tornado distribution in the upper figure. Some anomalies are seen (e.g., the Texas Panhandle) and may be related to low population density and nonsighting of tornadic events.

No state in the United States is immune to tornadoes, although, as noted, the frequency is highly variable. The tornado "season" varies over the country because the conditions that give rise to tornado watches are a function of the atmospheric circulation patterns, specifically the relative location of upper air streams as indicated by jet stream location. These streams of rapidly moving air are integrally associated with upper air westerlies, the extent of which is a function of the temperature gradient from equator to pole. Small gradients, during the Northern Hemisphere summer, result in the jet stream's being located at its most northerly position (figure 7-10). As a temperature gradient becomes larger, the jet stream

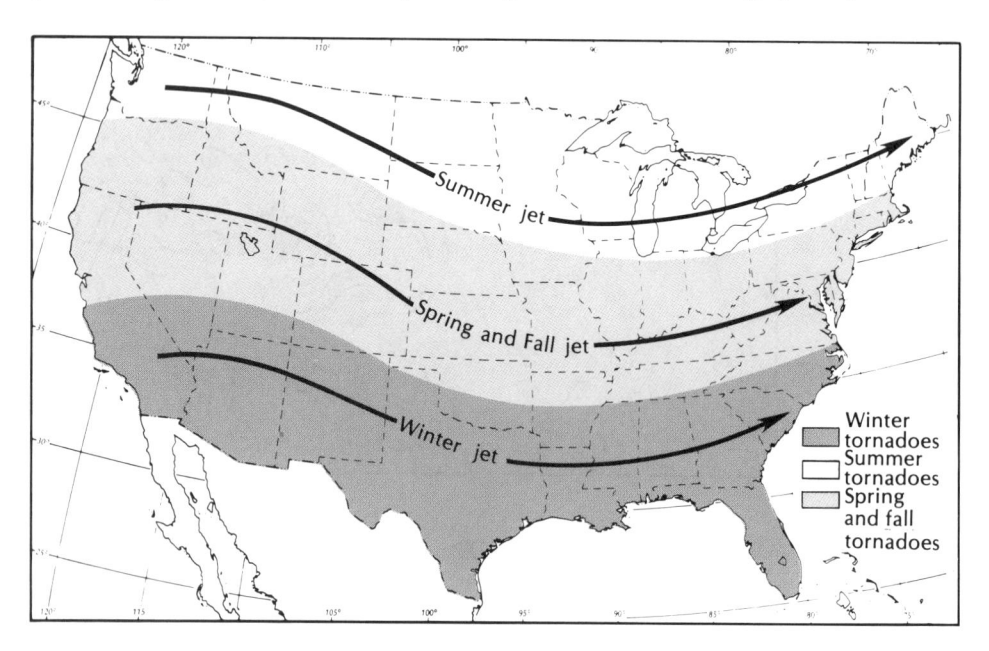

FIGURE 7-10 Migration of jet stream over the year causes areas of major tornado activity to follow a seasonal pattern. (From J. R. Eagleman, *Meteorology*, 1980, p. 246. Used by permission of Wadsworth Publishing Co.).

migrates equatorward until it attains its average winter position. The temporal distribution of tornadoes in the United States is closely related to this migration.

While tornados can form at any time of day, the majority (60 percent) occur between noon and sunset. Since only 19 percent of all tornadoes have been observed between midnight and noon, it is clear that their occurrence is concentrated during the warmest part of the day—afternoon and evening. This general observation does not apply equally to all regions in the United States. For example, in the northwest, tornadic activity seldom occurs at times other than the afternoon; on the other hand, tornadoes that occur between midnight and sunrise, although relatively rare, are most likely in the southern parts of the United States.

A method for assessing the relative severity of tornadoes was introduced by Fujita, who devised the *F-scale* (table 7-1). A rating on the scale is obtained by assessing the worst damage produced by a storm. This qualitative measure is categorized by providing approximate wind speeds. By far most tornadoes occur in the weak classes; only 2 percent of recorded tornadoes occur in the violent classes. This 2 percent, however, is responsible for more than 60 percent of the deaths from tornadoes.

Despite the large amount of ongoing research, the improved technology, and an increasing data base, there remain many unanswered questions about the outbreak of tornadic activity. As an example of the extent of damage and widespread occurrence, table 7-2 provides a summary of available data. It is an impressive array of numbers, and despite improved warning systems, a large decrease in losses is not anticipated in the immediate future.

TROPICAL CYCLONES

The *tropical cyclone* is the typhoon of the Pacific Ocean and the hurricane of the Atlantic; this storm is a large vortex that derives its energy from the latent heat of condensation from the warm oceans over which it originates. Satellite images of hurricanes (figure 7-11) show a vortex of clouds covering hundreds of square kilometers. A cross-section of the storm, as depicted in figure 7-12, shows that high winds and clouds may occur 480 km (300 mi) ahead of the center of the storm, a center characterized by the eye. In the vicinity of the eye, winds reach their highest

TABLE 7-1 Approximate speed corresponding to the Fujita intensity scale.

F-Scale	Category	Approximate Speed (mph)
0	Weak	72
1		73-112
2	Strong	113-157
3		158-206
4	Violent	207-260
5		261-318

TABLE 7-2 Selected tornado data.

Tornadoes and Deaths by Months (1953-1980)

	Tornadoes Total	Tornadoes Mean	Deaths Total	Deaths Mean
Jan.	416	15	90	3
Feb.	560	20	192	7
Mar.	1359	49	210	8
Apr.	2999	107	1063	38
May	4352	155	657	23
June	4046	145	446	16
July	2276	81	39	1
Aug.	1532	55	60	2
Sept.	1086	39	61	2
Oct.	659	24	62	2
Nov.	598	21	49	2
Dec.	476	17	96	3

Tornado Number and Deaths By State, 1953-1980

Number of Tornadoes

Rank	State	Number
1	Tex.	3344
2	Okla.	1477
3	Kans.	1212
4	Fla.	1155
5	Mo.	758

Number of Deaths

1	Tex.	371
2	Miss.	316
3	Mich.	231
4	Ind.	205
5	Ala.	202

Deaths per 10,000 mi^2

1	Mass.	120
2	Miss.	66
3	Ind.	56
4	Ala.	39
5	Ohio	36

velocity and may be in the order of 160-320 km/h (100-200 mph). The eye itself, which may be 24-32 km (15-20 mi) across, is an area of strange calmness and relative clearness.

Clues to the formation of these storms can be found in their time and place origin. They form over tropical oceans (outside a belt extending 5° N to 5° S) and have a distinctive seasonal distribution. Both factors point to a set of special conditions necessary to initiate hurricane development.

FIGURE 7-11 A well-formed hurricane over the western Atlantic Ocean (Source: NASA).

The characteristic vertical structure of the atmosphere in the trade wind belt is a relatively shallow layer of warm, moist air above which is a deep layer of warmer, dry subsiding air. This forms the Trade Wind Inversion, a characteristic that limits the vertical development of clouds (figure 7-13[a]). The inversion is sometimes interrupted by a low pressure trough, an easterly wave, which allows thunderstorm development behind the wave (figure 7-13[b]). Increased convection and the prevailing pattern of high altitude winds cause the trough to deepen and surface pressure to fall to form an isolated low pressure system. The trade wind pattern is interrupted, and winds circulate cyclonically around the low pressure. If the pressure continues to fall, winds accelerate and a severe tropical storm is born.

The change in status from tropical storm to hurricane requires a mechanism to stimulate vertical air motion and inflow of wind. A number of possible trigger

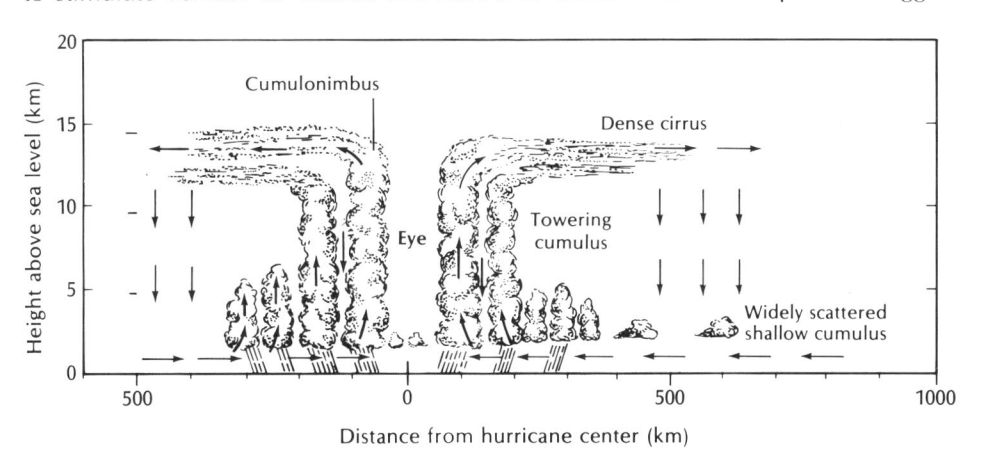

FIGURE 7-12 Schematic cross-section through a hurricane (From J. E. Oliver [1981, p. 157]. Used by permission of V. H. Winston and Sons.).

PHYSICAL AND DYNAMIC CLIMATOLOGY

FIGURE 7-13 The trade wind inversion and an easterly wave in the Atlantic Ocean (a) Normal pressure patterns with cross-section showing the trade wind inversion; (b) an easterly wave disrupts the normal pattern and ruptures the inversion, permitting development of tropical storms.

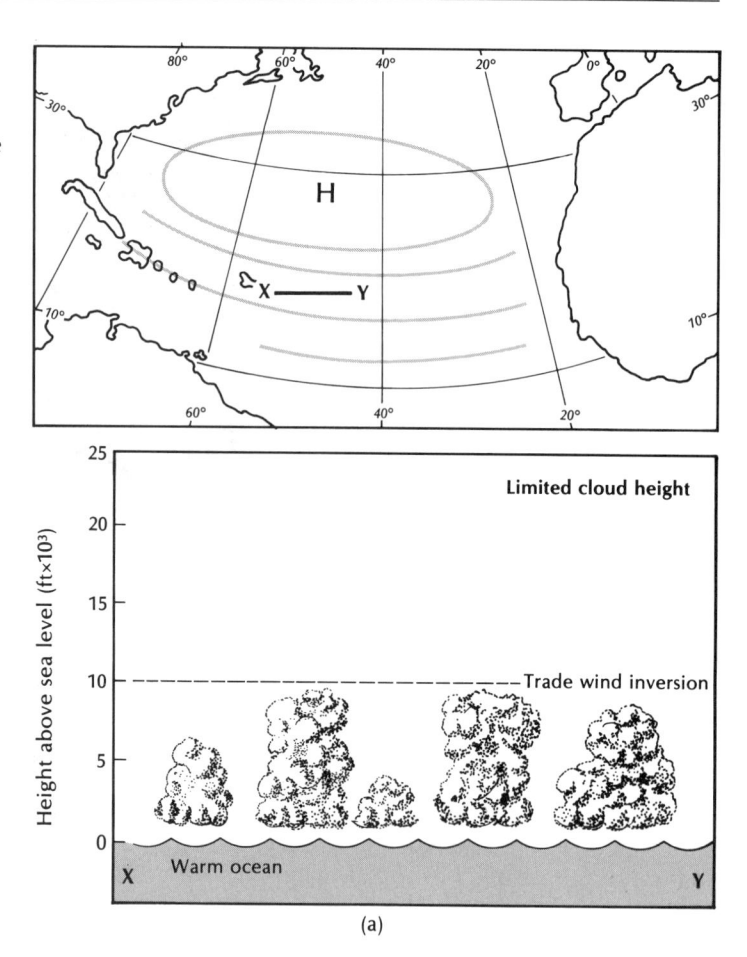

(a)

mechanisms exist, with an intruding high altitude low pressure system being the most often cited cause (figure 7-14). Derived from the remains of an upper troposphere cyclone wave in the westerlies, these abandoned waves act in two ways to promote instability. First, divergence is associated with the eastern side of the abandoned system; second, the low has a cold core so that the lapse rate is modified beneath it. Both the divergence and the modified stability enhance surface pressure differences enough to generate a hurricane.

When a hurricane does form, its size and potential for damage are variable enough to allow hurricanes to be categorized. As table 7-3 shows, this Saffir/Simpson scale ranges from a value of 1, the weakest, to a value of 5 for severe storms. The derivation of the scale rating is based upon pressure, wind speed, the size of the ocean waves generated, and the relative damage that the storm could produce.

The specialized conditions required for hurricane formation limits their origin to tropical oceans during the high sun season (figure 7-15). Over the central and western portions of the Pacific Ocean, where twenty storms may occur each year, the prime season is from June to October. In the eastern Pacific, southwest of Mexico, an average of three hurricanes may occur each year, but these die before striking any land. The areas of the United States most prone to hurricane damage are the Atlantic and Gulf coasts, where the hurricane season extends from June through November. Data applicable to hurricanes have been accumulated by the staff of the

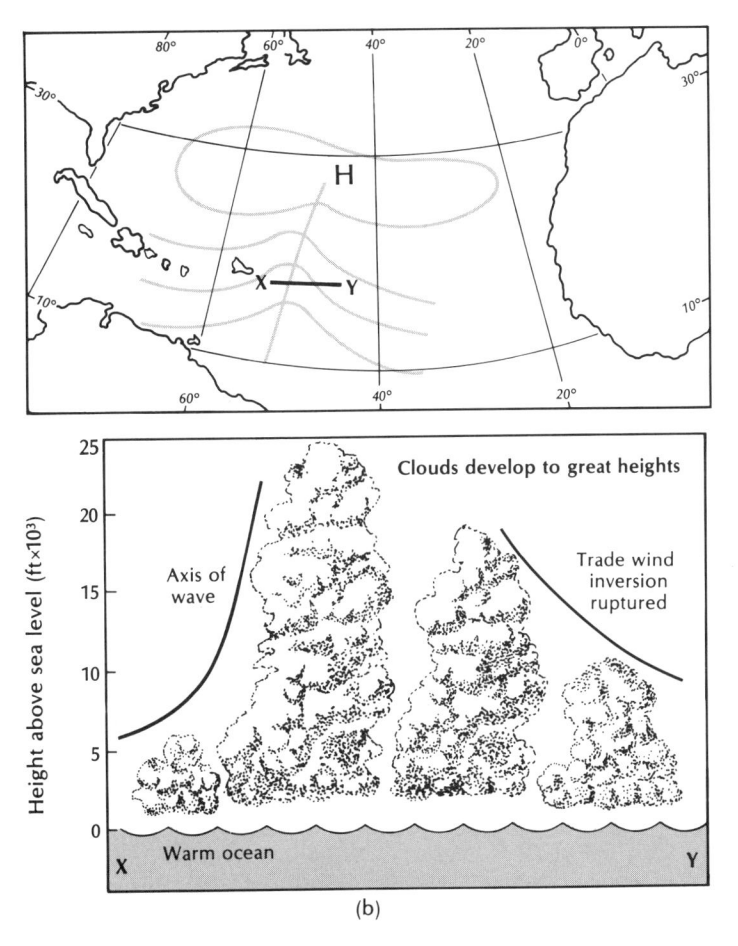

TABLE 7-3 Saffir/Simpson hurricane scale ranges.

Scale Number (Category)	Central Pressure (mb)	(in.)	Winds (mi/h)	Surge (ft)	Damage
1	≥980	28.94	74–95	4–5	Minimal
2	965–979	28.50–28.91	96–110	6–8	Moderate
3	945–964	27.91–28.47	111–130	9–12	Extensive
4	920–944	27.17–27.88	131–155	13–18	Extreme
5	<920	<27.17	>155	>18	Catastrophic

PHYSICAL AND DYNAMIC CLIMATOLOGY

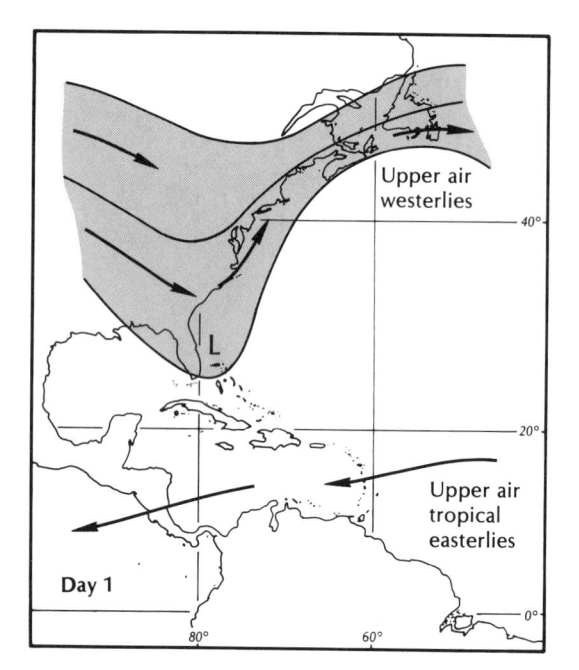

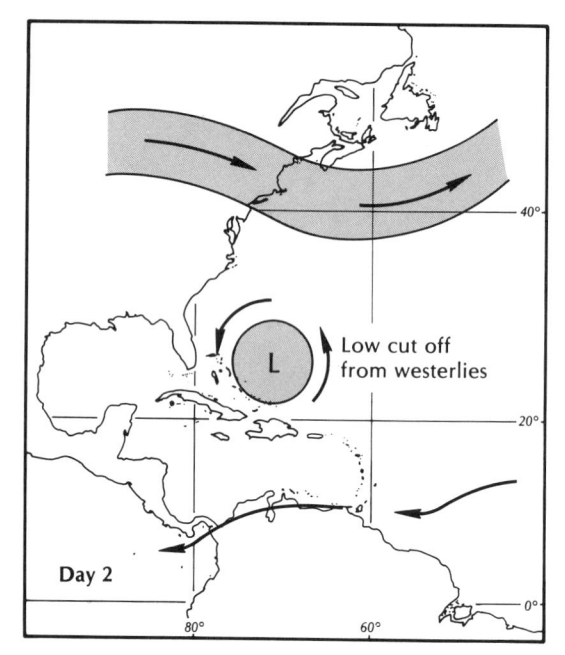

FIGURE 7-14 A cutoff from waves in the upper air westerlies creates favorable conditions for hurricane formation. (After *The Science and Wonders of the Atmosphere*, S. D. Gedzelman, Copyright © [1980, John Wiley & Sons]. Reprinted by permission of John Wiley & Sons, Inc.)

National Hurricane Center in Florida and provide the sources for the varied information given in table 7-4. Each component of the table is discussed below:

1. Major hurricanes are those classified as 3, 4, and 5 on the Saffir/Simpson scale. Of the hurricanes influencing the United States for the period 1900-1978, fewer than half can be considered as major, with the majority falling in the Scale 1 category.

2. For the period 1900-1978, more than 50 percent of hurricanes affecting the United States occurred during September.

3. In terms of losses, it is necessary to separate the storms into the categories "deadliest" and "costliest." It will be noted that the two do not coincide. This reflects a number of developments over time—largely the fact that deaths in recent years have been minimized because of early warnings, whereas costs have increased because of the development of the Atlantic and Gulf coasts.

The lengthy intervals between hurricanes at some coastal areas (as much as 60 years) often lead to a false sense of security by the inhabitants. Unfortunately, this illusion leads to the problem of lack of planning for hurricanes. Such a case is discussed in Applied Study 7.

152

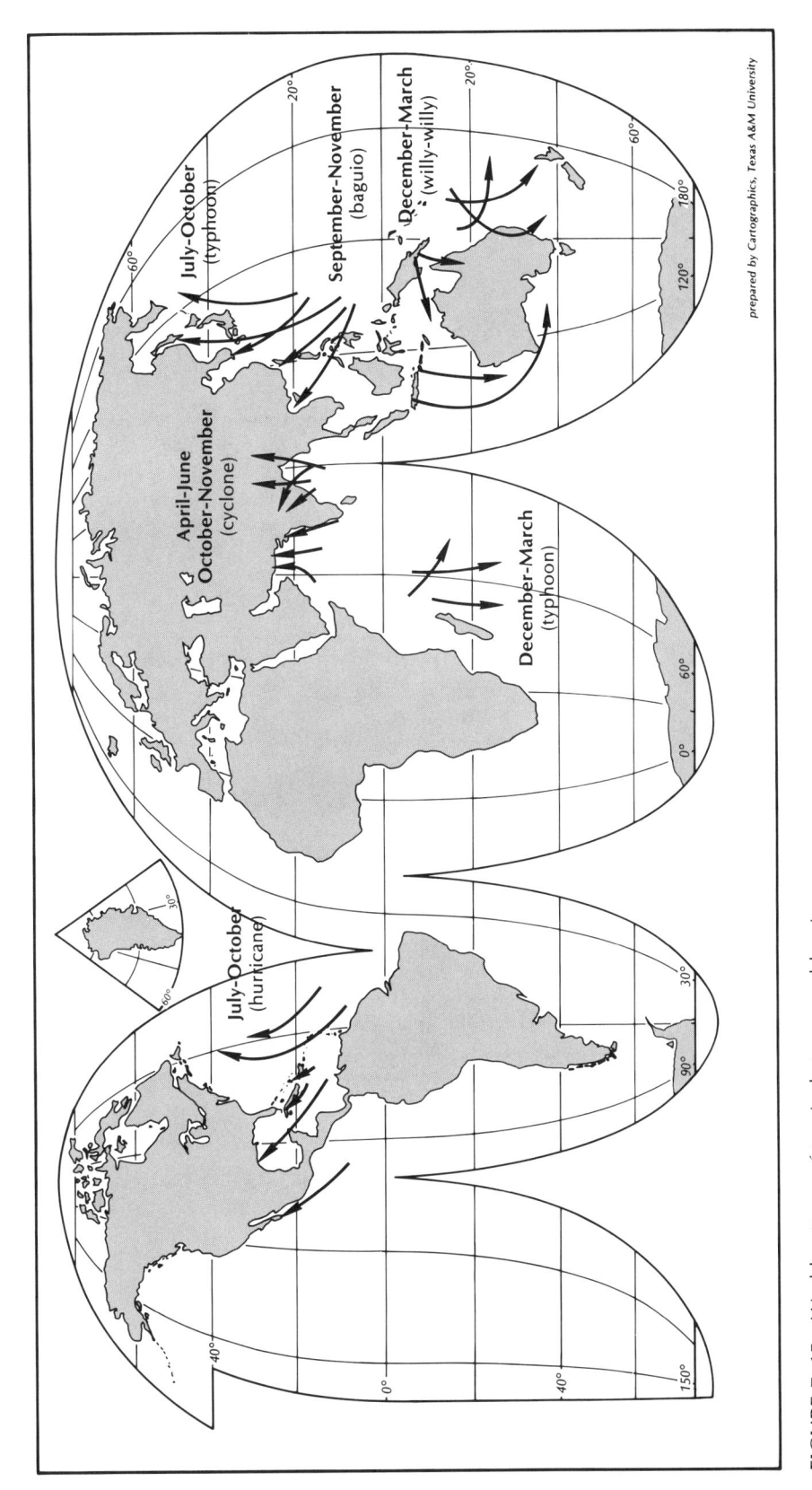

FIGURE 7-15 World patterns of tropical storms and hurricanes.

TABLE 7-4 Hurricane data.

Hurricanes Affecting United States (1900–1978) By Category
Saffir/Simpson Scale Number

	1	2	3	4	5	All
U.S.	47	29	38	13	2	129
Fla.	18	11	15	5	1	50
Tex.	9	9	7	6	0	31
La.	4	6	6	3	1	20
N.C.	9	6	3	1	0	19

Hurricanes by Month
(More Than 3 on Saffir/Simpson Scale)

	June	July	Aug.	Sept.	Oct.	All
U.S.A.	2	2	11	30	7	53
Fla.	0	1	1	14	5	21
Tex.	1	1	5	6	0	13
La.	2	0	3	4	1	10
N.C.	0	0	1	5	1	7

Deadliest and Costliest Hurricanes

Location (Name)	Year	Category	Deaths
1. Tex. (Galveston)	1900	4	6000
2. Fla. (Lake Okeechobee)	1928	4	1836
3. Fla. (Keys)	1919	4	900
4. New Eng.	1938	3	600
5. Fla. (Keys)	1935	5	408

Location (Name)	Year	Category	Cost ($ millions)
1. Agnes (Fla., NE U.S.A.)	1972	1	2,100
2. Camille (Miss./La.)	1969	5	1,420
3. Betsy (Fla., La.)	1965	3	1,420
4. Diane (NE U.S.A.)	1955	1	831
5. Eloise (Fla.)	1975	3	550

SUMMARY

The study of the nature and distribution of severe storms is a significant part of climatology. Such storms may be considered as systems that are responsible for energy exchanges and as climatic hazards that frequently result in losses of property and life.

  The severe storms that originate in middle latitudes are mostly associated with the formation of cumulonimbus clouds. High winds, lightning, hail, and tornadoes are associated with such clouds, the severity of which depends upon both local and regional controls. Middle latitude coastal areas are also influenced by hurricanes that have their origin in the lower latitudes. Their greatest impact is felt in coastal areas where wind, high seas, and torrential rains occur.

The climatic study of severe storms allows analysis of their distribution in both time and space, thereby permitting an estimate of the probability of their occurrence.

APPLIED STUDY 7

Storms and Planning

The hazard of storms is very real. Climatologists can provide much significant information concerning the probability and potential distribution of damage by storms. This study looks at the hazards of two storm types and the way people react to them. The first is hail; to the farming community in many areas of the country, this represents a distinctive problem. The question raised here is what can farmers do to decrease the risk of hail loss? The second hazard is hurricanes. This discussion examines the problem of indiscriminant development of areas most prone to hurricane damage. It is a significant planning problem.

To combat hail it is necessary for farmers to make some form of adjustment to minimize potential losses. On a local small scale, farmers in hail-prone areas could grow crops that are less susceptible to hail damage. However, there seems to be a general consensus among farmers that an increased net return by reducing losses to hail would be more than offset by lower yield and/or higher input costs of alternate crops. This does not appear a viable adjustment.

Another change that could be made in actual farming practice is that of orienting the direction of crop planting so that the row would be parallel to the most common direction of the hail swath. Such an orientation would permit rows of crops to protect each other and allow some of the hail to fall on non-hail-prone crops. Such a project is economically feasible but leads to other problems. Wind erosion, for example, would increase. Since this is a constant hazard for farmers, it takes precedence over hail damage, which may occur on only a few days each year.

Related to placing hail-prone crops in alternating rows is the idea of parcelling out the land so that the crop is grown in noncontiguous areas. A single storm would not then destroy a large part of that year's planting. There is some evidence that this is occurring in the Great Plains states. Whether it actually reflects hail concern or whether it represents the buying or leasing of parcels of land as they become available is not totally clear.

Currently, the most common method of planning for hail loss is through insurance coverage. Insurance is a means by which an individual can substitute small, certain payments for large, uncertain, and unpredictable losses. Unfortunately, the cost of the insurance is based upon the probability of hail's occurring, and, as indicated in the text, hail-prone areas can be readily identified. In some parts of the country the insurance rate can be as high as 25 percent to 30 percent of the crop value; as a result, many farmers do not purchase it.

Some farmers have turned to cloud seeding as a way of reducing hail loss. Cumulonimbus clouds are seeded with silver iodide, reducing the amount of supercooled water that is required for hail growth and, consequently, diminishing the amount and size of hail. Evidence of the relative success of this approach is not indicated by results of the National Hail Research Experiment. This experiment began after Soviet scientists reported that through cloud seeding they were able to reduce hailstorms by 60 percent. The five-year United States Program was begun in 1972 to check the Soviet claims. It halted after three years because no positive results were obtained. Despite this, there is an active cloud seeding program, financed by farmers, in wheat growing areas of the Great Plains.

Unlike the problem of hail, which concerns the protection of an individual farmer, hurricanes pose severe problems for large numbers of people living on some coastlines. A case in point concerns Hurricane Eloise, which occurred in the Florida Panhandle in September 1975. This storm destroyed many buildings, some of which were supposedly hurricane-proof. How and why a hurricane could be so destructive in a coastal area was the topic of a study by researchers at the National Hurricane Center in Florida. They discovered that the area around Panama City, in the Panhandle, had not experienced a major hurricane in this century. Most residents had never experienced a hurricane and believed they never would. They had forgotten, or never known about, the late nineteenth century storms in the area; in the period 1885–1896 nine hurricanes struck the Panhandle.

The study group suggested that going through a major hurricane is a learning experience, and that we should draw upon the experiences to learn basic lessons about hurricanes. Specifically:

1. Major hurricanes are relatively rare events at any one location, and coastal residents have a good chance of living many years without experiencing one. But none of the coastal areas of the eastern United States is immune, and residents must not assume that because a hurricane has not hit their area in recent years, one will not occur in the future. Major hurricanes can and will occur at any location along the coast from Brownsville, Texas, to Eastport, Maine; and people should be prepared to respond when hurricane warnings are sounded.

2. The development of the coastal sections of northwest Florida was poorly planned. The coastal area from Panama City to Pensacola had become a summer tourist region for many residents of the Southeast. Beach-front hotels, motels, and cottages mushroomed as developers rushed to capitalize on this market. Unfortunately, this type of development usually proceeds with little control. Local government struggles to keep pace, and effective planning is slow. This is what happened in northwest Florida.

This strip of coast has some of the finest beaches in the United States. Parallel to the beach are sand dunes up to 6 m (20 ft) high, around 15 m (50 ft) inland from the high-water line. The dunes are nature's method of beach preservation, and they also provide a measure of protection for inland areas. Developers literally leveled the dunes to provide guests staying in beach-front motels with instant access to the ocean. At one motel, sand dunes were leveled and replaced by the foundation of the motel. The sea wall for the motel was less than 30 m (100 ft) from the water. When Hurricane Eloise struck in September 1975, the storm surge smashed into the building and destroyed the sea wall. Water undermined part of the building, and it collapsed.

3. Building codes in northwest Florida were inadequate. Pilings were not required for most beach-front buildings. Floor-slab wings were not required for most beach-front buildings, and floor slabs were placed on top of sand dunes. Much to their horror, owners returning after Eloise had passed found their homes had toppled after being undermined.

4. In the Florida Panhandle area, the building codes that did exist were not enforced. The undermining caused by the storm's surge exposed the foundations of many buildings and provided a rare opportunity to inspect the construction practices of developers.

Another concern is how well high-rise buildings could sustain a major hurricane, particularly since people who were stranded in a hurricane might be tempted to take refuge in such tall and seemingly sturdy structures. For years, scientists had been anxious to inspect the effect of a major hurricane on a high-rise building located on the waterfront. When Eloise hit a high-rise condominium, they had an excellent opportunity. A 14-story building was less than 30.5 m (100 ft) from the water and appeared to survive the storm without major damage. The sea wall was destroyed, and undermining revealed the building was on pilings. Closer inspection, however, showed a very disturbing fact: Cement was missing from the top of many pilings, and metal rods were exposed. What if Eloise had moved more slowly and subjected the condominium to a more prolonged pounding from the sea? Would the building have collapsed? No one knows, but it is frightening to consider the fate of people stranded on the upper floors. It is evident that careful planning is required to reduce potential hurricane disasters.

TWO

Regional Climatology

CHAPTER 8: Regional Climates: Scales of Study
CHAPTER 9: Regionalization of the Climatic Environment
CHAPTER 10: Tropical Climates
CHAPTER 11: The Mid-Latitude Climates
CHAPTER 12: Climates of North America
CHAPTER 13: Polar and Highland Climates

8 Regional Climates: Scales of Study

DEFINITIONS

MICROCLIMATES
General Characteristics
The Role of Surface Cover
The Role of Topography

LOCAL CLIMATOLOGY

MESO- AND MACROCLIMATES
Synoptic Climatology
Regional Climatology

SUMMARY

APPLIED STUDY 8
Donora: An Air Pollution Episode

REGIONAL CLIMATOLOGY

Climates can be identified and analyzed over a broad range of areal units from the small vegetable garden up to the continental size regions. The climate over a plowed field is different from that over a field of clover or pasture. The climate of a city differs from that of a rural area, and the climate of a desert differs from that of a rain forest. Significant differences occur in the climate at all of these different scales. Climatologists must deal with these differences at all levels. Often the analytic approach used varies with the relative size of the study area. Before considering major world climates, it is desirable to examine some other scales of study briefly.

DEFINITIONS

There has been much confusion in identifying and naming the range of scale in climatic studies, largely as a result of the difficulty of separating the atmospheric continuum into discrete units. The identification problem has been compounded by the historic evolution of climatic studies. Researchers in different countries have used names for areas that, in other countries, are described by different terms. Thus we find such terms as *gelandeklima*, *ecoclimate*, and *topoclimate* referring to areas of somewhat similar size. At the same time, the dimensions suggested as suitable boundaries for identified scales frequently differ from one source to another.

In a discussion of scale in climatology, M. M. Yoshino derived a general consensus of definitions; the description given here follows his grouping. Figure 8-1 provides examples of the scales that he suggests. The major subdivisions include the following:

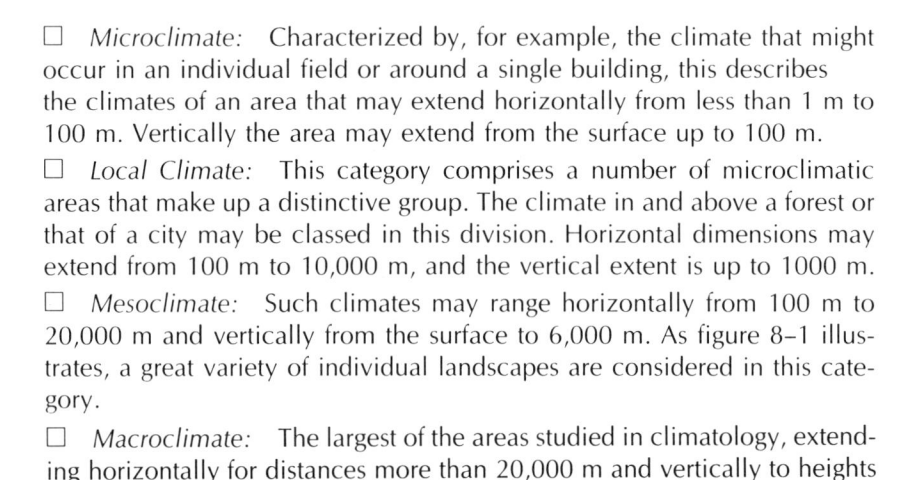

☐ *Microclimate:* Characterized by, for example, the climate that might occur in an individual field or around a single building, this describes the climates of an area that may extend horizontally from less than 1 m to 100 m. Vertically the area may extend from the surface up to 100 m.

☐ *Local Climate:* This category comprises a number of microclimatic areas that make up a distinctive group. The climate in and above a forest or that of a city may be classed in this division. Horizontal dimensions may extend from 100 m to 10,000 m, and the vertical extent is up to 1000 m.

☐ *Mesoclimate:* Such climates may range horizontally from 100 m to 20,000 m and vertically from the surface to 6,000 m. As figure 8-1 illustrates, a great variety of individual landscapes are considered in this category.

☐ *Macroclimate:* The largest of the areas studied in climatology, extending horizontally for distances more than 20,000 m and vertically to heights in excess of 6,000 m. Such areas can be continental in extent.

The dimensions provided in the listing are basic guides, with many area studies overlapping the identified groups. Perhaps the best way to illustrate the role of scale in regional studies is to examine specific examples.

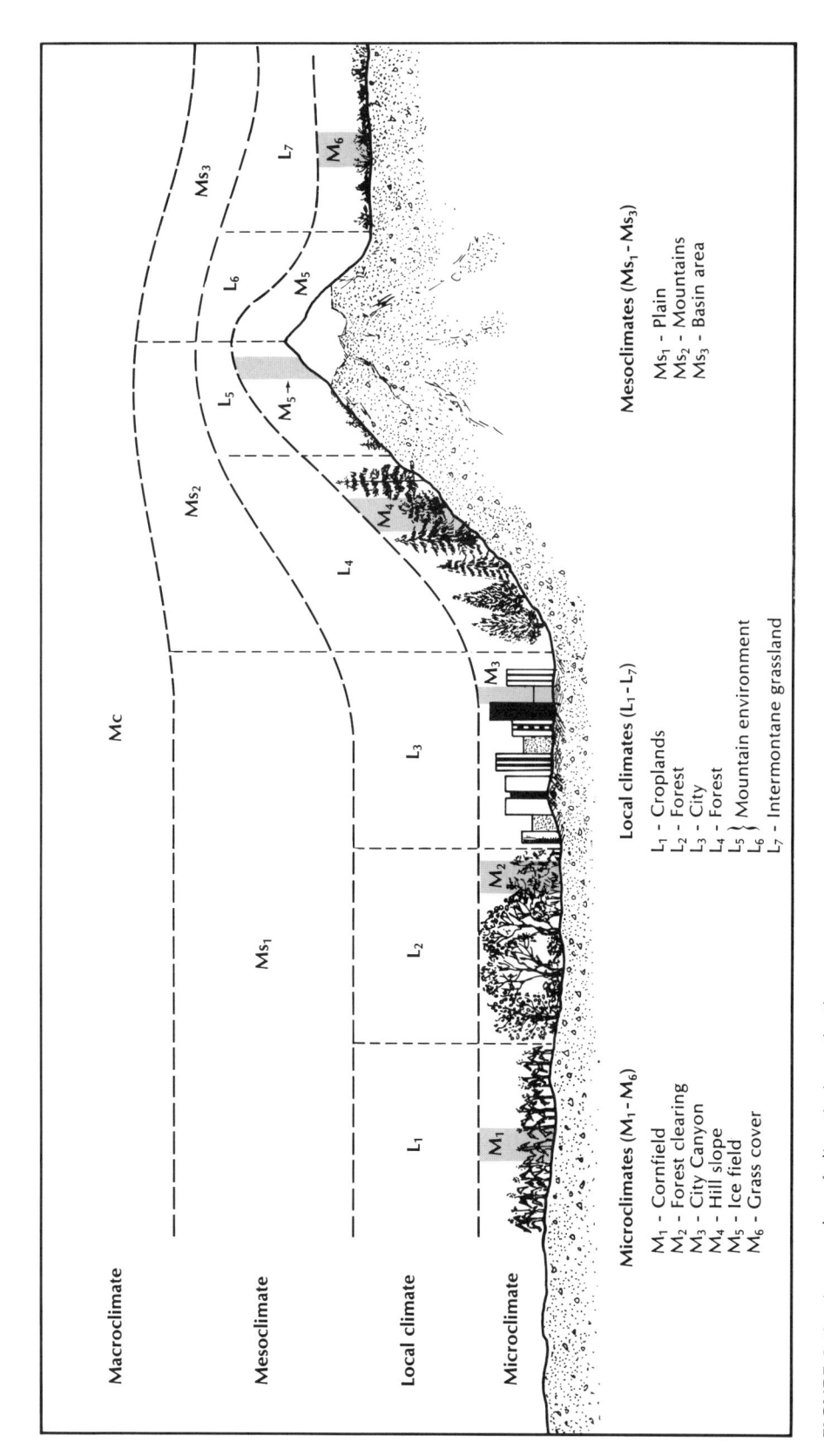

FIGURE 8-1 Area scales of climatic investigation.

MICROCLIMATES

Studies of these small areal units inevitably begin with field work. This preliminary step is required for accumulation of data and generally includes the measurement of variables. Such a procedure is needed because published climatological data are almost entirely composed of readings taken at the standard height of the instrument shelter. The microclimatologist is often concerned with the state of the atmosphere below that level. Furthermore, the stations that are used to report climatological data are widely spread, with perhaps one station being the representative of many square miles. To obtain data such that differences over small distances can be derived, it is necessary to emplace a fairly dense network of instruments within a small area.

Many studies of microclimatic environments have provided a generalized picture of climatic conditions near the ground. These findings indicate the types of conditions that will occur over a bare surface. Through analysis of surfaces covered by vegetation or synthetic materials it is possible to generate a more complete understanding of the variation of climatological processes that occurs at the microscale.

General Characteristics of Microclimates

Figure 8-2 shows the temperature characteristics at the interface of the earth and atmosphere. A large diurnal temperature range occurs at the near-earth levels; during the day interface temperatures have been found to be as much as 10°C warmer than the air only 1 m above the surface. At night, the situation is reversed; an

FIGURE 8-2 The daily course of temperature on a summer day at three elevations (Data from Geiger [1950]).

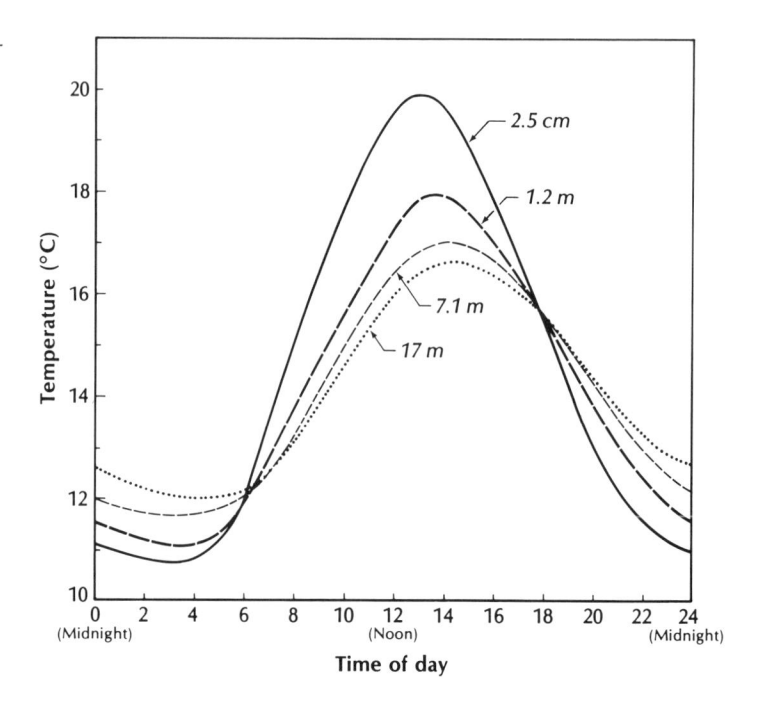

FIGURE 8-3 Idealized profiles of temperatures near the boundary layer in clear weather (After Oke, 1978).

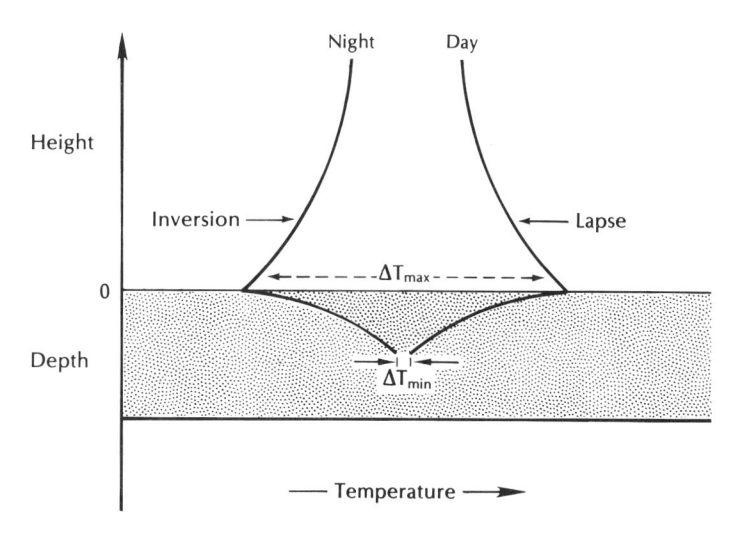

inverted lapse rate occurs, with temperatures at the surface cooler than those immediately above. Such a response is to be anticipated. During daylight hours incoming solar radiation warms the surface and heat diffuses into the atmosphere, raising the temperature of the air by smaller amounts at increasing altitude above the ground. After sunset, under clear-sky conditions, the surface cools rapidly by radiation and the atmosphere loses heat by diffusion to the cold surface.

Temperature changes in soils decrease with depth. As figure 8-3 illustrates, the amplitude of diurnal temperature change is greater near the surface and decreases with depth until equilibrium is attained. A similar pattern is obtained for the annual cycle, although, of course, it follows the seasonal rather than the diurnal cycle.

Figure 8-4 is an idealized representation of wind near the surface. At the interface between the air and the ground, a thin layer of air adheres to the surface; in this layer the flow is *laminar*: streamlines are parallel to the surface and lack the cross-stream component of convective currents. The depth of this layer depends upon the surface roughness and wind speed but is seldom more than 1 millimeter (mm) in thickness. Its significance lies in its role as an insulating barrier in which all nonradiative transfer is by molecular diffusion rather than the turbulent transfer typical of most of the lower atmosphere. The turbulent surface layer comprises a complex flow of swirling eddies extending up some 50 m. In this zone, a general increase of wind speed occurs with height.

FIGURE 8-4 Flow of air at the boundary layer changes from laminar to turbulent flow.

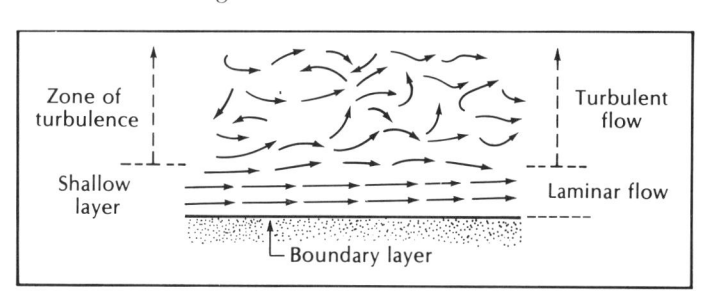

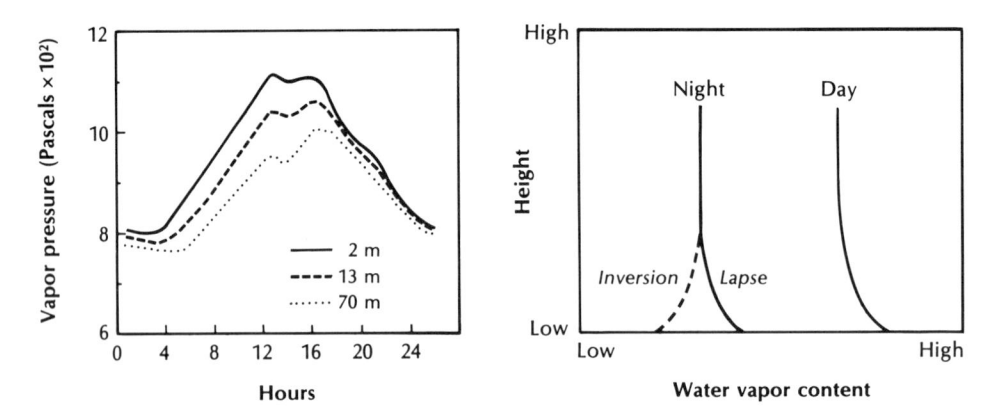

FIGURE 8-5 Generalized profile of water vapor concentration near the boundary layer (After Oke, 1978).

The exchange of moisture between the surface and the air above is reflected in humidity measurements at various levels. Figure 8-5 provides a generalized profile of water vapor concentration for day and night. During the day the concentration decreases away from the surface in a similar way to the temperature profile. At night, if the dewpoint of the air is not attained, the humidity profile is somewhat similar to that in daytime. This occurs because evaporation will continue during the night hours, but at a lower rate than during the day. If the dewpoint is reached, an inverted moisture profile will be found. The deposition of dew causes a lowering of near-surface moisture content; if the moisture is replaced by downward movement of air, dew formation will continue. Downward movement will occur through slight turbulence. If this turbulence does not occur, the near-surface air is not replenished with moisture and dewfall ceases.

Although it is possible to generalize about the nature of the microclimate above a bare soil surface, it is important to note that the actual characteristics will depend, at least in part, upon the type of soil surface exposed and the amount of water it contains. Sandy soils, for example, have a lower heat capacity and thermal conductivity than clays. This means that a sandy soil will heat up rapidly in its top layers during the day but at night will be cooler than less-sandy soils because it will undergo rapid radiational cooling. Organic matter in the soil reduces heat capacity and thermal conductivity, and the dark color will increase the absorption of solar radiation. At night, such soils will be relatively warm.

The presence of water in the soil or at the surface will greatly modify the exchanges of energy that occur. When water is present, incoming energy is used for evaporation, making less available for sensible heat. A comparison of the energy budgets for an irrigated field and one that is dry shows that the temperatures over a moist soil are usually lower than those over dry ground.

These general principles apply to all microclimates. However, modification of the bare surface through vegetative growth or human interference alters the intensity and rates at which ongoing processes occur. Such changes are described in the following section.

The Role of Surface Cover

The role of surface covering in creating microclimates is a response to the way in which incoming energy is disposed at the surface and the way in which the surface modifies air flow. The differences that exist can best be shown through illustrative data. Figure 8–6(a) shows, in schematic form, some modifications that

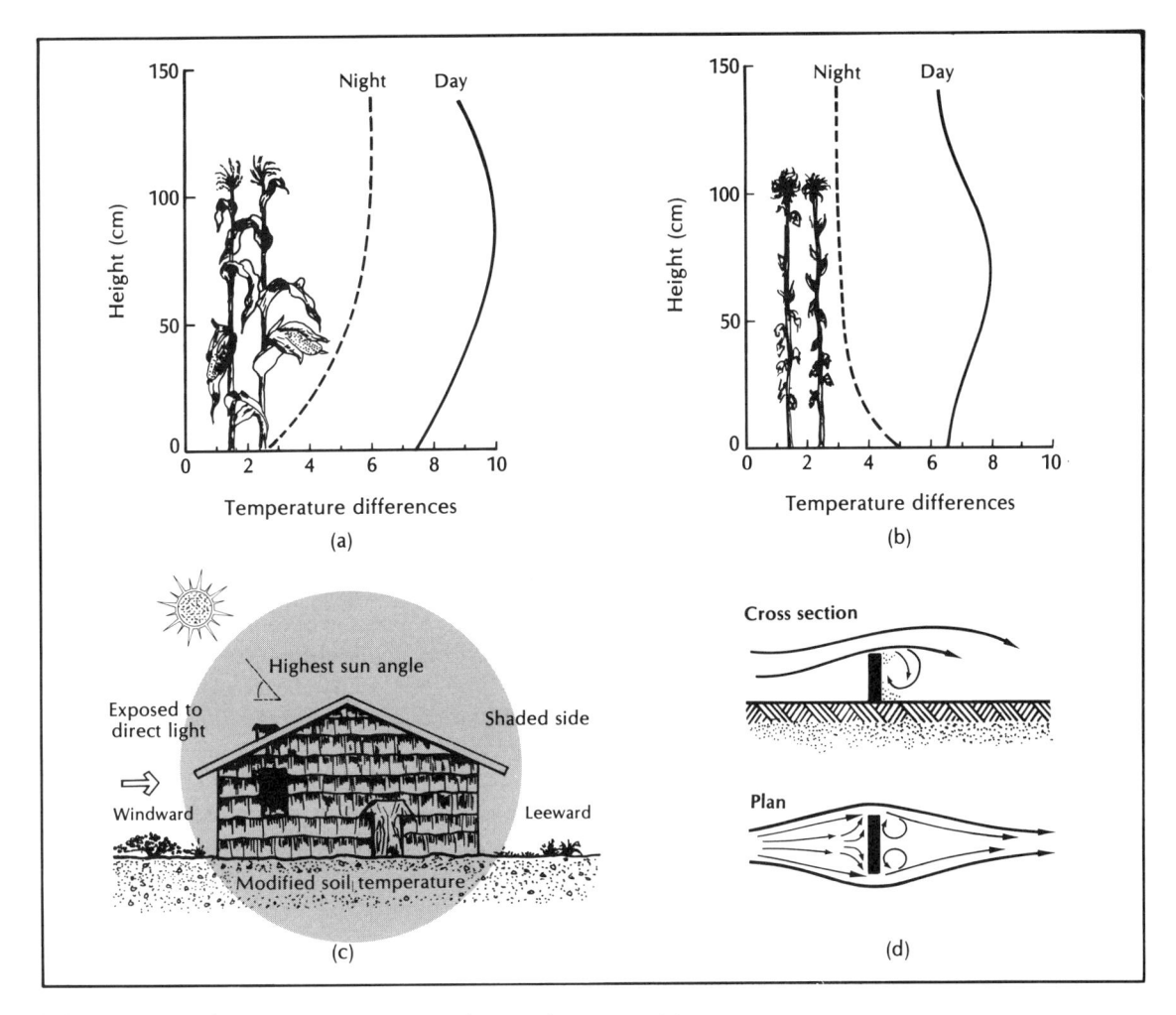

FIGURE 8-6 Schematic representation of microclimatic modifications caused by surface cover. (a) Temperature profiles in a crop whose growth shades the ground. (b) Temperature profiles for crop with essentially vertical growth. (c) Climatic sheath around a building. In the sheath many microclimatic variations occur. Examples show creation of windward-leeward sides, sunlit and shaded sides, and soil modification. (d) Airflow across a barrier at right angles to wind direction. Cross section shows how such an obstruction is used as a snow fence.

occur when plants cover the surface. Of particular significance is the creation of a canopy layer in which the microenvironment reflects the nature and extent of the canopy. Clearly, the relative continuity of the canopy plays a major role, so that plants with large leaves horizontal to the surface form a more effective canopy than those whose leaves are small and are aligned vertically. The canopy becomes most effective when a plant stand has grown to the extent the ground is shaded. This growth causes the highest temperature zone to move away from the surface to the canopy area. What in effect happens is that the top of the canopy, rather than the soil surface, becomes the energy exchange layer.

A building or a fence, for example, has a marked effect upon the microclimate. A building creates a climatic "sheath" (Figure 8-6[b]), in which temperature variations will occur as a result of shading, humidity anomalies are found, and even a rain shadow effect may occur. When the changes in the microenvironment of a single building are assessed, it is not surprising that cities consisting of buildings can create their own climate.

The modification of wind in the microenvironment is well demonstrated by effects of a vertical structure on air flow. As the diagram in figure 8-6(c) shows, the patterns change with the direction from which the wind is blowing. Such modification has been put to good use in the construction of snow fences to keep roads and other used areas clear of drifting snow.

The Role of Topography

In chapter 3, it was noted that topographic aspect is of importance in determining the distribution of temperature at a given location. The height of the snow line on the equatorward facing slopes compared to those facing poleward was stressed. This difference can be translated to the microlevel. Most people have seen the effect during winter when snow on one side of a small hill remains in place long after that on the opposite slope has melted.

Topographic variations also modify thermal patterns in small areas. One of the best-known effects of this is the creation of an inversion that, at times, can create a distinctive air flow. On cool, still nights air close to the surface becomes cooler than that above. If this cooling occurs in areas of uneven topography, the cold dense surface air will tend to flow downslope and accumulate in the bottom of valleys and depressions. Such air flow is called a *katabatic wind*.

If the temperature of the ground is at or just below freezing, the collection of cold air in the valley bottom causes a localized frost. Such an effect in citrus growing areas can cause appreciable damage to crops. Another negative effect of topographically induced air drainage is the creation of an air pollution potential. If the pool of cold air becomes deep enough, any effluent carried into the inversion layer will be trapped in it. Should the inversion remain in place for more than a few days, the stagnant air can become highly polluted. Such a case occurred in the well-known Donora, Pennsylvania, air pollution disaster, which is described in Applied Study 8.

LOCAL CLIMATOLOGY

Local climates consist of a number of microclimatic environments because they comprise a larger area. There is, of course, overlap in the organizational framework, but a further differentiation can be made because local climatic studies stress horizontal rather than vertical differences in climate. There are many examples of local climatic studies, and here the climate of a forest area is used to illustrate this level of analysis.

Forest climates differ from those of surrounding nonforested areas. The boundary layer of the forest is its canopy, for it is at this level that energy exchanges will occur. Some insolation is returned directly to space, the amount depending upon the albedo, or reflectivity, of the canopy layer. In some forests the quantity will vary enormously from season to season. Some energy will be trapped within the canopy layer and some will penetrate to the floor of the forest. As illustrated by the data in table 8-1, the amount of penetration is characteristically low. The evaluation of the data is further complicated in that the amounts involved in the disposition of radiant energy vary both with the state of the sky and the amount of foliage that exists (table 8-1).

Compared to the flow of air over open areas, wind inside the forest is reduced. The amount of decrease will depend upon the type and structure of the forest. For example, in a deciduous forest the wind velocity is reduced by as much as 60-80 percent at a distance of 30 m (100 ft) inside the forest. Similarly, in a Brazilian forest, the wind speed has been found to decrease from about 8 km/h (5 mph) to 1.6

TABLE 8-1 Variations in radiation receipts within forests.

	Daily totals of net radiation in and above a young pine forest (ly/day)					
Height (m)	10.0	5.0	4.1	3.3	2.1	0.2
7 July 1952						
total	566	555	223	36	—	35
percentage	100	98	39	6	—	6
9 November 1951						
total	291	—	104	—	14	—
percentage	100	—	35	—	5	—

	Solar radiation received on a horizontal surface at the 1-m level in an oak forest as a percentage of incident radiation above the canopy	
	Clear Sky	Overcast Sky
Foliaged	9%	11%
Defoliaged	27%	56%

Source: From Munn (1966).

km/h (1 mph) with the same distance. The flow of air is, of course, highly complex and varies in the vertical as well as the horizontal dimensions. Studies have shown that winds of 8-24 km/h (5-15 mph) above the canopy are often less than 3.2 km/h (2 mph) at the surface.

The forest environment also modifies local moisture conditions. Evaporation from the forest floor is relatively low because of reduced insolation and wind velocity. This effect is counterbalanced by the fact that with the profuse vegetation, high transpiration occurs. As such, the humidity within a forest depends upon the density of the forest and the rates of transpiration that occur. It is found generally that the relative humidity in the forest may be from 2 percent to 10 percent higher than that of nonforested areas, with the highest humidities occurring during the high-sun season. Note that comparisons of humidity using relative humidity values are not always meaningful, for the modified thermal environment directly influences the water-holding capability expressed by using relative humidity.

The thermal differences that occur result from a combination of the factors already outlined: shelter from direct rays of the sun, heat modification through water transfer, and the "blanket" effect of the canopy. The essential result of the interaction of these factors is that temperatures inside a forest are moderated, the maximum is lower, and the minimum higher than those of nonforested areas experiencing a similar climatic regime. The amount of variation is seasonal: the main difference in summer being as much as 2.8°C (5°F) within a low-altitude, middle-latitude forest. Exceptions to such moderating influences do occur; the Forteto oak forests of the Mediterranean, for example, experience higher temperatures than neighboring, nonforested areas. Such trees transpire slowly, with the result that the "usual" hygric and thermal conditions of the forest are modified.

Forest temperatures vary vertically as well as horizontally. For the most part, temperature increases with height during the day and lapse conditions prevail at night, reflecting the role of the canopy in the energy exchange.

MESO- AND MACROCLIMATES

Mesoscale climates are those that are frequently identified with a distinctive geographic region. In such a region the physical controls of climate are similar and are not modified by major differences within the region. Thus a mesoscale climatic study might concern the climates of areas such as the Central Valley of California, the lands in the vicinity of the Great Lakes, or the Mississippi Delta lowlands.

Climatologists have completed extensive research on mesoscale climates. The type of research has evolved over time as newer methodologies became available. Early studies often comprised an inventory approach of the climate of a region. Classic works on such locations as the Paris and Thames basins provided the basis for the understanding of much regional climatology.

More recently, mesoscale climatic studies have often concerned scales of motion rather than scales of area. This development corresponds to the meteorological concept of mesoscale phenomena, which includes analysis of such features as severe storms, mountain-valley winds, and the like. The climatological equivalents are seen in studies ranging from the cause of the Sahel drought to precipitation variations resulting from circulation patterns in the Midwest of the United States.

While the distinction between various scales of climate is not always clear, the identification of *macroclimates* is aided by the concept of *filtering*, the averaging of circulation features over successively longer time periods. As the time scale becomes longer, the smaller scale patterns are filtered out and only the general characteristics remain. Thus a mesoscale feature such as a hurricane or a land-sea breeze would not appear on a macroscale representation.

The climatic analysis of macroclimates can follow two methods of approach. First, the analysis can consider the major surface climatic characteristics, ranging from temperature and precipitation patterns to statistical analysis of other elements. Such an approach provides the descriptive climatology of large climatic regions or even continents. Second, the analysis can deal with the dominant circulation patterns that influence the climate and provide the key to understanding its cause. This process is accomplished through analysis of the state of the atmosphere as inferred by pressure maps showing patterns and winds at various levels of the atmosphere. Alternatively, the patterns can be depicted as cross-sections through the troposphere and stratosphere. This approach obtains a synopsis, or condensed view, of the atmosphere at a given time and is referred to as the *synoptic climatology*.

Synoptic Climatology

A good example of synoptic climatology concerns the relationship between surface conditions and midtropospheric waves over the United States. To gain a three-dimensional view of this relationship requires the somewhat difficult step of passing from the surface pressure chart to the upper air contour chart. This transition is compounded by the change in units that apply. The surface chart shows air pressure (the variable unit) reduced to sea level (the fixed unit). The upper air chart shows conditions by means of constant pressure charts in which the variable unit is height. In fact, the constant pressure chart can be thought of as a sheet of rubber that is distorted out of flat plane (the pressure surface) and on which are drawn lines that are a measure of the distortion.

Figure 8–7 provides an example of a surface synoptic, an 850-mb, and a 500-mb chart over Western Europe. The 1000-mb (surface) chart shows the low pressure systems over Scandinavia and the southeast Mediterranean Sea together with the associated frontal systems. Over the Atlantic, a broad anticyclone is seen. When this situation is viewed on the 850-mb chart (middle) the low pressure over Scandinavia is seen to deepen and the Atlantic high to become even more ridge-like. At 500 mb (bottom chart) the situation is dominated by the Scandinavian low. This example shows how the pressure patterns, although related in terms of the circulation that exists, vary with elevation. Furthermore, the use of the pressure surface in the charts is illustrative of the use of the constant pressure charts in synoptic analysis.

REGIONAL CLIMATOLOGY

FIGURE 8-7 Synoptic charts at three levels over Western Europe (After Giles, [1976]).

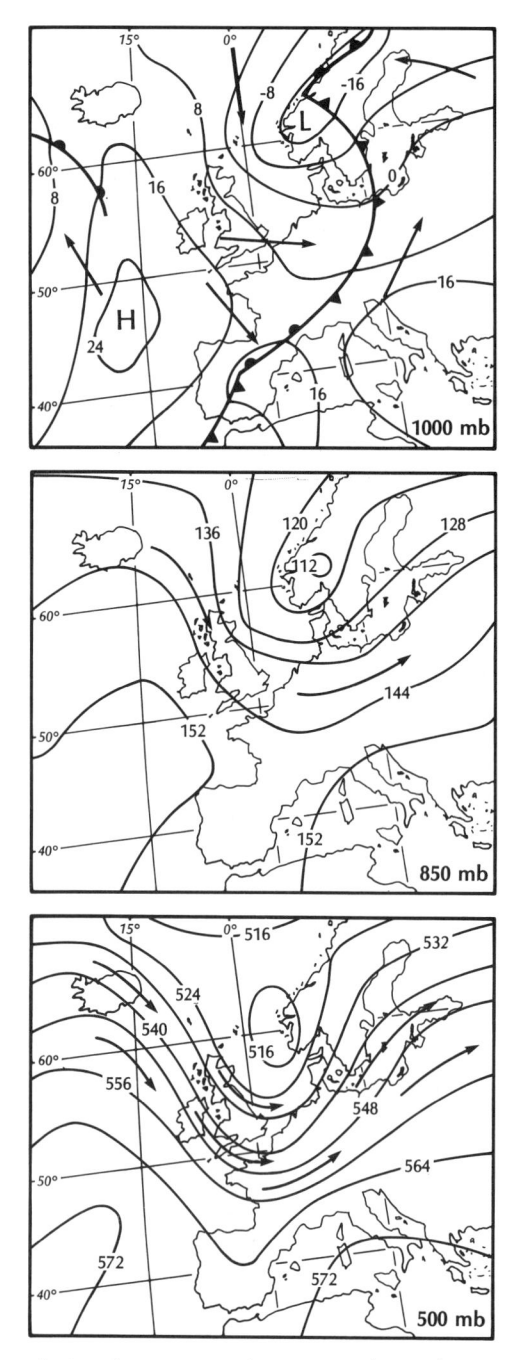

Given these relationships, some discussion about the units used to show the pressure charts and air movements that occur is required. The heights on the upper air charts are given in geopotential meters (gpm). *Geopotential* is a measure of the potential energy a body has due to earth's gravity. It thus depends upon height above an arbitrary base—in this case, sea level. Since gravity varies over the earth, horizontal surfaces are not necessarily level surfaces, so a certain amount of energy is required to move air particles along horizontal surfaces over the earth. In reality, there is little difference between geopotential meters (gpm) and the ordinary geometric meter (m); thus in the illustrations, the interval in gpm's can be considered as presenting meters.

To gain insight into air motion at various levels, wind-flow arrows have been added to the isobaric maps in figure 8-7. Note that in the surface map the winds circulate around the high and low pressure centers, crossing the isobars at an angle. On the 500-mb chart, the winds move parallel to the isobars. It can be seen that surface pressure conditions are but one expression of conditions aloft. In fact, the sinuous curves of the 500-mb chart are part of the prevailing Rossby regime that encircles the earth in middle and high latitudes.

The normal circulation of the upper air flows associated with the Rossby regime is illustrated in figure 8-8(a). Essentially, the air flows in an west-east direction with the flow interrupted by wavelike patterns. The highest velocity winds in the flow occur in a corridor of rapidly moving air called a *jet stream*, which represents the boundary between cold polar air and warm tropical air aloft; at the surface, the location of the junction is marked by fronts. The normal patterns, as illustrated, show that the jet stream is located in its typical winter position across the southern parts of the United States and that the wave patterns are shallow.

Compare these normal conditions with those showing the 1976-1977 winter season (figure 8-8[b]). There are remarkable differences that can be assessed by comparing—

1. the amplitude of the wave pattern,
2. the location of the high-pressure ridge over the Pacific Ocean,
3. the southerly extent of the low pressure normally off the northern Canadian islands,
4. the intense high pressure in the polar area, and
5. the deepening of the low-pressure system off the Aleutians.

In all, the pattern is quite different from the normal, with the result that the winters in many places in the United States varied appreciably from what was normally expected. A comparison of Figure 8-8(a) with 8-8(b) permits an estimate to be made of the ways in which various locations felt the effects of the different circulation patterns of 1976-1977. Of immediate note are the following:

1. The high-pressure ridge that extends over the eastern Pacific Ocean causes the basic jet stream path to be far north of its usual path. This means that the West Coast of the United States does not fall in the path of the frontal systems that bring rain in from the Pacific. The area is dry.
2. The same ridging effect means that warm, moist air would extend into Alaska. Usually dominated by cold air, the Alaskan area would have experienced a warm or mild winter.

FIGURE 8-8 Schematic diagrams illustrating upper air circulation patterns in winter: (a) the "normal" pattern; (b) circulation that resulted in the severe winter over much of the United States in 1976–1977.

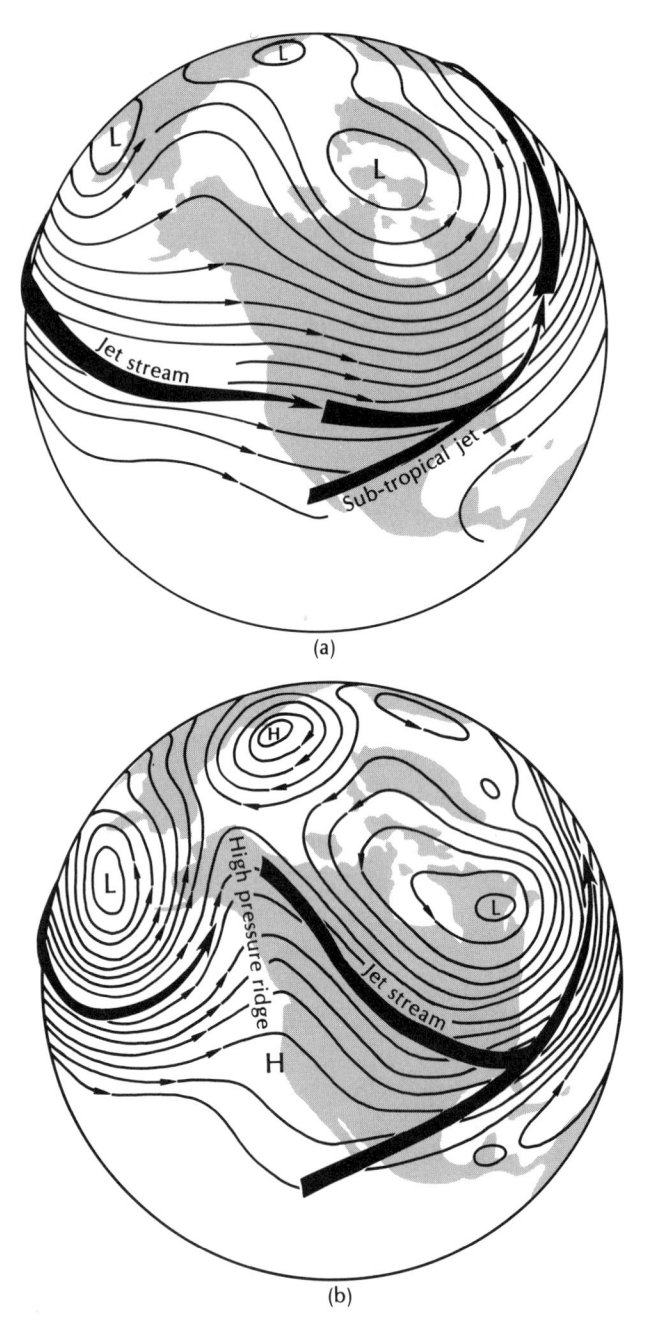

3. The exaggerated jet streams axis would be maintained by movement around the low pressure over Labrador, and cold Arctic air would stream across much of the central United States.

These conditions, hypothesized from the observed circulation patterns, actually occurred. Figure 8–9 shows the temperature departure (in °F) from normal

REGIONAL CLIMATES: SCALES OF STUDY

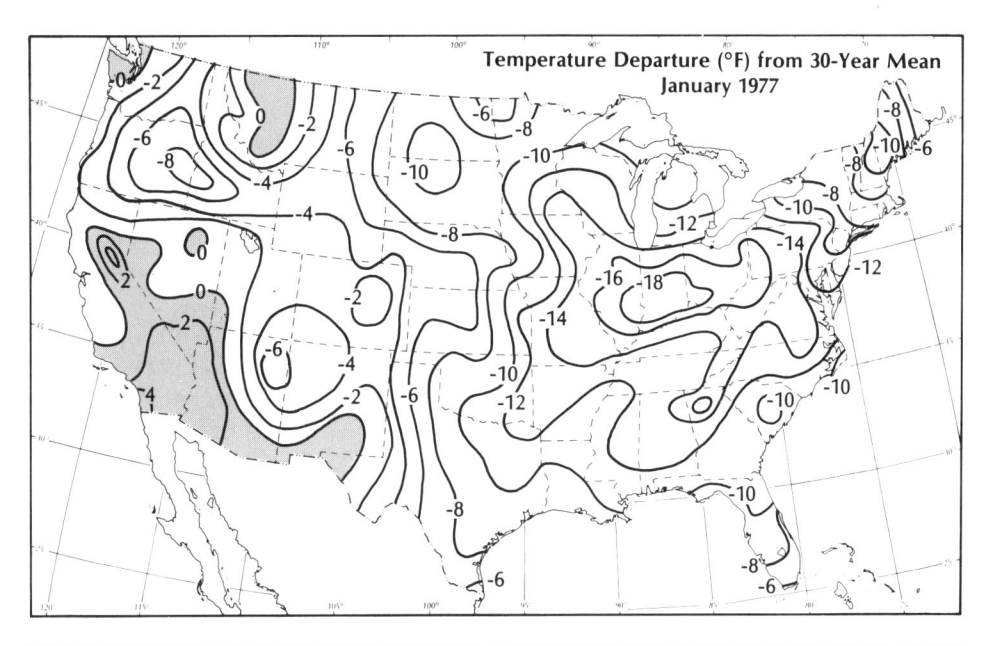

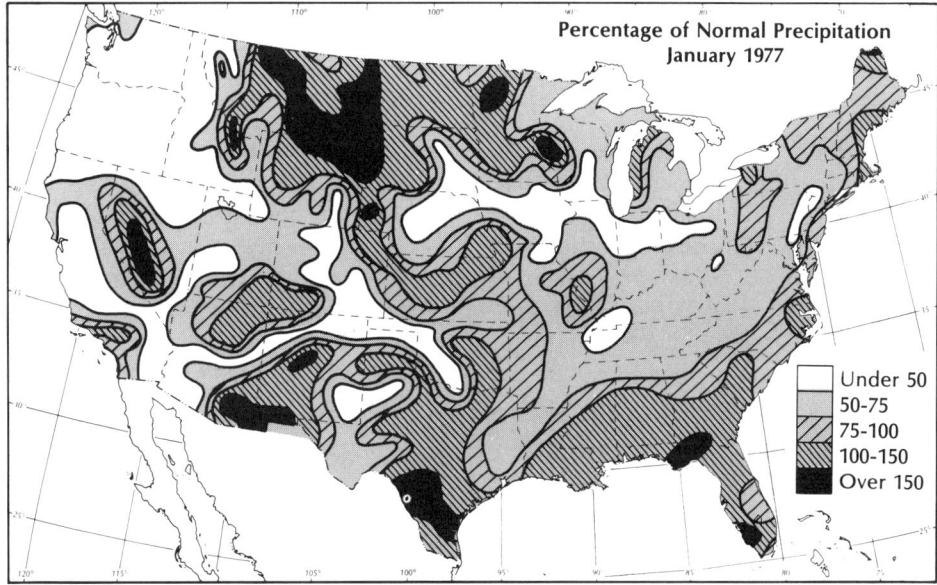

FIGURE 8-9 Unusual circulation patterns of winter 1976–1977 resulted in significant departures from the normal: (a) January temperature departures (NOAA); (b) percentage of normal precipitation for the same month (NOAA).

during January 1977. Look at the departure in the Midwest, where average temperatures were 18°F less than normal with dire results.

Figure 8-9(b) shows percentage of normal rainfall in January 1977. As anticipated, large parts of the west are unshaded, indicating rainfall less than 50

175

percent of normal. The lack of winter rainfall continued an already dry period and furthermore bode ill for the coming year, since much of the river water that flows from the Rockies is fed by melting snow. During this time the western mountains failed to receive their winter replenishment, making prospects for sufficient summer irrigation needs extremely dim.

The synoptic climates that produced the unusual January conditions can be clearly analyzed. There is an important question, however, of why such conditions occurred. In recent years much research has been undertaken to understand why such circulation patterns occur, and a number of possible explanations have been suggested. One that is receiving much attention is the role of the oceans in determining such circulation patterns. For many years little was known about the nature of the climate over the oceans and the ways in which they influenced global climates. The main problem was lack of observed data over great expanses remote from land and shipping lanes. In recent years the availability of satellite data and the development of international research programs have permitted data to be gathered and theories to be propounded. In fact, much of the abnormal weather that has occurred in recent years has been attributed to conditions occurring in and over the oceans.

Some of the ideas—such as the role of El Niño, pools of warm or cold water in the Pacific Ocean, and the concept of the Southern Oscillation—are discussed in some detail in chapter 16. In that chapter, the topics are treated in the context of modeling and predicting climate.

Regional Climatology

As noted earlier, one approach to describing the climates of the world involves the descriptive climatology of large regions. Historically, the boundaries of such regions have been defined in terms of the surface conditions, and a number of classification schemes have been implemented. Such classification procedures are considered in the next chapter. The remaining chapters of part two will describe and analyze the climates of the region. The treatment is in terms of both the traditional descriptive regional climatology, in which the major surface climatic characteristics are utilized, and the dynamic analysis that uses the principles of synoptic climatology. Of necessity, the scale used to provide an overview of the entire climate of the world is at the macrolevel.

SUMMARY

To comprehend the incredible variety of climates that exist on earth, climatologists work at a number of areal scales. Microclimatic studies deal with energy and matter exchanges in a small area, with concentration upon the exchanges that occur at the earth-atmosphere interface. Local climates represent the next size scale, and their study incorporates the characteristics and processes that occur over a variety of surfaces and in a more extensive vertical area.

Mesoscale climatic studies generally deal with an area in which the physiographic features are somewhat similar and where the dynamic cause of the existing climate can be explained in terms of circulation patterns. The largest scale of investigation concerns macroclimatic regions. Such regions are identified in a variety of ways and their grouping permits the climates of the world to be studied in their entirety. Although many significant data are lost and while extensive generalizations must be made, climatic study at the macroscale allows a basic understanding of the variety of world climates.

APPLIED STUDY 8

Donora: An Air Pollution Episode

Air pollution is defined as a change in the concentration of material (or energy) in the air, such that the modified content adversely affects the well-being of organisms. Air pollution is not new; long before the earth was inhabited, air pollution occurred. Gases from fetid marshes, dust from deserts, and ashes spewed from volcanoes all influenced the quality of the air. However, the difference between these natural pollutants and those associated with the modern urban-industrial world lies in the rate at which they are produced and the amounts involved. The atmosphere has self-cleansing mechanisms—dispersal by the wind and cleansing by rain—that could easily handle the isolated cases in early times. What modern society has done, is, at times, to add pollutants more rapidly than the rate at which they are removed. Since modern industrial societies tend to concentrate their activities, high air pollution levels are frequently localized and are analyzed in terms of local climates.

It follows, then, that the amount of pollution at any location is a function of the amounts added and the ability of the air to remove them. In relation to the latter, the key here rests with air stability (See Appendix 1, Note 4). In unstable conditions both vertical and horizontal movement of air occurs, and the pollutants are dispersed. Conversely, when stable conditions exist there is no tendency for the air to move, and calm, windless conditions prevail. If pollutants are added to the air under stable conditions, they remain until a dispersion mechanism occurs: under such conditions, the added pollutants accumulate to attain high levels. If such conditions persist, the pollution content becomes extremely hazardous to health, and an air pollution episode is identified.

Stable atmospheric conditions are often associated with inversions, which result from the accumulation of cold air below a layer of warmer air. Such conditions can occur in upper levels of the troposphere or at ground level. In this study, air pollution episodes associated with ground level inversions are considered.

In December 1930, the heavily industrialized Meuse River Valley in Belgium was blanketed for a week by a dense fog. Industrial pollutants added to the air remained in place with the fog; by the end of the week, hundreds of people had fallen ill from respiratory ailments, and 60 died. In December 1962, London experienced a thick, sulfurous fog that lasted five days. The pollution-

REGIONAL CLIMATOLOGY

laden fog was credited with contributing toward almost 4000 deaths. Such air pollution episodes ultimately led to an awareness of the problems of air pollution and eventually to protective legislation. In the United States, the death toll of air pollution is not so evident as in these examples. That it can be killing, however, is aptly demonstrated by the air pollution episode that occurred in Donora, Pennsylvania, in 1948.

Donora is located about 28 miles south of Pittsburgh on the Monongahela River. It is a sheltered location; bluffs up to 450 ft rise from the river and enclose it on the north, east, and south sides. To the west, high hills complete the encirclement. In October 1948, the weather over a large area, including Pennsylvania, New Jersey, and New York, was dominated by an anticyclone to give clear, calm conditions. Days were warm and the clear nights were cool.

FIGURE 8-10 The development of valley inversions: (a) On cold, clear nights the ground chills the air above, which flows downslope to form a pool of cold air. Disposition of stack effluents depends upon height of stack and level of inversion. (b) Topographic setting and low sun angle inhibit heating of inversion layer. (c) Even light regional winds do not create mixing conditions; air becomes heavily polluted.

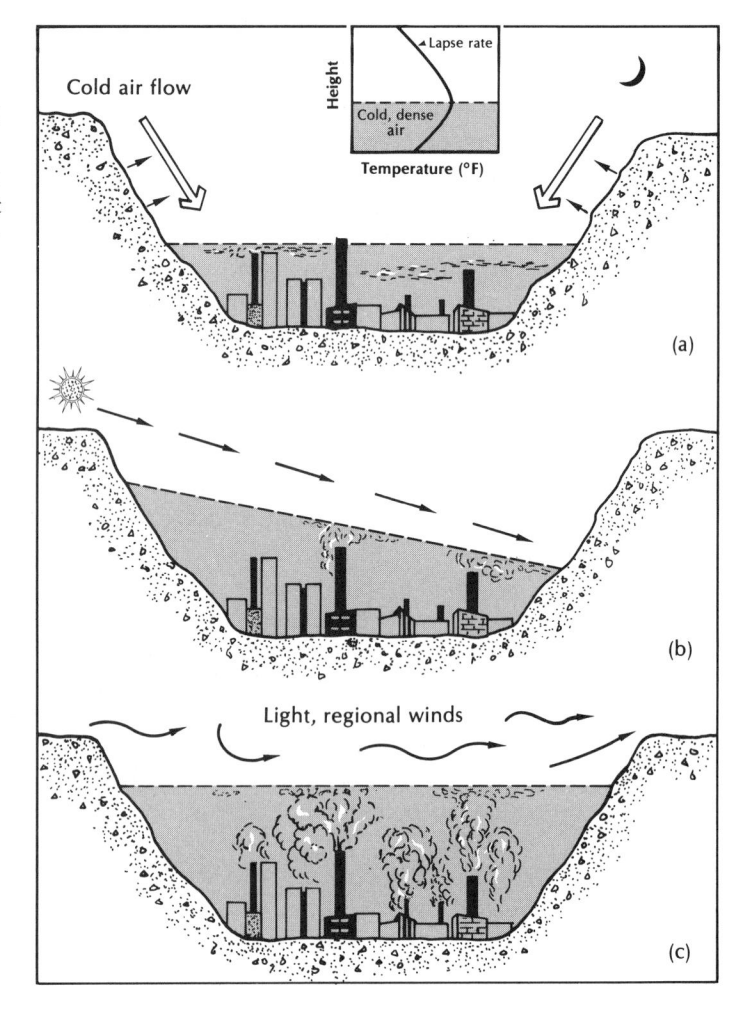

At night, the layers of air next to the ground were chilled; in the hilly area in the Donora region, the dense layer of cold surface air flowed downhill under gravity. A pool of cold air formed in the enclosed valley (figure 8–10a). In the calm conditions, no wind mixed the cold pool; the low sun angle and the shadow of the mountains prevented its heating by direct sunlight. The cool air remained and, on the next night, deepened as more flowed down the hillsides (figure 8–10b). This continued for six days.

In the industrial town, most people were employed in the steel plants and in factories that produce wire and sulfuric acid. These factories have stacks about 100 ft high, well below the level of the surrounding hills. The factories continued to pour out pollutants to accumulate in the stable air, which became laden with sulfur dioxide and particulates (figure 8–10c). As time progressed, hospitals and doctors' offices reported a sharp increase in illness, eventually amounting to six thousand cases. Some people did not recover from breathing the polluted air, and 20 people died. Only when a low pressure system moved across the area, bringing wind and rain, did the conditions improve.

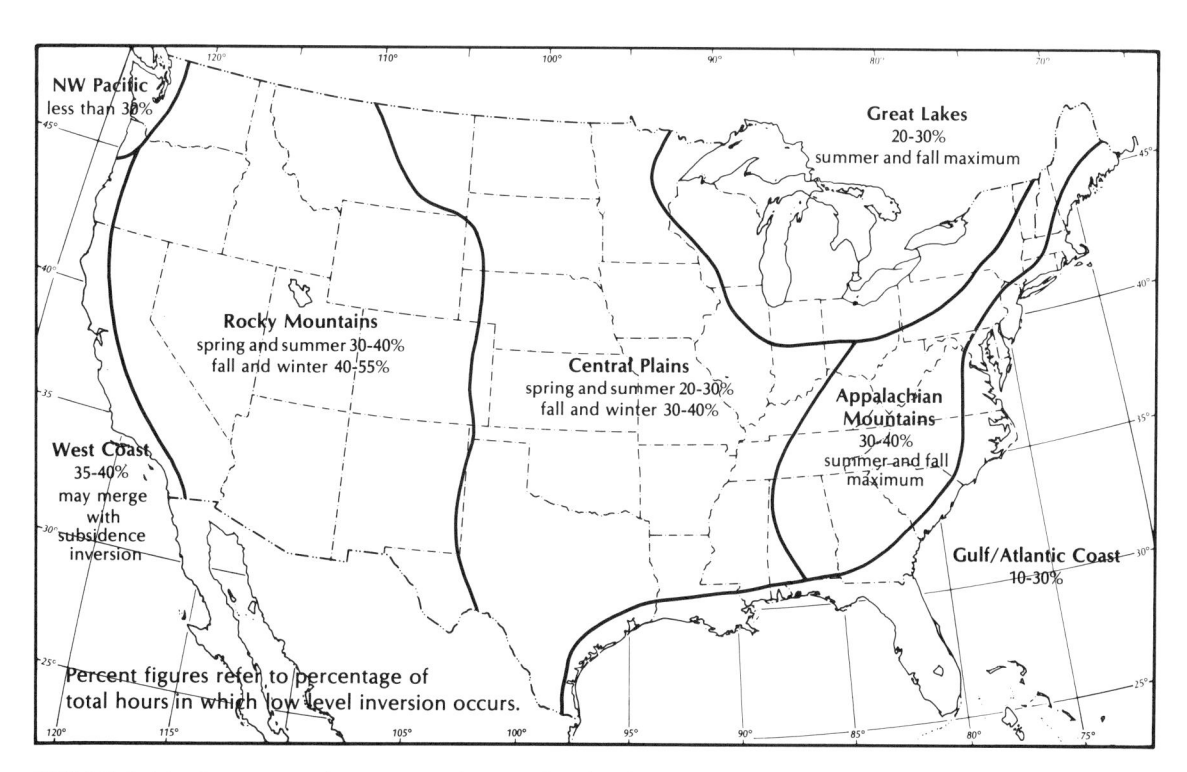

FIGURE 8–11 Frequency of summer and winter low-level inversions over the United States (From *Climate and Man's Environment*, John E. Oliver. Copyright © [1973, John Wiley & Sons]. Reprinted by permission of John Wiley & Sons, Inc.).

That Donora should be the location of a deadly air pollution episode might have been anticipated. When assessing the air pollution potential of the United States, the prime consideration is the percentage of time that inversions prevail. As figure 8-11 (page 179) shows, these stable conditions occur most frequently in the mountainous areas of the United States, reaching their maximum in the Rockies and the Appalachians.

The local conditions of the mountainous areas are ideal for the formation of low-level inversions and show how the study of local climatology can have far-reaching conclusions.

9 Regionalization of the Climatic Environment

THE RATIONALE
APPROACHES TO CLASSIFICATION
EMPIRIC SYSTEMS
The Köppen System
The Thornthwaite Classification
GENETIC SYSTEMS
The Model
CLIMATIC REGIONS AND ENVIRONMENTAL CHARACTERISTICS
Climate and the Distribution of Vegetation
Soil Classification and Climate
SUMMARY
APPLIED STUDY 9
Climatic Regions and Development: The Transamazon Highway

In the previous chapter the concept of areal scale in regional systems was examined. The classification of climates on a formal basis began with the macroscale climatic regions, which cover large geographical areas, some being subcontinental in size. This chapter is concerned with schemes for classifying the atmospheric environment of the earth into areas of similar climatic characteristics. The following chapters will examine each of these major climatic regions in some detail.

THE RATIONALE

To produce a useful classification of any data it is necessary first to group together those items that present the greatest number of common characteristics and then to subdivide those groups on a uniform basis until a satisfactory degree of subdivision is reached. In most areas of science, attempts have been made to produce classifications based upon the most fundamental characteristics possible, rather than on elements that might be more easily observed but of less intrinsic importance. Sometimes items don't lend themselves easily to grouping or classifying. Attempting to group similar regional environments on the earth is much like trying to group students on the basis of height or weight. There are no clearcut divisions.

The climatic elements of a region do not distinguish a region by their presence or absence but by the difference in the character of the elements. Spatial heterogeneity is a basic assumption in regional study, but this is not to imply that the differences that occur from place to place are not without order. There is a rather systematic variation in climatic elements from place to place, sometimes rather abruptly and in other cases over considerable distance. Since change with time and place is an integral part of the earth's environment, it is appropriate that a classification of climates should be based upon the patterns of variation that occur.

Throughout this book there have been repeated references to the annual and diurnal periodicities that exist. The motions of the earth in space give rise to periodic climatic, hydrologic, biological, and geological events. If the basic pattern of energy fluctuation is considered, we find that it is responsible for the largest share of the periodicity and spatial variation in the earth environment. The changing intensity and duration of solar radiation bring about changes in the atmosphere and seas; influence migratory and hibernation habits in animals; and set the life cycle of much of the biota including humans. The changing intensity of solar radiation between the Northern and Southern hemispheres produces a shifting of the general circulation of the atmosphere, which in turn is responsible for periodic changes in the hydrologic cycle over the earth. Thus, an energy flow that varies systematically through time and space results in an environment that also varies systematically through time and space. It is the purpose of this chapter to consider the major regions that result from periodic patterns of energy and moisture and to show how other environmental variables are related to these seasonal patterns.

Attempts to classify regional differences date back thousands of years. During the period of Greek civilization, the earth was divided into three broad temperature zones—the torrid, temperate, and frigid zones. The use of the word *temperate* to describe mid-latitude weather may not have been a particularly good choice, but nevertheless the classification persisted through the centuries. Since the recording of that very early division of the earth, classification schemes for organizing regional differences on the earth have appeared from necessity and with ever greater frequency.

The division of the earth into three temperature zones as formulated by the Greeks centuries ago is appropriate. The *tropics* are essentially winterless areas, as they are not directly affected by the outreach of cold polar air; they are dominated by air currents that originate in the warm areas between 30° N and 30° S. The *polar regions* are areas that lack a warm summer with the incursion of tropical air currents; the *mid-latitudes* are characterized above all else by the very marked summer and winter seasons, since these areas vacillate between the dominance of tropical and polar air currents.

A second element that varies markedly over the earth is the seasonal pattern of moisture. Some areas have a nearly equal probability of rain every day of the year. Most of the earth is subjected to a seasonal probability of rainfall as the primary circulation shifts back and forth with the solar energy supply. Some desert areas have nearly equal probability of having no rain on any day of the year. This breakdown gives perennial precipitation regimes, seasonal precipitation regimes, and dry regions. Such basic divisions of global climate using temperature and precipitation form the basis of a number of well-known classification systems.

APPROACHES TO CLASSIFICATION

Of basic importance in the classification procedure is the selection of the variables used in the delineation process. The use of temperature and precipitation as the major variables has been determined historically. Reliable data have been available only over the last 100 years (much less for most climatic stations), and many early records consist only of temperature and precipitation. Much of the early trend-setting work in classification was limited to the use of these two variables. Furthermore, much of the early work was carried out by plant physiologists and plant geographers who found a correlation between vegetation and the two parameters. Mid-nineteenth century researchers such as Dove, Linsser, and Supan were all influential in the development of climatic classification, so that it is not surprising to find, in view of the botanical training of these men, that the distribution of climate and natural vegetation should be treated simultaneously. As an example of the influence of plant geographers on climatic classification, many climatic regions are identified by plant association, a procedure that is still followed by some writers today; it is not unusual to find a climate type described as savanna, taiga, or tundra. It will be noted that both plant and animal names are given to climatic regions. While it would be quite unusual to find a climate described as "yak" or "penguin" climate in modern literature, the correlation between climate and vegetation is still prevalent; and climates of the world are still described in terms of natural vegetation distribution.

It is evident that, in relating the distribution of natural vegetation to the distribution of climate, the effect of climate is being measured, instead of the climate itself. It is assumed that a given climate gives rise to a distinctive vegetation association. To identify the climatic type, it is first necessary to determine the vegetation and then infer the climate. If climate can be so identified—that is, by expressing it as the result of the distribution of one selected component of the environment—then equally useful climatic regions may be identified through other measures of the effects of climate. It becomes possible to devise climatic schemes using factors ranging from the human response to climate to the effect of climate on rock weathering. Such systems would be based upon the observed effects of climate, and the criteria used to delimit their boundaries established by best-fit properties. Systems derived through such methodology may be collectively termed *empiric classifications*. The use of the qualifying term *empiric* connotes identification through the observed effects of climate.

The empiric systems essentially concern themselves with identifying similar climate types. An equally valid approach to climatic distribution employs the study of why climatic types occur in distinctive locations. Systems that attempt to deal with the question "why" must concern the cause of climate variation. As such, they may be termed collectively *genetic classifications*. As in the empiric systems, there are a number of methods of examining the causes of climates, so that the basis of genetic systems can vary appreciably.

Much controversy concerns the relative merits of the genetic and empiric approaches to classification. The view presented here is that two great groups of climatic divisions exist: those proceeding from actual observation and those proceeding from explanation. If it is accepted that there are two approaches and that both are valid, then the genetic versus empiric argument need not be of concern. What is important is which classification should be selected for a given distributional problem. A discussion of the climatological implications of the migration of the polar front may not use an empiric system; a genetic approach may not prove of great value in the discussion of specific temperature requirements for plant types. In effect, the approach used and the classification selected depend on the purpose to which they are to be put.

Because it is possible to observe the effects of climate on a whole range of environmental phenomena, there are many bases that can be utilized in the formulation of an empiric system. The following illustrates some of the innumerable interrelationships that can be examined.

1. The human response to climate
2. Climatic requirements for crop growth
3. Water needs and precipitation effectiveness related to vegetation
4. Study and identification of climatic analogs, for example, agricultural analogs
5. Vegetation distribution related to climatic controls
6. Geomorphic processes acting under different climatic conditions

7. Climate and soil-forming processes
8. Continentality and oceanicity as climatic determinants

A little thought could provide many other relationships, and for each there is probably a climatic classification available.

EMPIRIC SYSTEMS

Of the many classifications that have been devised, it is inevitable that one or two develop into what might be termed "standard systems." Such classifications become standard as a function of their wide usage; since contact with climatic classification occurs mostly in introductory courses, it follows that the most widely used systems are those that facilitate an orderly description of world climates. To achieve any prominence such a system must, of course, be acceptable conceptually.

One system that has developed along these lines is that formulated by Köppen. In its various forms, it is probably the most widely used of all climatic systems. A second scheme that, largely as a result of its innovative methodology, is also widely used is that devised by C. W. Thornthwaite.

The Köppen System

Wladimir Köppen made one of the most lasting and important contributions in the field of climatic classification. He was trained as a botanist, and in the early stages of his work he was strongly influenced by the work of de Candolle and Supan. The systems he formulated range from a highly descriptive vegetation zonal scheme to a classification in which boundaries are defined in relatively precise mathematical terms. The Köppen system has been considerably modified over time.

Beginning with his doctoral dissertation (Leipzig) in 1870 and continuing up to his death in 1940, Köppen proposed, modified, and remodified his system. By 1951, it had become so established that F. Kenneth Hare reported to the Royal Meteorological Society of Canada that some regarded the system "as an international standard, to depart from which is scientific heresy." Such rigorous interpretation was not probably intended by Köppen, to whom the scheme was never completely satisfactory. The evolution of the system shows that Köppen was not so concerned with the precise boundaries as he was with attempting to use simple observations of selected climatic elements to provide a first-order world pattern of climates.

Köppen's early work was completed at a time when plant geographers were first compiling vegetation maps of the world. His early publications (1870, 1884) were concerned with temperature distribution in relation to plant growth, and it was not until 1900 that any of his publications were really concerned with world climatic classification. The 1900 system, which did not obtain much notice, is a highly descriptive scheme making use of plant and animal names to characterize climate. In 1918, Köppen produced a system that is substantially the one in use at the present time. Boundary values have changed, new symbols have been introduced; but the framework of the present system was clearly evident. The scheme demonstrates Köppen's major contribution to the systematic treatment of the climates of the world. He recognized a pattern underlying world climatic regions and introduced a quantitative method that allows any set of data to be located within the system. The classification is considerably enhanced by the introduction of a unique set of letter symbols that obviates the necessity of long descriptive terms.

The classification is based essentially on the distribution of vegetation. Köppen's assumption was that the type of vegetation found in an area is very closely related to the temperature and moisture characteristics of the region. These general relationships were already known at the time Köppen's classification was produced, but he attempted to translate the boundaries of selected plant types into climatic equivalents. The Köppen system is based on monthly mean temperatures, monthly mean precipitation, and mean annual temperature.

Köppen recognized four major temperature regimes, one tropical, two mid-latitude, and one polar (table 9-1). After identifying the four regimes, he assigned numerical values to the boundaries (table 9-2). The *tropical climate* is delimited by a cool month temperature average of at least 18°C (64.4°F). This temperature was selected because it approximates the poleward limit of certain tropical plants. The two *mid-latitude climates* are distinguished on the basis of the mean temperature of the coolest month. If the mean temperature of the coolest month is below −3°C (26.6°F), it is *microthermal*; if the temperature is above −3°C, the climate is *mesothermal*. The fourth major temperature category is the *polar climate*. The boundary between the microthermal and polar climates is set at 10°C (50°F) for the average of the warmest month, which roughly corresponds to the northern limit of tree growth. A fifth major regime, the *dry climates*, was based not upon temperature criteria but upon lack of moisture. Dry climate boundaries are obtained using derived formulas. Figure 9-1 shows the distribution of the identified climatic types.

TABLE 9-1 Köppen's major climates.

	A	Tropical rainy climates
	B	Dry climates
	C	Mid-latitude rainy climates, mild winter
	D	Mid-latitude rainy climates, cold winter
	E	Polar climates

Principal Climatic Types According to Köppen's Classification

Af	Tropical rainy	Cw	Mid-latitude wet-and-dry, mild winter
Aw	Tropical wet-and-dry	Cf	Mid-latitude rainy, mild winter
Am	Tropical monsoon	Dw	Mid-latitude wet-and-dry, cold winter
BS	Steppe	Df	Mid-latitude rainy, cold winter
BW	Desert	ET	Tundra
Cs	Mediterranean	EF	Ice cap

Source: From Hidore, 1972.

TABLE 9-2 The Köppen classification.

A Temperature of the coolest month above 18°C (64.4°F)

Subcategories:

- f: rainfall in driest month at least 6 cm (2.4 in.)
- m: rainfall in driest month greater than $10 - r/25$, but less than 6 cm when r = annual rainfall in cm

  OR

  rainfall in driest month greater than $3.94 - r/25$, but less than 2.4 in., when r = annual rainfall in inches
- w: rainfall in driest month less than 6 cm (2.4 in.), but insufficient for m̲. and dry season in low sun period
- s: rainfall in driest month less than 6 cm (2.4 in.), but insufficient for m̲. and dry season in high sun period
- w': maximum rainfall in autumn
- w": two rainfall maxima, with intervening dry periods
- i: annual temperature range less than 5°C (9°F)
- g: warmest month precedes summer solstice

B Evaporation exceeds precipitation for the year.

Subcategories:

BS (Steppe): Derived by the following, when r = annual rainfall in cm and t = annual average temp. in °C

70% of rainfall in summer six months:	$r = 2(t + 14)$
70% of rainfall in winter six months:	$r = 2t$
Even rainfall distribution or neither of above:	$r = 2(t + 7)$

OR, when r = annual rainfall in inches and t = annual average temp. in °F

70% of rainfall in summer six months:	$r = .44t - 3.5$
70% of rainfall in winter six months:	$r = .44t - 14$
Even rainfall distribution or neither of above:	$r = .44t - 8.5$

The value r̲ is the BS/humid boundary. When the derived r̲ is greater than the value on the right of the equation the climate is humid; when less it is B̲. If B̲, then determine if BS by dividing the answer by 2. If, after dividing, r̲ is greater than value on right, climate is BS; if less, climate is BW (desert).

BW (Desert): Derived, as indicated above

TABLE 9-2 *continued*

Subcategories of BW:

- h: average annual temperature above 18°C (64.4°F)
- k: average annual temperature below 18°C (64.4°F)
- k': average of warmest month below 18°C (64.4°F)
- n: high frequency of fog
- s: 70% of rainfall in winter six months (summer dry season)
- w: 70% of rainfall in summer six months (winter dry season)

C Coolest month temperature averages below 18°C (64.4°F) and above −3°C (26.6°F); warmest month is above 10°C (50°F).

Subcategories:

- f: at least 3 cm (1.2 in.) of precipitation in each month; or, neither w or s
- w: minimum of 10 times as much precipitation in a summer month as in driest winter month
- s: minimum of 3 times as much precipitation in a winter month as in driest summer month, and one month with less than 3 cm (1.2 in.) of precipitation
- x: rainfall maximum in late spring or early summer; dry in late summer
- n: high frequency of fog
- a: warmest month over 22°C (71.6°F)
- b: warmest month under 22°C, but at least four months over 10°C (50°F)
- c: only one to three months above 10°C
- i: mean annual temperature range less than 5°C (9°F)
- g: warmest month precedes summer solstice
- t': hottest month delayed until autumn
- s': maximum rainfall in autumn

D Coolest month temperature averages below −3°C (26.6°F) and warmest month over 10°C

Subcategories:

- d: coldest month below −38°C (−36.4°F)

Other subcategories same as for C.

E Warmest month temperature averages less than 10°C (50°F)

Subcategories:

- ET: average temperature of warmest month between 0°C (32°F) and 10°C (50°F)
- EF: average temperature of warmest month below 0°C (32°F)

Source: From Hidore, 1972.

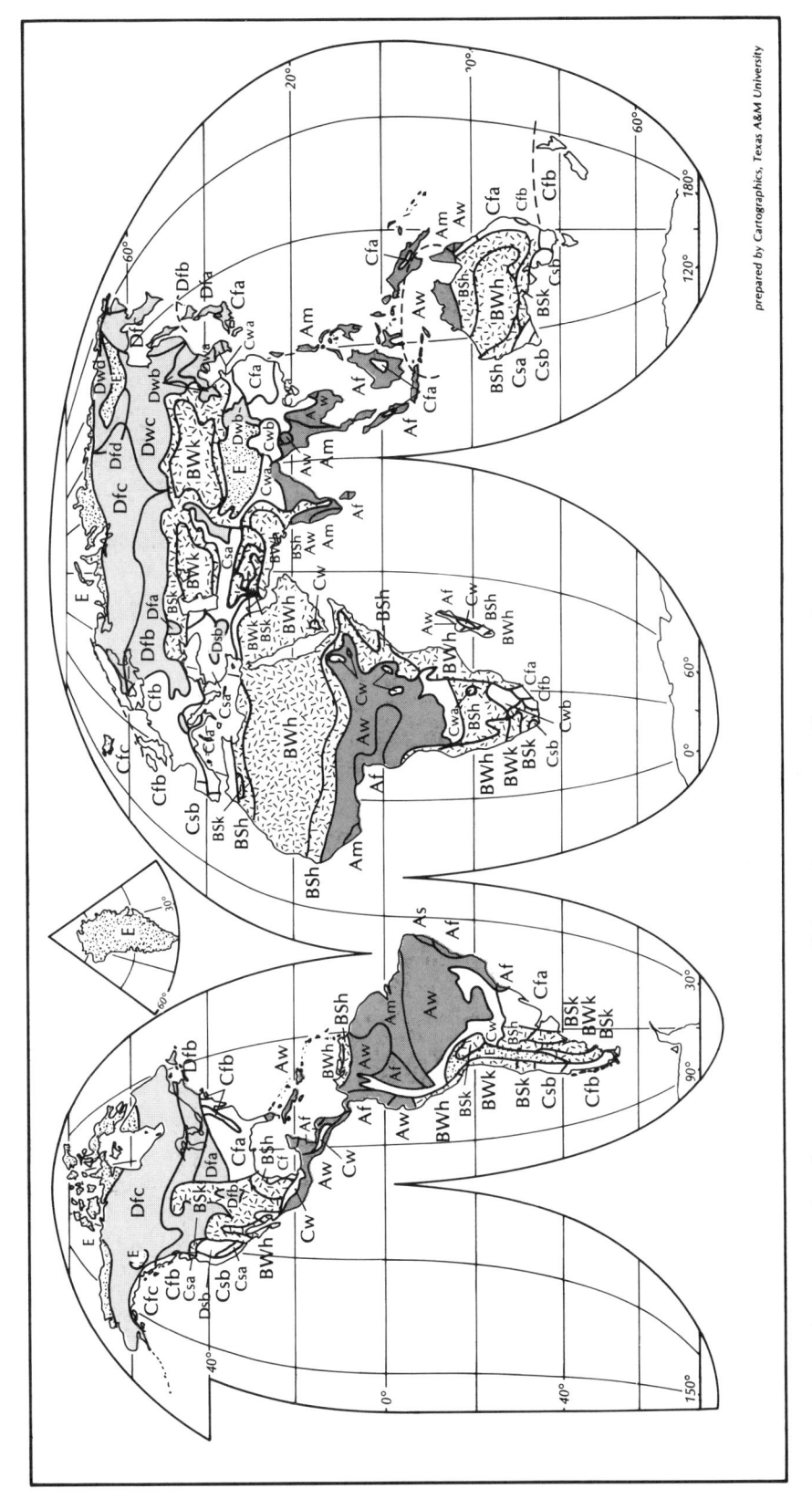

FIGURE 9–1 The Köppen classification (From J. F. Griffiths and D. M. Driscoll, *Survey of Climatology*. Used by permission of Charles E. Merrill Publishing Co.).

The system has been subjected to criticism from two aspects: (1) There is no complete agreement between the distribution of natural vegetation and climate. This is to be expected since factors other than average climatic conditions (soils, for instance) affect the distribution of vegetation. (2) The system is also criticized on the basis of the rigidity with which the boundaries are fixed. Temperatures at any site differ from year to year as does rainfall, and the boundary based on a given value of temperature will change location from year to year. In spite of the criticisms and the empiric basis of the classification, it has proved quite usable as a general system.

The Thornthwaite Classification

In a paper published in 1931, C. W. Thornthwaite proposed a climatic classification that was a marked departure from preexisting systems. Unlike most classifications available at the time (based upon temperature zonation for first-order grouping), Thornthwaite based his system on the concepts of moisture and thermal efficiency. Many authors prior to Thornthwaite had suggested that the relationship between precipitation and evaporation could provide a useful measure of precipitation effectiveness, but few had utilized the concept because of lack of evaporation data. Faced by the same shortage, Thornthwaite produced a precipitation-evaporation index that could be determined empirically from available data. Using this index, Thornthwaite determined humidity provinces; these formed the first-order division of his classification scheme. It differs from the Köppen system in that boundaries between provinces are not related to any practical vegetation or soil criteria; instead, they are based upon regular arithmetic intervals of derived values. The 1931 system has been adequately described and analyzed by a number of authors; despite some adverse criticism, it may be considered as a major contribution to the process of climatic classification.

Of more significance at present (because it has superseded the earlier system) is Thornthwaite's 1948 classification, which is a radical departure from the 1931 system because it makes use of the important concept of evapotranspiration. The earlier system had been concerned with the loss of moisture through evaporation, whereas the new approach considers loss through the combined process of evaporation and transpiration. Plants are considered physical mechanisms by which moisture is returned to the air. The combined loss is termed *evapotranspiration,* and when the amount of moisture available is nonlimiting, the term *potential evapotranspiration* is used.

As with any widely used system, the 1948 classification has been subject to criticism. Many of the criticisms relate to the empiric formula used to express evapotranspiration and to the way in which the water budget of a station is manipulated. Such criticisms are dealt with in a later section of the chapter.

The two major aspects of the system are the uses of precipitation effectiveness and temperature efficiency. Precipitation effectiveness was designed as an indicator of net moisture supply, taking into account both the actual amount of precipitation and the estimated consumption of moisture by evaporation. The precipitation effectiveness is determined by calculating the ratio of the *precipitation to evaporation* (P/E) ratio for each month of the year and summing them to form the *precipitation effectiveness* (P-E) index. Temperature efficiency in this classification is used as an indicator of the energy or heat supply relative to evaporation rates. The T-E index is calculated in the same fashion as the P-E index, using temperature and evaporation data.

On the basis of the precipitation effectiveness index, nine moisture provinces were established (table 9-3). The boundary of each progressively more humid region was established at a doubling of the P-E index. On a comparable scale of the temperature efficiency index, nine major temperature provinces are recognized. As in the case of the P-E index, each progressively warmer province is bounded by an index double that of the preceding province. Thornthwaite used many of the same

TABLE 9-3 Thornthwaite's 1948 classification of climate.

Nine Divisions Based upon Moisture Efficiency			Nine Divisions Based upon Temperature Efficiency		
Climatic type		Moisture index	TE index (cm) / (in.)		Climatic type
A	Perhumid	100 and above	14.2 / 5.61	E'	Frost
B_4	Humid	80 to 100	28.5 / 11.22	D'	Tundra
B_3	Humid	60 to 80	42.7 / 16.83	C'_1	Microthermal
B_2	Humid	40 to 60	57.0 / 22.44	C'_2	
B_1	Humid	20 to 40	71.2 / 28.05	B'_1	
C_2	Moist subhumid	0 to 20	85.5 / 33.66	B'_2	Mesothermal
C_1	Dry subhumid	−20 to 0	99.7 / 39.27	B'_3	
D	Semiarid	−40 to −20	114.0 / 44.88	B'_4	
E	Arid	−60 to −40		A'	Megathermal

Subdivisions Based upon Seasonality of Precipitation

	Moist climates (A, B, C_2)	Aridity index
r	Little or no water deficiency	0–16.7
s	Moderate summer water deficiency	16.7–33.3
w	Moderate winter water deficiency	16.7–33.3
s_2	Large summer water deficiency	33.3+
w_2	Large winter water deficiency	33.3+

	Dry climates (C_1, D, E)	Humidity index
d	Little or no water surplus	0–10
s	Moderate winter water surplus	10–20
w	Moderate summer water surplus	10–20
s_2	Large winter water surplus	20+
w_2	Large summer water surplus	20+

Aridity index = water deficit/water need.
Humidity index = water surplus/water need.

alphabetic symbols that Köppen used in his classification. To the two indexes of moisture and temperature are added a letter designation for rainfall distribution through the year. The initial classification yields 32 different climatic types.

Many other classifications of climate have been proposed and numerous modifications of these schemes have been suggested (table 9-4). In classifying continuous variables, it must be realized that any classification is going to be arbitrary and also that there are an infinite variety of climatic regions found on the earth. No two square miles of earth's surface are likely to have the same atmospheric conditions.

GENETIC SYSTEMS

In comparison to the empiric classifications, genetic systems are generally less formally defined and often less well developed. As a result they are less widely used in general descriptive climatology. The types that have been proposed are based either upon identified physical determinants or upon air mass dominance. The scheme described here, and used as a basis for classification of regional climates in the following chapters, is an air mass approach first formulated by Hidore (1969).

The Model

On the basis of seasonal patterns of radiant energy and precipitation, the earth's environments can be identified as belonging to nine basic types, as shown schematically in figure 9-2. It can be noted in the figure that each row and column of three has the same seasonal characteristics of either temperature or moisture. Thus for each of the three types of tropical systems there is a corresponding mid-latitude and polar system with a similar seasonal moisture pattern.

Figure 9-3 illustrates several characteristics of four of the regional systems. Each of these four areas tends to lack any strong seasonal variation in energy or moisture. Some differences between seasons do exist, of course. In the polar ice caps, radiation increases in the summer season; but even so, it remains cold throughout. The increase in radiant energy simply is not sufficient to bring about real warming, and tropical air currents don't penetrate that far poleward. Each of the four systems is influenced primarily by one kind of air mass, and each is associated with one of the four semipermanent pressure zones in the atmosphere.

Four of the remaining five types of regions have a marked seasonal variation in either temperature or moisture, but not both. Two have pronounced seasonal moisture regimes and two pronounced seasonal temperature regimes. These four environmental groups are situated between the major pressure zones and are affected by the seasonal migration of the primary circulation. They are subjected to air currents that are considerably different from one season to the next (figure 9-4). One of the nine systems, the mid-latitude seasonal rainfall regime, represents the maximum in seasonal variation of weather systems. Periodic invasion of air currents developing in each of the major types of source regions brings a variety of weather.

TABLE 9-4 Chronological development of some world climatic classifications.

Author(s)	Year	Base of System
de Martonne	1909–	Based 9 first-order divisions upon temperature and precipitation criteria; numerous subdivisions named for local areas in Europe; considerable attention given to desert limits, but most boundaries derived nonquantitatively
Penck	1910	Physiography and world climate; 3 main types of climates significant in determining weathering and erosion: humid, arid, and nival; each subdivided into 2 parts
Köppen	1918–	Based upon vegetation regions, with specific climatic values for boundaries between regions
Vahl	1919	5 zones appraised by temperature limits, a function of the data for warmest and coldest months; subdivision by precipitation expressed as % of number of wet days in a given humid month
Passarge	1924	Recognition of 5 climatic zones subdivided into 10 regions, emphasizing vegetation distribution
Miller	1931–	5 temperature years based upon vegetation; each temperature zone divided into 3 possible moisture zones: areas with rain in all seasons, marked seasonal drought, or drought all year
Thornthwaite	1931	Use of precipitation effectiveness and temperature efficiency to construct 5 humidity provinces and 6 temperature provinces; 30 major climatic types
Philipson	1933	Based upon temperature of warmest and coldest months and precipitation characteristics; 5 climate zones with 21 climatic types and 63 climatic provinces
Blair	1942	5 main zonal climates: tropical (T), subtropical (ST), intermediate (I), subpolar (SP), and polar (P); 14 types and 6 subtypes distinguished using letter notation; based on precipitation and temperature data and related to vegetation types

TABLE 9-4 (Continued)

Author(s)	Year	Base of System
Gorsczynski	1945	10 "decimal" types associated with 5 main zonal climates; emphasis on continental vs. marine climates and definition of aridity
von Wissman	1948	Related to the Köppen approach; 5 temperature zones subdivided by precipitation distribution and temperature regimes
Thornthwaite	1948	9 primary climates based on a moisture index, 9 divisions based on temperature index; evapotranspiration introduced as an index
Creutzberg	1950	Annual rhythm of climate based on identification of *Isohygromen* (lines of equal duration of humid months) and *Tag-Isochione* (lines of equal daily snow cover duration); 4 major zones differentiated, subdivided by monthly moisture values
Geiger-Pohl	1953	Modification of the Köppen system
Brazol	1954	Human comfort zones; use of wet and dry bulb temperatures to establish comfort months; 12 ranked classes ranked from No. 12 (lethal heat) to No. 1 (glacial cold)
Trewartha	1954	Major modification of the Köppen system
Budyko	1958	Distribution of energy in relation to water budgets
Peguy	1961	Modification of de Martonne's system
Troll	1963	Climates differentiated by thermal and hygric seasons
Carter and Mather	1966	Modification of the 1948 Thornthwaite system
Papadakis	1966	Agricultural potential of climatic regions; crop-ecologic characteristics of a climate based upon empirically derived threshold values; 10 main climate groups recognized, each divided into subgroups that are subdivided
Hidore	1966	Based on dominance of air masses; 10 primary climates identified by annual distribution of air mass frequency
Terjung	1968	Thermal classification of the earth's climates
Oliver	1970	Genetic classification based on air-mass frequency

REGIONALIZATION OF THE CLIMATIC ENVIRONMENT

FIGURE 9-2 Schematic representation of the nine basic types of climate found on the earth. The seasonal patterns of temperature and moisture, the types of air masses most common to the area, and the location in the general circulation are shown.

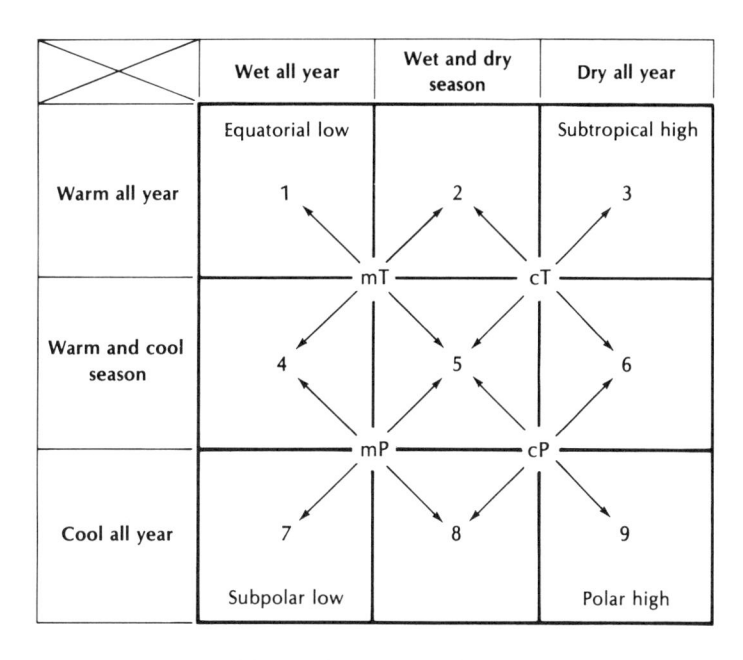

	Wet all year	Wet and dry season	Dry all year
Warm all year	Equatorial low — 1	2	Subtropical high — 3
	mT		cT
Warm and cool season	4	5	6
	mP		cP
Cool all year	7	8	9
	Subpolar low		Polar high

Using the model, it is now possible to identify the climatic types that occur within the model (figure 9-5) and to list their characteristics (table 9-5). When the identified types are located on a world map, a highly generalized distribution results, with large areas grouped as a single climatic type. Given that the distribution is based upon only nine major climatic types, such might be expected. The resulting map (figure 9-6) does, however, provide a useful guide to the classes identified within the model. For each of the regions, it is possible to subdivide the climates to whatever scale is needed by using a selected criterion. Thus, in figure 9-6 the middle latitude wet climate has been divided with summer conditions as a criterion.

FIGURE 9-3 Four of the climates are dominated primarily by single types of air masses. Each of these is found in the core area of one of the semipermanent pressure zones.

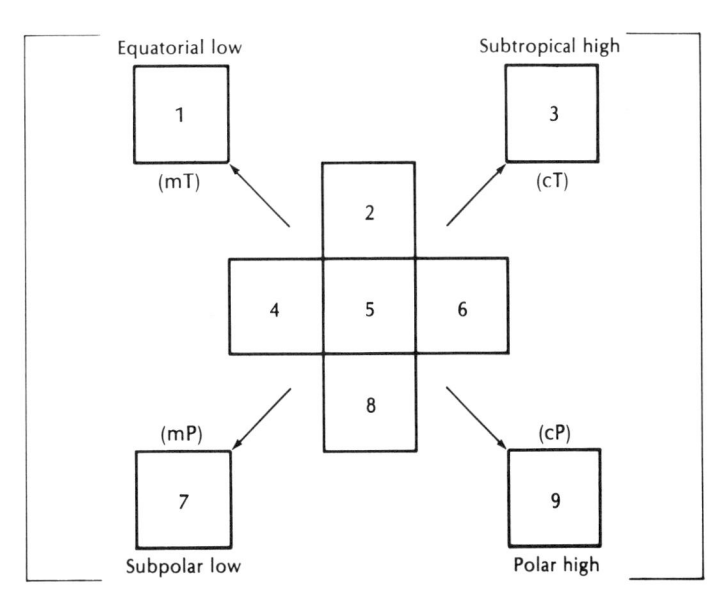

195

REGIONAL CLIMATOLOGY

FIGURE 9-4 Four of the climates have marked seasonal change in temperature or moisture associated with a seasonal change in air mass control. The other climate, that shown in the center of the diagram, is subject to weather associated with all four basic kinds of air masses. It has both wet and dry seasons and summer and winter seasons.

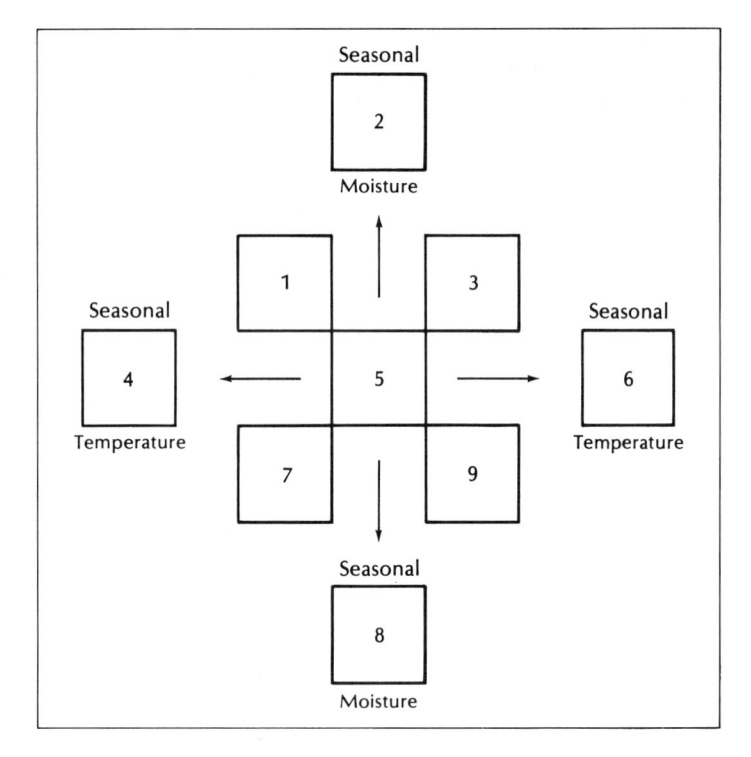

Each type of climate is detailed in the chapters that follow. It will be noticed that mountain regions are not identified. Two points must be remembered with regard to mountain regions: (1) the seasonal patterns of energy and precipitation will be the same as those of the surrounding lowlands; and (2) mountain regions may be, and probably are, found in each of the regions differentiated in the previous paragraphs. The altitudinal variations that occur in mountains are similar whenever they are found.

CLIMATIC REGIONS AND ENVIRONMENTAL CHARACTERISTICS

Climate, per se, does not exist as an instantaneous environmental factor. It is a concept derived from an orderly examination of atmospheric conditions over a long time period. As a result, one cannot "see" climate. Its impact, however, is clearly seen in other environmental systems, particularly in the vegetation and soils that exist in a given locale. Thus, before concluding the discussion of climatic classification, it is useful to consider the role of climate in the world grouping of vegetation and soil systems.

Climate and the Distribution of Vegetation

Among the factors that determine the distribution of flora are moisture availability, radiant energy, soils, parent material, slope, and other biota. Clearly

FIGURE 9-5 Climates differentiated on the air mass model.

	Wet ←————————→ Dry			
Hot	Tropical wet (equatorial)	Tropical wet and dry	Tropical desert	Group I
	Mid-latitude wet	Mid-latitude summer or winter dry	Mid-latitude desert	Group II
Cold	Polar wet	Polar wet and dry	Polar desert	Group III

climatic factors are a major determinant in the distribution of individual species and communities. At the biome level, the implications of climate are most readily visible. Where the climatic conditions are fairly similar from year to year in terms of temperature and moisture, a fairly distinct climax plant formation has evolved. In the classification of climates four such regions were identified: the tropical rainy, the tropical desert, the polar rainy, and polar desert or ice caps. A recognized biome associated with the tropical rainy climate is the tropical rain forest (see Applied Study 9). The relationship between the rain forest and climate is close enough so that until recent years the distribution of rain forest has been used to map the areas with a tropical rainy climate. It is now clear that the association between the two is not that perfect. In the tropical deserts is found the desert biome, which covers large areas on the earth's surface. Associated with the polar rainy is to be found the coniferous forest (taiga); and in the vicinity of the ice caps little or no vegetation is found, as a result of very harsh conditions. In the intervening areas between these four very different biomes, classification is more difficult and often confused by the intermix of

TABLE 9-5 Classification of climates by air mass control.

Type		Air Masses
Group I	Tropical air masses dominate	
	1. Tropical wet	mT or mE air masses dominates
	2. Tropical wet and dry	mT/cT seasonally
	3. Tropical dry	cT dominates
Group II	Tropical/polar air masses seasonally	
	4. Mid-latitude wet	mT or mP dominates
	5. Mid-latitude wet and dry	
	5S summer dry	cT/mP
	5W winter dry	mT/cP
	6. Mid-latitude dry	cT/cP seasonally
Group III	Polar air masses dominate	
	7. Polar wet	mP dominate
	8. Polar wet and dry	mP/cP seasonally
	9. Polar dry	cP dominate

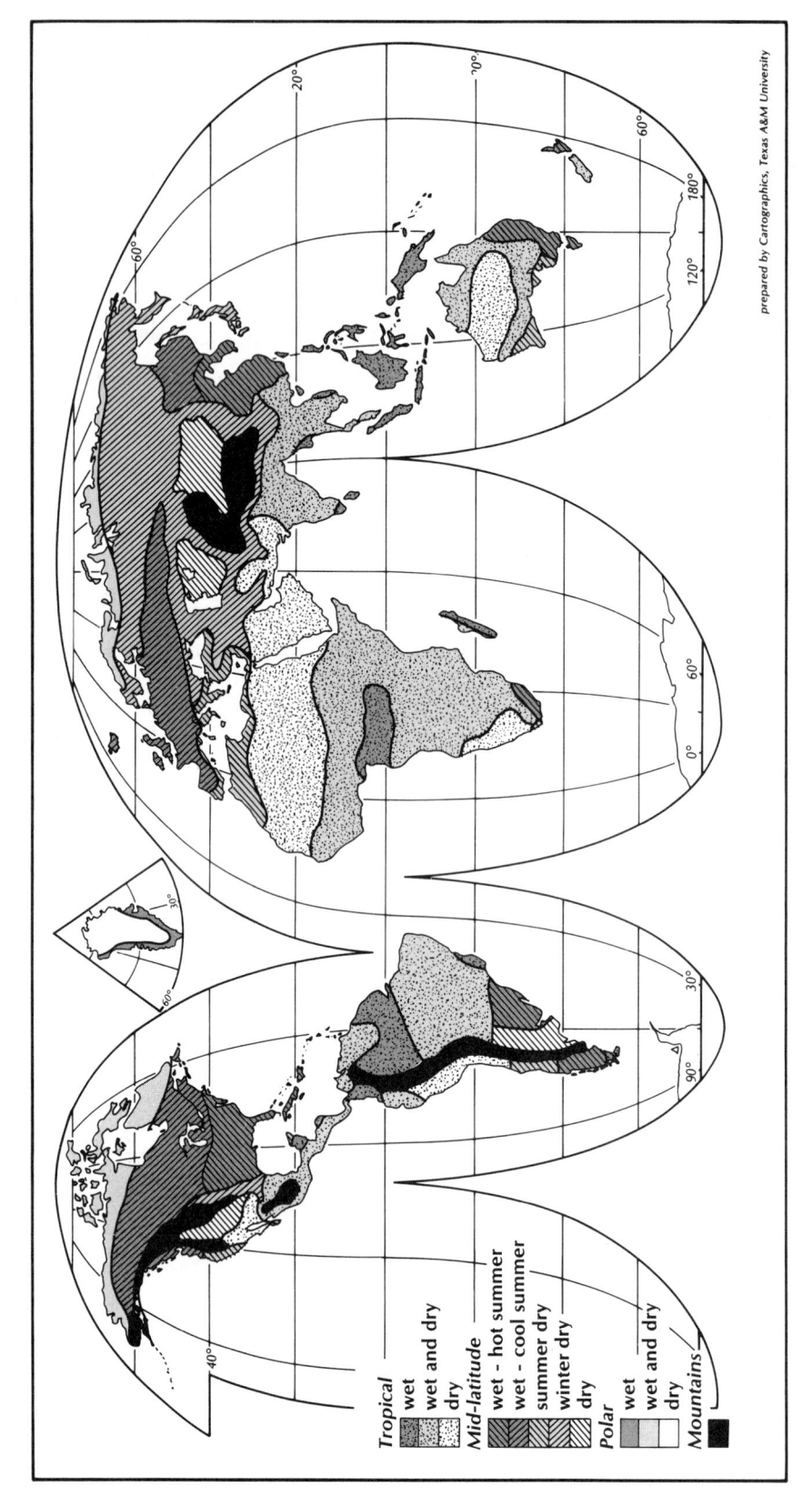

FIGURE 9–6 World climatic regions differentiated using the air mass model.

TABLE 9-6 Formations associated with tropical wet and dry regimes.

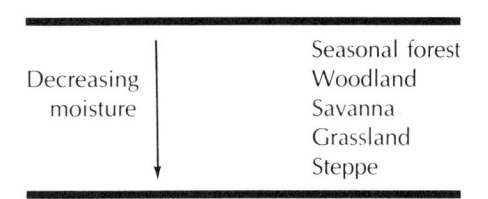

Decreasing moisture ↓	Seasonal forest Woodland Savanna Grassland Steppe

communities and species. In these regions the biomes present transitions from tropical to polar and humid to dry. For instance, table 9-6 lists some of the different biomes found in the tropical seasonal rainfall regime. Since the communities form a continuum over the landscape, placing the boundaries is generally difficult. A brief description of some of the more widely recognized biomes follows.

Rain forests There are many definitions of a forest because of the many varieties of forest communities. Here, *forests* include those formations in which trees are a prevalent plant form. As already suggested, this classification covers a multitude of different ecosystems, and here it is possible to mention but a few.

Rain forests are formations in which evergreen trees are the dominants and where the canopy is more or less continuous. Rain forests exist where moisture is abundant, if not on a year-round basis, at least during the greater part of the year. Most of the vegetation in these communities is found in the canopy. The foliage of the trees themselves is concentrated in the crowns; hence the canopy and the foliage of the lianas, epiphytes, and parasites are concentrated in the canopy. The understory consists largely of young trees of the dominant species and a sparse ground cover of shade-tolerant or shade-demanding shrubs and lower forms of vegetation. These forests are very limited in extent.

Seasonal forests The majority of the world's forests occur where there is a dry season long enough to affect a seasonal change in the forest community. The seasonal forest may include evergreen, semideciduous, deciduous trees, or some combination of these. Where a mixture exists it is not usually a random mixture of individuals of each species but mixed stands of one type or the other. Local differences in soil characteristics or other site characteristics often determine which community will persist. Since the seasonal forests exist where there is seasonal precipitation, the character of the forest is closely associated with the length of the rainy season. As the length of the rainy season decreases, the density of the canopy decreases. The most dense forests, or true jungles, are found in the seasonal forests where the dry season is long enough to spread the canopy and allow sunlight to reach the ground, but where there is not such a long dry season that edaphic drought can occur with any regularity.

Deserts A desert is characterized by a discontinuous plant cover. The total vegetal coverage is usually small. Some desert plants, such as the creosote bush, produce their own population control devices. They produce hormones through their roots or leaves that inhibit the sprouting of other individuals within the proximity of the parent plant. This controls spacing of the plants and allocates water and nutrient resources in a fashion that increases the probability of survival of some individuals of the species.

Many of the animals found in the deserts are the same as those found in the grasslands but in greatly reduced numbers. There are some animals, including the kangaroo rats, which have adapted specifically to the deserts; but such species are relatively few in number.

Savanna Savannas are tropical grasslands with scattered trees or clumps of trees. Isolated trees are often found right at the desert edge. It is this scattering of drought resistant trees that gives the tropical grasslands a distinctly different appearance from the mid-latitude grasslands. Fires are a recurrent phenomenon in the savannas, and both the grasses and trees are fire-tolerant. The varieties of species of trees and grasses are few compared to the tropical forests. Though the species are fewer, because of the necessary adaptation to drought and fire, they are at the same time very hardy and respond rapidly after a disturbance. Acacia and baobob trees are among the species that spread through the savannas.

Grasslands The world's grasslands consist of formations made up of communities in which the herbs and shrubs are dominant. The communities are dominated by grasses and legumes, some of which reach considerable size. Grasses ranging upward of 3 m (10 ft) in height are not uncommon in the more humid grasslands. Perhaps the tallest well-known member of the grass family is bamboo, a phenomenally fast-growing plant that has been observed to grow as much as 20 cm (8 in.) per day. Although they lack the upper story of trees, the grasslands are still multistoried, with usually more than one story of herbs or shrubs and some of the lesser structural forms closer to the ground.

Grazing animals live in their greatest numbers in the grasslands; among their attributes are their tendencies to live in groups and depend upon speed for defense against predators. Some creatures have developed the abilities to leap high above the grasses for easier traveling and to see over the grass to watch for predators. The jack rabbit of the American and Australian plains is a good example. Many small animals burrow for shelter and concealment. Among this group is the prairie dog of the Great Plains of North America, which exhibits both the group social structure and the construction of extensive underground towns.

Very few areas of natural grassland remain on the face of the earth. The grasslands have proved to be the most useful of the biomes when measured in terms of agricultural purposes. The natural communities of the grasslands have been either burned off or plowed up and replaced by the simpler communities of the domestic cereals such as corn, rice, wheat, and barley.

The grasslands are found where there is a seasonal moisture regime. They occur on all of the continents except Antarctica, and convergent adaptation has led to similar species in each and strange forms in some. Grasslands have been subdivided by secondary structural characteristics; grasses have been divided between steppe and prairie, tall and short, and sod and bunch grass. The grasslands may border forests, seasonal forests, woodlands, or deserts.

Taiga The dominants of this forest are the needle-leaved evergreen trees. The members of the spruce, pine, and fir families are most common. Under a mature coniferous forest there is very little understory as a result of the dense shade. The ground often is covered with a fairly thick layer of undecomposed and partially decomposed needles. This forest is associated with cool, moist conditions poleward to the limits of tree growth. The forest extends equatorward considerable distances along mountain ranges, where favorable temperature and moisture conditions prevail.

Tundra Vegetation of the tundra consists largely of grasses, sedges, lichens, and dwarf woody plants. The tundra is associated with the seasonal rainfall areas around the Arctic Ocean, and at high altitudes in mountains. It exists where there is a short summer season that is too cool for trees to thrive. The vegetation of the tundra has distinct characteristics that allow it to survive the cold temperatures, wind, and the very long physiological drought. Almost all tundra plants are low growing and compact to escape the wind, reduce evaporation, and conserve heat. Perennials predominate, and many reproduce asexually by runners, bulbs, or rhizomes. Being perennial, the plants store food through the winter, and the buds are sheltered either underground or close to the surface.

Disturbed formations Some scientists maintain that the grassland, savanna, and brush formations are disturbed formations that have persisted because of repeated razing by fire. The sharp boundaries that exist, the variety of formation boundaries, and the lack of woody plants have led to this hypothesis. At this time the debate seems to be a long way from being resolved. The suggestion that grasslands are a product of disturbance does not in any way change the fact that disturbance is a very real factor in formation structure. Disturbance affects formations in several ways. It tends to sharpen the boundary between formations, at least in some areas, and often favors the intrusion of one formation into another. Local disturbances within a formation may cause an alternative formation to become established. The effects of a disturbance vary, depending upon the type of formation. Some communities in mid-latitude grasslands may completely re-establish a climax community in less than 50 years. The tundra, where ecological processes are very slow, may not recover from a disturbance for centuries.

Soil Classification and Climate

Many different classifications of soils have appeared over the years, each new system generally an improvement over the preceding ones. Until recent years those most widely used were variations of a scheme proposed by V. V. Dokuchaiev around 1870. Additions and changes were made to the original proposal until around 1950, when soil scientists decided that the old system was not adequate to include all of the soils that were being classified during the rapid expansion of soil science following World War II. In 1960 *Soil Classification, A Comprehensive System, 7th Approximation* was published. It is based upon characteristics of the soil

TABLE 9-7 Soil category levels.

Level	Category
1	Order
2	Suborder
3	Great group
4	Sub-group
5	Family
6	Series

profile and has six levels of categorization, given in descending order of generalization in table 9-7. The system allows for the inclusion of soils worldwide. At the order level, which is the most general category, all of the soils included in the order must have some basic similarities, basically color and organic content. These two elements are often, but not always, a reflection of climate.

In the classification of vegetation over the earth, plant formations are most closely associated with climatic regimes where the weather conditions tend to be fairly persistent through the year. This is also the case with soils. Associated with the tropical rainy regimes and the rain forests are the *oxisols* formed largely by the laterization process. They tend to be stained red by iron and aluminum oxides and to be low in organic content. In the deserts are found the *aridisols*, light in color from the calcification process and low in organic matter. The coniferous evergreen forest and the cool moist climate have combined to produce the *podosols*, a light colored soil accumulation of mineral and organic matter in the lower horizon. On the ice cap, soils are largely absent because of the severe cold and lack of weathered regolith.

As with the classification of climate and vegetation, the classification of soils in the regions of more variable climate is more difficult. There are locations in midlatitudes, for instance, where the laterite process prevails in the summer, and the podzolization process in the winter.

SUMMARY

The classification of climate is a difficult task and one that has a long history. Generally, however, two main approaches to classification have been used. The *empiric approach*, typified by the Köppen classification, is based upon observed distributions of the environment; *genetic classifications* attempt to define climates by their cause. The classification followed here uses air masses and air mass interaction as its genetic base. Nine primary climates are identified by the system, with other variations introduced in some of the units.

The relationship between climate and other environmental variables is evident, and in many ways the distribution of vegetation and soils is related to climatic realms.

APPLIED STUDY 9

Climatic Regions and Development: The Transamazon Highway

The regionalization of climates provides a general guide to conditions that exist over different parts of the earth. The generalized view of the nature of the climate and its associated vegetation has led to many misconceptions about a region's potential for human use. One climatic region about which many incorrect ideas occur is the wet equatorial climate. This climate gives rise to an incredible luxuriance of plant growth. On viewing the vegetation cover, it is easy to understand why early colonists and explorers assumed that very high agricultural yields would be obtained with introduction of "westernized" development methods. Such a conclusion is far from the truth; for in the hot, wet regions, physical problems place severe limitations on development.

One of the great paradoxes of the equatorial rain forest is that profuse vegetation exists on what is potentially one of the world's most infertile soils. The soils result from the process of laterization (giving soils such as oxisols and ultisols) caused by high inputs of precipitation and high temperatures over long time periods. Although the forests produce vast amounts of litter, it is soon leached out of the soil because organic materials decompose rapidly in such climates. Intense weathering leads to rapid decomposition of minerals that are also washed from the system. The result is a thick, infertile soil largely devoid of organic materials and minerals, made up largely of oxides of iron and aluminum.

That luxuriant natural vegetation can thrive on such a soil is the result of the ecological balance between the amount of nutrients produced by the plants and the rate at which they are used. Figure 9-7 illustrates how nutrients derived from the forest litter are rapidly taken up by other plants so that there is a minimum leakage of nutrients from the system. Also note the other ways in which the forest system is adjusted to the prevailing conditions. If vegetation is cleared over wide areas, the recycling mechanism is obliterated. Exposed to the sun and direct rainfall, the denuded surface alternately resembles hard rock and a sea of red mud. Soil temperatures rise to desertlike conditions to inhibit growth of seedlings, while direct rain on the unprotected surface will lead to much erosion.

Such a setting is the background for a development project in the Amazon Basin of Brazil. The key to the project was the construction of a road through the equatorial rain forest—the Transamazon Highway (see figure 9-8). The purpose of the construction was threefold: (1) to provide a settlement area for people from the overpopulated, drought-stricken northeast part of Brazil, (2) to settle an area that represents one-half of Brazil's territory but has only 4 percent of its population, and (3) to provide access to mineral and timber reserves. The construction of the road, completed in 1975, has largely failed on all three counts. While some of the problems are economic and others political, there is little doubt that the prevailing climate and its ecosystem response accounted for many of the problems that occurred. A few examples illustrate this fact.

REGIONAL CLIMATOLOGY

FIGURE 9–7 Immediate decay of forest litter allows living vegetation to recycle available nutrients rapidly. By maintaining this tight cycle, the forest can exist on soils of limited fertility. (From *Climate and Man's Environment*, John E. Oliver, Copyright © [1973, John Wiley & Sons]. Reprinted by permission of John Wiley & Sons, Inc.).

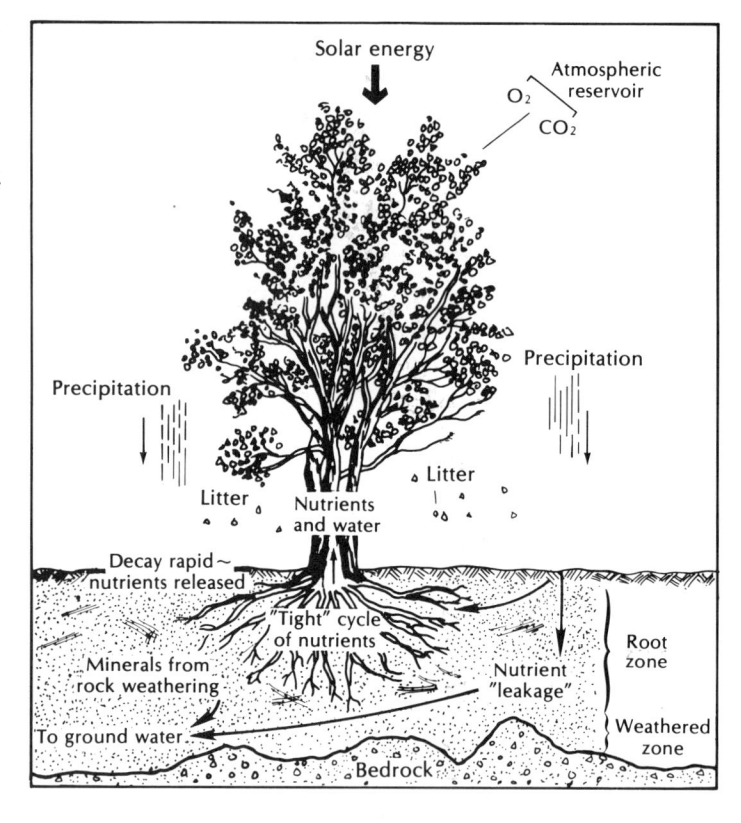

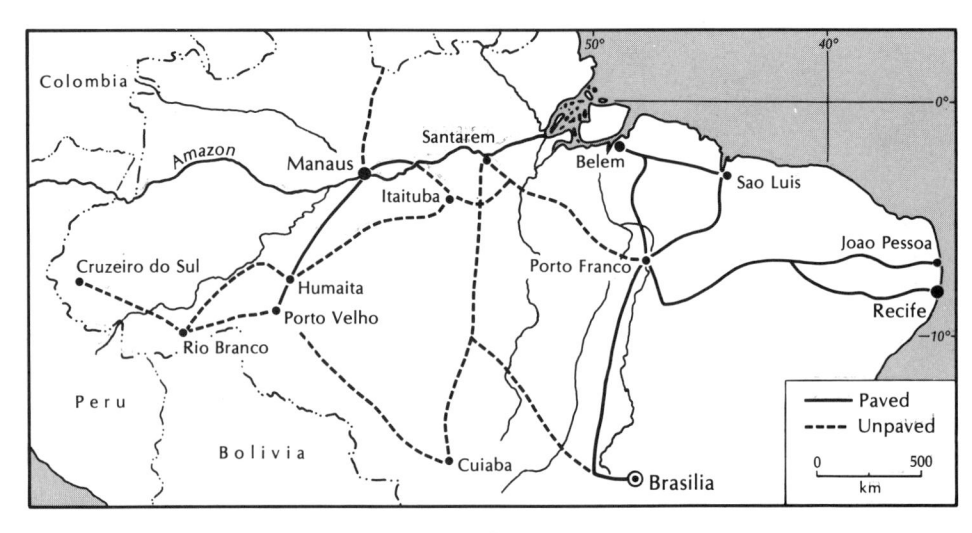

FIGURE 9–8 Highway development, proposed and actual, in the Amazon Basin.

Erosion The exposed road undergoes extensive lateral erosion during the wet season and cannot be repaired until the dry season—or a long dry spell—occurs. Side roads are cut off, and because of blocked drainage rice along the highway often rots in the fields because it cannot be harvested and dried. Accidents and breakdowns are frequent on the degraded road surface.

Agriculture The opening of lands to settlers for farming has not progressed well, and farming is characterized by very low yields. Poor crop selection, infertile soils, pests, and erosion contribute to the failure.

Health There are severe health problems for people moving to the region. Surprisingly, however, these are not related to "exotic" diseases; rather they are largely caused by diseases brought in by the settlers themselves. Malaria is widespread, contaminated drinking water results in gastroenteritis, and diarrheal diseases are the leading cause of infant mortality.

Timber Only a few small sawmills have been established. The most important wood, mahogany, has been cleared from a zone on either side of the highway, and further exploitation is limited by access and transport costs.

Minerals No sizeable mineral deposits have been discovered; where mining does occur, the Transamazon Highway is used in a very limited way.

The potential development of the equatorial forests has other far-reaching problems. Because of its enormous vegetation cover, the Amazon Basin represents one of the main photosynthesis regions of the world. It has been estimated that it provides 20 percent of the world's oxygen supply. Just how a diminished forest would affect the world's atmospheric composition is a question of some importance and is the topic of ongoing research.

10 Tropical Climates

GENERAL CHARACTERISTICS
Radiation and Temperature
Precipitation

CLIMATIC TYPES
The Tropical Wet Climate
The Tropical Wet and Dry Climate
The Tropical Deserts
Coastal Deserts
Desert Storms

SUMMARY

APPLIED STUDY 10
Desertification

REGIONAL CLIMATOLOGY

*T*he tropics have long had a certain mystique for mid-latitude peoples. The rich flora and fauna of certain tropical areas have encouraged some to envision great economic development. In the past, diseases such as malaria and yellow fever, fears of the debilitating effects of the tropical climate, and fears of ferocious insects, animals, and people have retarded development. The diseases can now be controlled, and the climate has been found tolerable and—with air conditioning—even comfortable indoors. But many of the constraints on growth in the tropics are inherent in the very nature of the environmental systems found there.

GENERAL CHARACTERISTICS

Radiation and Temperature

There are several attributes that characterize tropical climates, even though there is no single tropical climate as such. The most important of these are associated with the energy balance. There are several aspects of the energy balance that distinguish these climates from those further poleward. One is that the influx of solar energy is high throughout the year, although it does vary with the seasons. In this sense these are indeed winterless climates. Intensity of solar radiation is high all year because the solar beam is nearly always at a high angle, and there is very little variation in the length of the day from one part of the year to the next. The *photoperiod*, the relative lengths of day and night to which plants must adjust, varies between 11–13 h from winter to summer. Contrary to popular notion, though solar radiation is relatively high all year it is not necessarily greater in the tropics than in mid-latitudes, particularly in the summer. In the equatorial lowlands, for instance, clouds can and usually do reflect more than half of the total solar radiation. In addition to the large amount reflected by clouds, the high humidity and smoke from widespread burning further reduce solar radiation near the ground. So much solar radiation is reflected and scattered that a light-skinned person has difficulty in acquiring a tan. The ultra-violet radiation has been largely filtered out. At no time during the year do the tropics receive as many hours of sunlight as mid-latitude locations receive in the summer.

Although there is little difference in the length of the photoperiod in the tropics, the variations that exist are important biologically because even very small differences in length of day are sufficient to stimulate flowering or dormancy in plants. Evidence indicates that many tropical plants are more sensitive to slight changes in day length than are plants found in higher latitudes. The common poinsettia is a tropical shrub whose flowering is triggered by reduced day length. In the tropics these shrubs do not bloom at Christmas but much later in the year. They are forced to bloom before Christmas by controlling exposure to light and reducing the exposure artificially in early autumn.

Annual temperatures average about the same throughout the tropics, with slightly higher averages in the drier areas due to more intense radiation at the surface. The annual range in temperature is fairly closely associated with the length of the dry season. Where there is no dry season, the annual range in mean monthly temperatures may be as little as one or two degrees. With dry weather occurring in the winter months the mean temperatures drop, and the annual range increases.

Another significant aspect of the energy balance of tropical climates is that the primary energy flux is the diurnal one (figure 10-1). The variation in temperature from day to night is greater than the variation from season to season (figure 10-2). For this reason the tropics are not necessarily places of continuous great heat. Nights throughout the tropics can be rather cool. After arriving in the village that was to be home for one of the authors for a year in tropical West Africa, the first items that he and his family went looking for were blankets, for one night without a cover was enough. The diurnal radiation and temperature cycle is more important than the annual temperature cycle as a regulator of life cycles.

Precipitation

In the tropics rainfall, rather than temperature, determines the seasons; and it is the amount and timing of rainfall that form the chief criterion for distinguishing the various climates. Contrary to popular belief, only a very small portion of the tropics has a year-round rainy season. For the continent of Africa, for instance, the area with substantial rainfall in each month is less than 10 percent of the total land area. The largest portion of the tropics has a marked seasonal regime of rainfall that governs the biological productivity of the system. The remaining areas of the tropics are deserts, where rainfall is infrequent throughout the year. The seasonal pattern of precipitation is certainly the most critical aspect of moisture in the tropics. It is, in

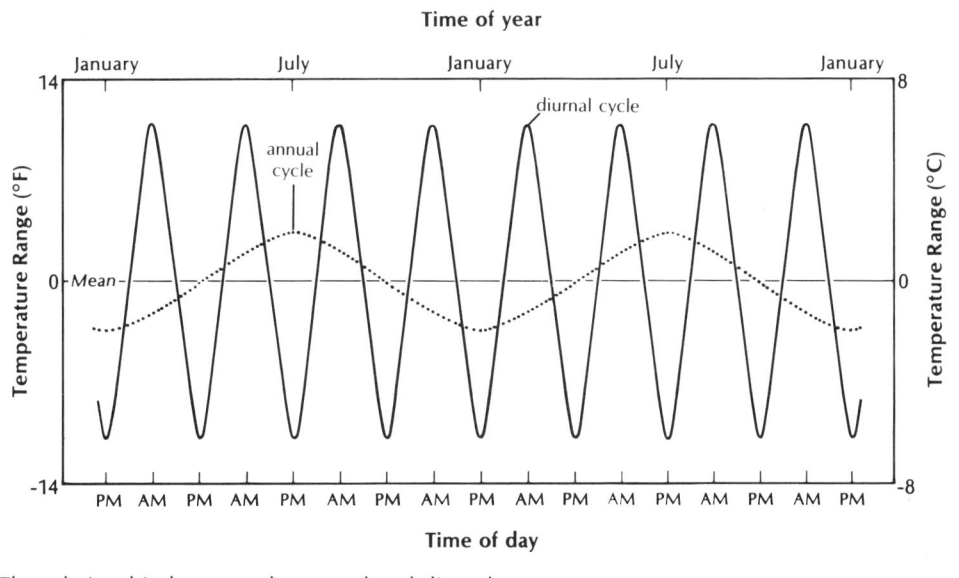

FIGURE 10-1 The relationship between the annual and diurnal energy cycles in the tropics. The diurnal range is of greater magnitude than the annual range.

REGIONAL CLIMATOLOGY

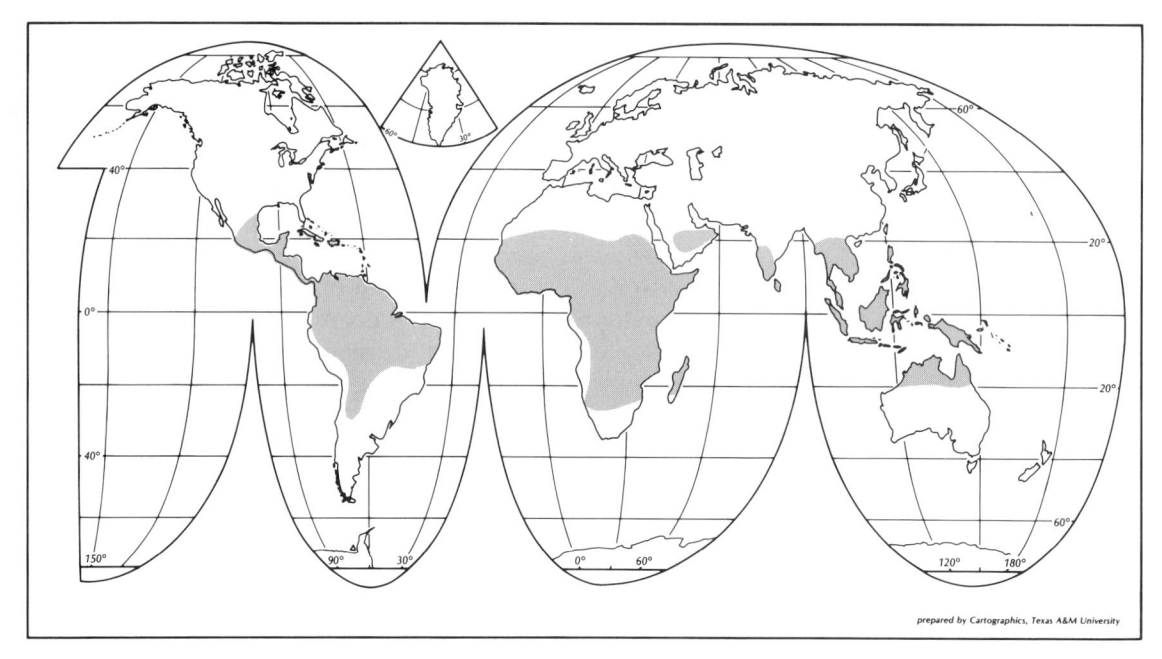

FIGURE 10-2 Area where the diurnal energy cycle is dominant. The shaded areas have a diurnal temperature range exceeding the annual range.

fact, the seasonal moisture pattern that distinguishes the major tropical environments—the rain forest, the savannas, and the desert—from each other.

The notion of the tropics as being monotonous, seasonless regions is far from the truth. Like so many other things, everything is relative. Compared to the mid-latitudes, the tropics lack the snow and ice storms, the tornadoes, and the hurricanes. But change is an attribute of the tropics as well as anywhere else, and to the resident just as much a conversation item.

CLIMATIC TYPES

The Tropical Wet Climate

Climates dominated by mT air

Köppen type	Af
Thornthwaite types	AA′r
	BA′r
	CA′r

The term *wet climate* is commonly associated with the luxurious evergreen forest typical of the wet tropical lowlands. The distribution of this climate is generally more limited than most people imagine because it is associated with a rather limited set of climatic conditions. It is characterized by fairly even and high temperatures averaging between 20°C–30°C (68°–86°F), a high frequency of precipitation year-round, and total rainfall of over 200 cm (80 in.) or more.

Part of the reason for the restricted area is that the region must lie within the tropical convergence zone throughout the year, which is not likely to happen over an extended amount of land because the convergence zone is continually shifting north and south. This migration is related to the passage of the overhead sun, as schematically illustrated in figure 10–3. The actual amount of movement is given in figure 10–4, which shows the average extreme positions of the ITCZ. Where the migration of the ITCZ is greatest, the rain forest is the smallest in extent. The rain

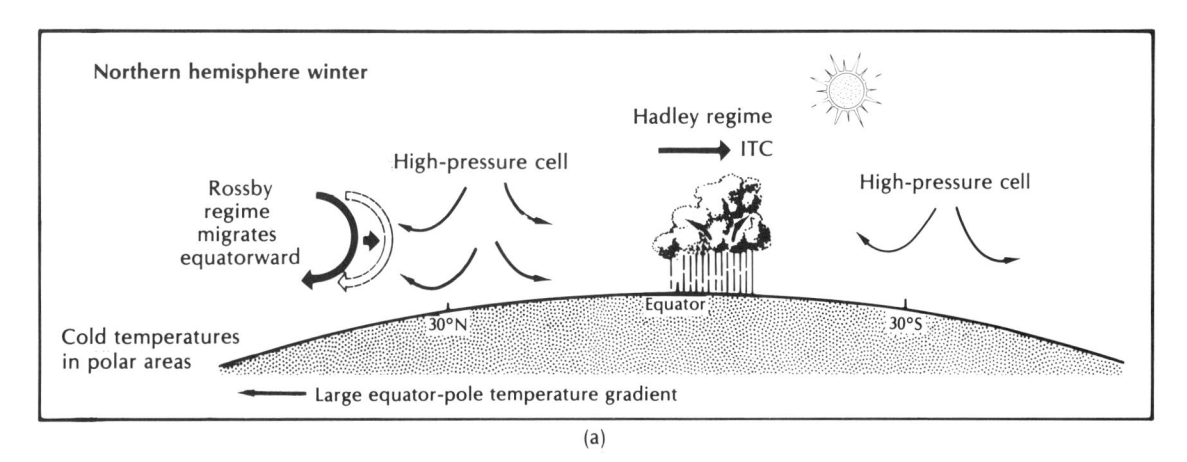

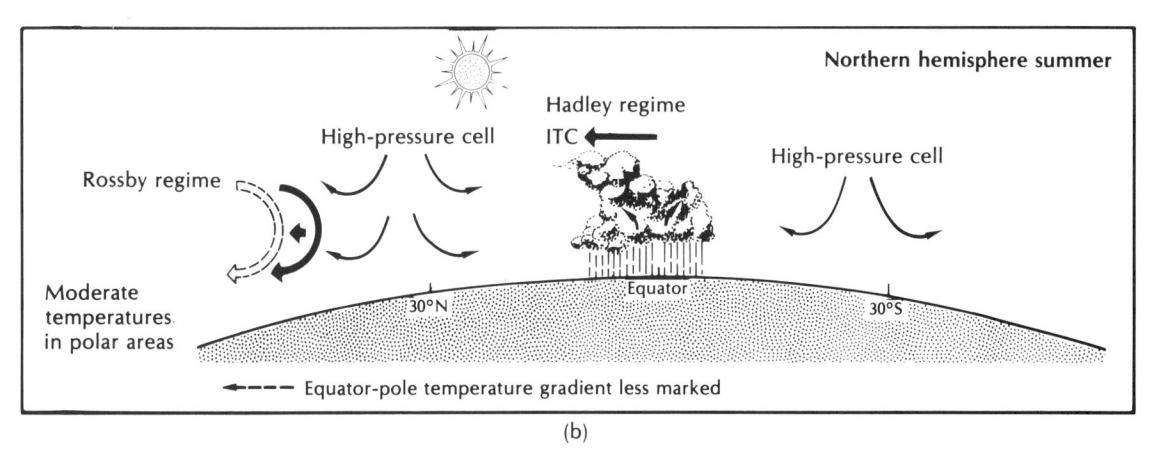

FIGURE 10-3 Schematic diagram showing the extent and migration of the Rossby and Hadley regimes in (a) winter and (b) summer.

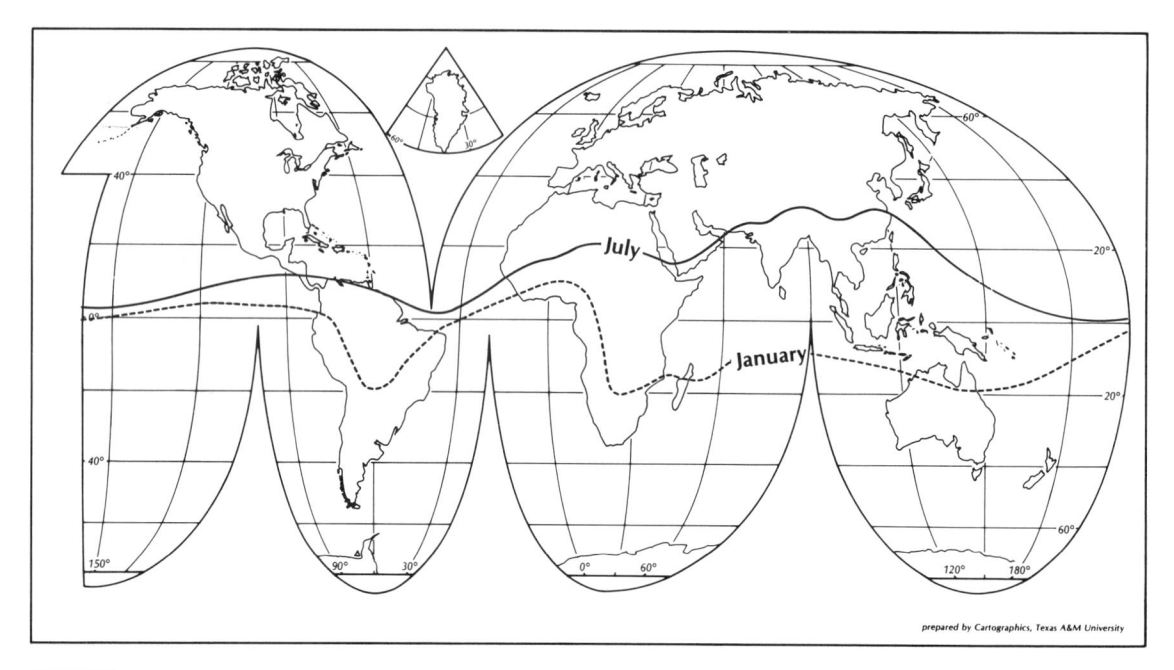

FIGURE 10-4 Average extreme positions of the ITCZ. (From J. Griffiths and D. Driscoll, *Survey of Climatology*, Charles E. Merrill Publishing Company, by permission.)

forest is also restricted to low elevations, usually below 1000 m (3300 ft), because at higher altitudes temperatures are considerably lower. As altitude increases, the forest changes character: trees become shorter in height, the number of species decreases, and highland vegetation replaces the rain forest.

The fairly even temperatures of the rain forest are due primarily to—

1. the equal or nearly equal periods of daylight and darkness,
2. the uniformly high intensity of radiation throughout the year,
3. the constantly high humidity.

The monthly average temperatures vary from 24°C to 30°C (75°–86°F), with an annual range of only 3°C (5°F) or so. Belem, Brazil, for instance, experiences a range from the warmest to coolest month of only 1.6°C (3°F). Since there is greater variation in radiation within the day than from day to day, the diurnal range may be two to three times the annual range.

Associated with the convergence of tropical maritime air are high humidity and cloudy skies. Since humidity is so high in the daytime, when nocturnal cooling occurs, early morning fogs and heavy dews are common. The dewpoint is reached while temperatures are in the range of 15°–20°C (59°–68°F).

While the primary circulation determines the regional pattern of atmospheric circulation and the seasonal regime, the thunderstorms and easterly waves

FIGURE 10-5 The diurnal distribution of rainfall at Kuala Lumpur, Malaysia. The midafternoon maximum is typical of many wet tropical locations. (Data from Ooi Jin-bee [1959]).

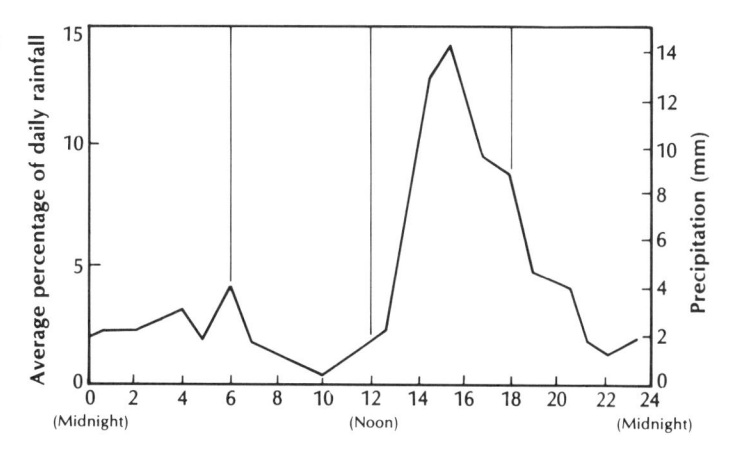

account for most of the day-to-day weather. As might be expected from the degree of surface heating, local thermals, cumuloform clouds, and thunderstorms predominate. The cumulus clouds build during the daytime hours into towering cumulus and thunderstorms when the air becomes unstable. This process provides a distinctive diurnal rainfall, such as that shown in figure 10-5. Precipitation can be expected on more than 50 percent of the days. Duitenzorg, Java, averages 322 days a year with thunderstorms. Since the precipitation is largely from thunderstorms, it is of high intensity and short duration. Rainfall amounts of 50-65 mm (2.0-2.5 in.) per hour have a return period of less than two years. Maximum hourly rainfalls of 95-120 mm (3.7-4.7 in.) have been recorded in the East Indies. Though rainfall intensities for 1-h periods are not significantly higher in the tropics than in mid-latitudes, sustained periods with high-intensity rainfalls are significantly more common in the tropics. The greatest 24-h rainfall on record occurred at Cilaos, La Reunion, on March 16, 1952, when 1870 mm (73.62 in.) of rain fell (table 10-1). Thunderstorms in the tropics develop their maximum turbulence at higher temperatures than in mid-latitudes; therefore, hail is fairly infrequent, but it does occur on occasion. In 1922 Mount Cameroon erupted in West Africa. The eruption produced a storm that left a snow cover on the summit, and at Debundscha at the foot of the mountain 572 in. of rain were recorded. Not all rainfall is convectional, however, as there are periods of prolonged showers.

TABLE 10-1 Precipitation extremes established in the tropical wet climates.

Highest 12-h total	Belouve, La Reunion 28-29 Feb. 1964	1340 mm (52.76 in.)
Highest 24-h total	Cilaos, La Reunion 16 Mar. 1952	1870 mm (73.62 in.)
Highest 5-day total	Cilaos, La Reunion 13-18 Mar. 1952	3854 mm (151.73 in)
Highest number of rain days in a year	Cedral, Costa Rica 1968	355 days

Average annual rainfall is not as important a factor as is the frequency of precipitation but ranges upwards to 2 m (6.6 ft). The highest amounts are found on mountain slopes such as Mount Waialeale, Hawaii, where the annual average is 11.8 m (39 ft).

As with almost all generalizations, there are exceptions to the rules about types of climate actually found at some locations along the equator. Some extremely dry environments are found where the humid tropics should be expected. Herman Melville wrote of "The Encantadas" (Galapagoes).

> But the special curse, as one may call it, of the Encantadas, that which exalts them in desolation above Idumea and the Pole, is that to them change never comes; neither the change of seasons nor of sorrows. Cut by the equator, they know not autumn and they know not spring; while already reduced to the lees of fire, ruin itself can work little more on them. The showers refresh the deserts but in these isles, rain never falls. . . . Nowhere is the wind so light, baffling, and in every way unreliable, and so given to perplexing calms, as at the Encantadas (p. 182).

A shack, which was inhabited by a woman stranded on one of the islands, utilized a system of collecting dew to provide fresh water:

> and here was a simple apparatus to collect the dews, or rather doubly distilled and finest winnowed rains, which in mercy or in mockery, the night skies sometimes drop upon these blighted Encantadas. All along beneath the eave, a spotted sheet, quite-weather-stained, was spread, pinned to short, upright stakes, set in shallow sand. A small clinker, thrown into the cloth, weights the middle down, thereby straining all moisture into a calabash placed below. This vessel supplied each drop of water ever drunk upon the site by the Choles. Hunilla told us the calabash would sometimes, but not often, be half-filled over-night. It held six quarts, perhaps (p. 231).

The Tropical Wet and Dry Climate

Climates with seasonal mT and cT air	
Köppen types	Aw
	Am
	BSH
Thornthwaite types	BA'w
	CA'w
	DA'w

The distinguishing feature of this climate is a very pronounced periodic moisture pattern. Atmospheric humidity, precipitation, soil moisture, and stream flow change through the year in a rhythmic pattern. These environments are marginal to the tropical rain forest environments and in most cases poleward of them. They occupy much of the area from the boundary with the tropical rain forest to the tropics of Cancer and Capricorn, a much larger area than that occupied by the rain forest itself. Areas with this type of environment are found north and south of the Amazon Valley in South America, across Africa north and south of the Congo Basin, and in much of southeast Asia and parts of the Pacific islands. Their location in the general circulation is such that they experience the seasonal migration of weather types associated with both the equatorial convergence zone and the subtropical divergence zone.

The seasonal pattern of moisture is due to the migration of the tropical convergence zone. The rainy season is concurrent with the high sun and the subsequent presence of the convergence zone. The dry season is a product of the more stable air originating from the subsidence in the subtropical highs. The seasonality increases poleward as might be expected. In a traverse away from the equator, the low sun precipitation begins to decrease first, with the high sun precipitation remaining as high as in the tropical convergence zone. Midway between the equatorial convergence zone and the tropical desert, the winter precipitation drops to near zero. From that point poleward, the high sun precipitation declines until it is no longer significant and the tropical desert is reached.

This climate has the most pronounced seasonality of precipitation of any climatic type. An example of the extremes that are reached is found in Asia. Rangoon in Burma has a 3-month winter average of 25 mm (1 in.) and a 3-month summer average of 188 cm (74 in.). This average increases northward to Akyab, where the averages are 3.8 cm and 426 cm (1.5 in. and 167 in.). The maximum in the Asian region is reached at Cherrapunji, India, where in two winter months they receive an average of only 2.5 cm (1 in.) of precipitation but in the two summer months, 528 cm (207 in.). Cherrapunji recorded a 5-day total of 405 cm (159 in.) in August 1841; a 1-month total of 915 cm (30 ft) in July 1861; and in 1 year 26 m (85 ft) in the period August 1860 to July 1861. Usually the most intense storms occur at the onset and end of the rainy season, when instability is greatest.

The average annual precipitation is extremely varied and cannot be used as a criterion for distinguishing the region. On the wet margins where topography is favorable for orographic intensification, the totals run over 1000 cm (400 in.); on the dry margins it drops to less than 25 cm (9.8 in.). It is, in fact, quite likely that the highest annual average precipitation totals to be found on the earth occur where there is a strong seasonal pattern of precipitation. Annual variability is higher there than in the tropical rain forest because the precipitation total in any year is subject to the extent of migration of the general circulation and the length of time the zone of convergence stays poleward. The farther a site is from the heart of the convergence zone, the lower the annual average precipitation, the shorter the rainy season, and the greater the annual variability. The seasonal weather is, of course, in marked contrast. During the rainy season it is warm, humid, and there are frequent rainstorms. During the dry season more or less desert conditions prevail (table 10–2).

The lag of convergence and the rainy season behind the migration of the sun produces three seasons in many parts of the wet and dry tropics: the cool season, the hot season, and the rainy season. The cool season is during the winter months, when

TABLE 10-2 Data for tropical wet and dry stations.

	Jan.	Feb.	Mar.	Apr.	May	June	July	Aug.	Sept.	Oct.	Nov.	Dec.	Yr.
Calcutta 22°32' N 88°22' E: 6m
Temp. (°C)	20.2	23.0	27.9	30.1	31.1	30.4	29.1	29.1	29.9	27.9	24.0	20.6	26.94
Precip. (mm)	13	24	27	43	121	259	301	306	290	160	35	3	1582
Cuiaba, Brazil 15°30' S 56°03' W: 165 m
Temp. (°C)	27.2	27.2	27.2	26.6	25.5	23.8	24.4	25.5	27.7	27.7	27.7	27.2	26.5
Precip. (mm)	216	198	232	116	52	13	9	12	37	130	165	195	1375
Dakar, Senegal 14°39' N 17°28' W: 23 m
Temp. (°C)	21.1	20.4	20.9	21.7	23.0	26.0	27.3	27.3	27.5	27.5	26.0	25.2	24.3
Precip. (mm)	0	2	0	0	1	15	88	249	163	49	5	6	578
Darwin, Australia 12°26' S 131°00' E: 27 m
Temp. (°C)	28.2	27.9	28.3	28.2	26.8	25.4	25.1	25.8	27.7	29.1	29.2	28.7	27.6
Precip. (mm)	341	338	274	121	9	1	2	5	17	66	156	233	1562

solar radiation is at a minimum. The hot season follows, when temperatures rise as the sun moves higher in the sky and solar radiation increases. The onset of the wet season brings an increase in cloud cover, slightly reduced radiation, and cooler temperatures.

Generally where moist air of oceanic origin exists, precipitation occurs; and where the dry continental air prevails, precipitation is largely absent. Often there is a sharp boundary between the two types of air. The Hadley cells north and south of the equator and the convergence zone between them shift north and south, following the migration of the vertical rays of solar energy. This migration produces the seasonal pattern of precipitation that characterizes so much of the tropical region. The system moves through 20° of latitude through the year from about 5° S to 15° N.

The Tropical Deserts

Climates dominated by cT air

Köppen type	BWh
Thornthwaite types	EA'h
	EB'd
	EA'd
	DA'd

World deserts cover more than one-fourth of all the land area of the continents. Because they cover more area than any other single climatic type there is concern that the current deserts may be expanding (see Applied Study 10). They are found at all latitudes between 50° N and 50° S, with the largest found in the tropics (table 10-3). The heart of the tropical deserts lies near the tropics of Cancer and Capricorn primarily toward the west sides of the continents. They are less common on the east side of the land masses because the trade winds carry considerable amounts of moisture onshore in these latitudes. There are a number of characteristics that distinguish the tropical deserts. They are—

1. low relative humidity and cloud cover,
2. low frequency and amount of precipitation,
3. high mean annual temperature,
4. high mean monthly temperatures,
5. high diurnal temperature ranges,
6. high wind velocities.

The basic controls in the tropical deserts are the upper air stability and subsidence. As a result, divergence is a basic characteristic of the tropical desert, and general stability and low relative humidity prevail. Relative humidity averages 10-30 percent in the interior areas and slightly higher in coastal locations; it has been measured at as low as 2 percent. With relative humidity as low as it is, cloud cover is also low. The Sahara averages 10 percent cloud cover in winter and 4 percent in the summer.

Precipitation is very low in amount and very sporadic in distribution, both temporally and spatially. Low annual precipitation is the basic characteristic of these climates, but there are occasionally heavy downpours. Arica, in Chile, averages 5 mm (0.2 in.) of precipitation per year. Iquique, Chile, experienced a 14-year period with no rainfall, and Wadi Halfa in the Sahara Desert experienced a 19-year period with no rainfall. Iquique, on another occasion, received 100 mm (4 in.) of rain in 1 day. One rain may bring 125-250 mm (5-10 in.) of precipitation; then for a period of years no precipitation at all is recorded. One station in the Thar desert with an annual average of 100 mm rainfall received 850 mm in 2 days. At Dakhla, in southern Egypt, where the average is 100 mm, one 11-year period elapsed with no rain. General rains over large areas of the desert are infrequent, but they do occur. In

TABLE 10-3 Major deserts of the world.

Desert	Approximate Area km² (thousands)	mi² (thousands)
Sahara	9100	3500
Central Asia	4510	2200
Australian	3400	1300
North American	1300	500
Patagonian	680	260
Indian	600	230
Kalahari-Namib	570	220
Atacama	360	140

February of 1980 a widespread rainstorm struck eastern Saudi Arabia with paralyzing results. The major cities were flooded since there are no storm sewers to carry off the water, and the major highways from Riyadh to Dhahran were blocked by flood waters. There is a greater frequency of precipitation on the equatorward margins of the deserts in the summertime and on the poleward side in the winter because the major mechanism for precipitation on the equatorward margin is the ITCZ and on the poleward side the mid-latitude cyclones. The precipitation that does fall is primarily convective on the equatorward side, with only occasional cyclonic precipitation. As latitude increases, the frequency of cyclonic precipitation increases also.

Seasonal weather in the deserts is associated more with temperature than with precipitation (table 10–4). With the winter season come slightly cooler days and much cooler nights. The highest average annual temperatures of any climate are found in the tropical deserts. They vary between 29°C–35°C (84°F–95°F). At Lugh Ferrandi, in Somalia, the temperature averages 31°C (88°F), considerably higher than in any of the other tropical climates. The major factor controlling the average annual temperature is latitude, and the stations with the highest averages are those closest to the equator. The highest temperature yet recorded is 58°C (136°F) at El Azizia, near Tripoli in northern Africa, on 13 September 1922. In the United States the highest recorded temperature is 56.7°C (134°F) in Death Valley, California. Winter averages are below those of other parts of the tropics as earth radiation at night rapidly cools these areas. Averages are as low as 15°C–20°C (59°F–68°F). The lower temperatures of the winter season give the tropical deserts the highest annual range found in the tropics. Aswan, Egypt, located on the Tropic of Cancer, has an annual range of 19°C (34°F).

Diurnal ranges in the deserts are the highest of any of the climates, and, as in other tropical climates, they greatly exceed the annual range. The average diurnal range is from 14°C to 25°C (25°F–45°F), but on occasion it is much greater. At Birmilrha in the Sahara south of Tripoli the temperature went from 37.2°C down to −0.6°C (99°F–31°F) in one 24-h period, a range of 37.2°C (68°F). At the oasis of In-

TABLE 10–4 Data for tropical desert stations.

	Jan.	Feb.	Mar.	Apr.	May	June	July	Aug.	Sept.	Oct.	Nov.	Dec.	Yr.
Al-Hofuf, Saudi Arabia 25°22′ N 49°35′ E: 145 m														
Temp. (0°C)	14.0	15.9	20.6	25.4	30.8	33.5	34.6	33.9	31.3	26.8	20.9	16.1	25.3
Precip. (mm)	23	8	16	16	1	0	0	0	0	1	1	6	49
Marrakech, Morocco 31°37′ N, 08°00′ W: 458 m														
Temp. (°C)	11.5	13.4	16.1	18.6	21.3	24.8	28.7	28.7	25.4	21.2	16.5	12.5	19.9
Precip. (mm)	28	28	33	30	18	8	3	3	10	20	28	33	242
Alice Springs, Australia 23°38′ S, 133°35′ E: 570 m														
Temp. (°C)	28.6	27.8	24.7	19.7	15.3	12.2	11.7	14.4	18.3	22.8	25.8	27.8	20.8
Precip. (mm)	43	33	28	10	15	13	8	8	8	18	30	38	252

Salah in the Sahara on 13 October 1927, the temperature dropped from an afternoon high of 52.2°C to −3.3°C (126°F–26°F) the following morning, a range of 55.6°C (100°F).

Coastal Deserts

Along coastal sections of the tropical deserts a rather distinctive atmospheric set of conditions, which differs from some of the stereotypical aspects of a tropical desert, prevails. These desert areas are characterized by cool temperatures, shallow temperature inversions, cold water offshore, considerable fog and stratus cloud cover, but little rain (table 10-5). In fact, the areas with the lowest annual rainfall of all the deserts are to be found here. Areas where this type of climate can be found are Baja California, Ecuador, Peru, and along the Sahara and Namib deserts of Africa. The primary factor in producing these conditions is the onshore flow of air that has been chilled by crossing a cold current flowing along the coast. One effect of this onshore flow of air is to cool the area during the summer months, dropping the annual mean temperature by 5°C–10°C and greatly reducing the annual range in temperature. At Iquique, Peru, the relative humidity averages 81 percent in August, yet the rainfall averages only 28 mm (1.1 in.) per year. During one 20-year period, there was no measurable precipitation in 14 years. Caloma, Peru, averages 48 percent relative humidity in August, and no measurable precipitation has been recorded. These coastal deserts have extremely low precipitation, even for desert areas, and a much higher percentage of cloud cover. Humidity often hangs in the 80–90 percent range, and extensive fog is common. In the vicinity of Lima, Peru, in July and August fog often covers the city for days at a time. When conditions are well developed, the fog will pass over the city and penetrate for miles into the valleys of the Andes to the east of the city. These are unusual climatic areas that do not cover large areas of the earth's surface.

TABLE 10-5 Comparison of temperature and relative humidity characteristics of tropical and west coast deserts.

	Average of Warmest Month T(°F)	R.H. (%) 0700	1400 h	Average of Coolest Month T(°F)	R.H. (%) 0700	1400 h	Mean Annual Temp. (°F)	Annual Temp. Range (°F)
Tropical deserts								
In-Salah (Sahara)	98	36	25	56	63	37	77	42
Riyadh (Arabian)	93	47	31	58	70	44	75.5	35
Laverton (Gr. Australian)	87	36	24	52	60	43	74.5	35
West Coast deserts								
Arica (Atacama)	72	74	61	60	83	74	66	12
Walvis Bay (Kalahari)	66	91	73	58	83	65	62	8
Villa Cisneros (Coastal Sahara)	72	88	63	63	75	51	67	9

Desert Storms

Severe storms are not frequent in the deserts, primarily because of the lack of moisture to supply the energy. However, deserts tend to be very windy, particularly in afternoons and during the hottest months. Common features of the air over deserts are a steep lapse rate and instability in the air near the surface. *Dust devils* are the most visible form of the turbulence. Developing in clear air with low humidity, they can usually be observed by the debris they carry and have been known to reach a height of nearly 2 km. On occasion they will sustain velocities high enough to blow down shacks or blow screen doors off their hinges. These dust devils are very common in deserts, and it is not unusual to see a multitude of them at one time under the right atmospheric conditions. The dust devil results when there is an intense thermal at or near the ground surface, and surrounding air moves inward to replace the rising air. As the air moves in toward the thermal, the radius of curvature decreases and velocity increases.

Other storms of the tropical deserts are the dust storms and sand storms. *Dust storms* occur as a result of deflation. One of the readily visible aspects of much desert surface is the lack of fine sediments. Deflation tends to keep the surface swept clean. Winds blowing out of the deserts are often dust-laden. On the south side of the Sahara the dry northerly wind is known as the *harmattan*. Since it is a dry wind, it is cooling and welcome although it may limit visibility and leave household furnishings covered with dust. These same winds blow out of the north side of the Sahara Desert. Occasionally they travel across the Mediterranean Sea, bringing disaster to agricultural crops. They are important enough in the climate of Mediterranean countries to warrant local names, such as *sirocco* in Italy and Yugoslavia and *leveche* in Spain.

In the sandy parts of the deserts it is the ability of the wind to move sand that is significant. Only a small part of the tropical deserts is covered with sand dunes. Something less than 30 percent of the Sahara is covered with sand dunes, and in the Sonoran Desert of North America less than 2 percent is covered by dunes. *Sand storms* occur only when wind velocities reach a high enough value to move the sand. This value is called the *threshold value* and depends on the size of the sand particles, wetness of the sand, and other, less important variables. For medium-sized dry sand (0.25 mm) the threshold velocity is 5.4 m/s and is exceeded some 30 percent of the time. Sand drift increases rapidly as wind velocities increase above the threshold. An average rate of 12.0 l/m width was observed for winds 5.5–6.4 m/s and a rate of nearly ten times that for wind velocities of 10 m/s. The increase is rapid because the power of the wind increases as the cube of the wind velocity above the threshold. However, it should be noted that the frequency of winds decreases rapidly as velocity increases.

Two different kinds of sand storms develop in the tropical deserts. The first is the result of surface heating and the resultant turbulence. It is not unusual for sand temperatures at the surface to reach 85°C (180°F). This type of sand storm is chiefly a daytime phenomenon and most pronounced in the hottest months. At Al-Hofuf, in Saudi Arabia, the percentage of the hours of sand drift occurring between 0600 and 1800 hours ranges from 76 percent in February to 91 percent in June. The total number of hours per day when winds exceed the threshold velocity also has a seasonal pattern, increasing from 3 h per day in February to 9.4 h in June. As a result of the increasing number of hours of high velocity winds in summer, sand movement increases as well, increasing from 116 l/m/day in February to 406 l/m/day in June. In summer the high frequency of afternoon sand storms makes travel and other outdoor activity particularly difficult during the afternoon hours.

The second kind of dust storm is brought about by a low pressure disturbance passing through the area. These storms often generate higher velocity winds than does diurnal heating, and they last for longer periods. One such sand storm in the Nafud lasted for a period of 43 h, during which sand blew constantly and most of the time from the same direction. The wind averaged 10.7 m/s (24 mph) and exceeded 15.6 m/s (35 mph) for hours at a time. It was just such a regional sand storm that halted the American attempt to rescue the hostages from Iran in 1980.

Both types of sand storms have the ability to move huge amounts of sand over a period of time. In the Nafud, the wind moves an estimated 80 m^3 of sand across each meter width of the sand field each year. The dunes themselves move at a fairly high rate of speed as the sand is continually moved from place to place. In the Nafud, dunes with an average height of 10 m (33 ft) are moving at a rate of 15–19 m/yr. The main highway across Saudi Arabia and the only railroad go through the Nafud, and both are sporadically closed as the dunes encroach upon the right of way.

SUMMARY

Tropical climates cover a large portion of the earth's surface between latitudes 25° N and S. In most of this region temperatures change as much, or more, from day to night as from summer to winter. Radiation intensity varies but remains relatively high all year. Within the tropical realm climatic regions are distinguished on the basis of the precipitation regime. Some sites have a high probability of precipitation on every day of the year, and others have an equally low probability of precipitation on any given day of the year. In between these extremes are the vast areas with a rainy season and a dry season. In the wet-and-dry tropics there is a wide range in both the length of the rainy season and in the amount of precipitation received during the rainy season. The monsoon regions of Asia represent the extreme example of seasonal precipitation, and it is there that some of the greatest amounts of annual precipitation are to be found. The driest areas of the world are also found in the tropics. The coastal margins of the tropical deserts include stations with the lowest known annual total precipitation.

APPLIED STUDY 10

Desertification

In many parts of the world, particularly on the margins of the deserts, a process of environmental degradation is taking place that is commonly referred to as *desertification*. The process consists of the breakdown in the vegetal cover and soil until the land is no longer productive, and erosion by water and wind produce relatively sterile land. Desertification does not involve the outward expansion of the climatic conditions responsible for the core areas of the world deserts. Rather, it is the alteration of the land to desertlike character by human activity.

Lands on the fringes of the desert are particularly susceptible to these processes because of the nature of precipitation on the dry margins of the wet-and-dry tropics. Precipitation in the wet-and-dry lands is subject to considerable fluctuation from year to year, as is illustrated by the rainfall data for Khartoum, Sudan, shown in figure 10-6. Precipitation during the period from 1941 to 1974 varied from less than 50 mm to more than 300 mm. The amount of seasonal rainfall translates into human carrying capacity, because the people are semisubsistence farmers or herders (figure 10-7).

Even the total annual precipitation is often misleading because the intraseasonal distribution varies from year to year. In central Sudan most of the rainfall is concentrated in the months of May through September. In 1973 there was 175 mm of rainfall during the growing season but none in the months of June

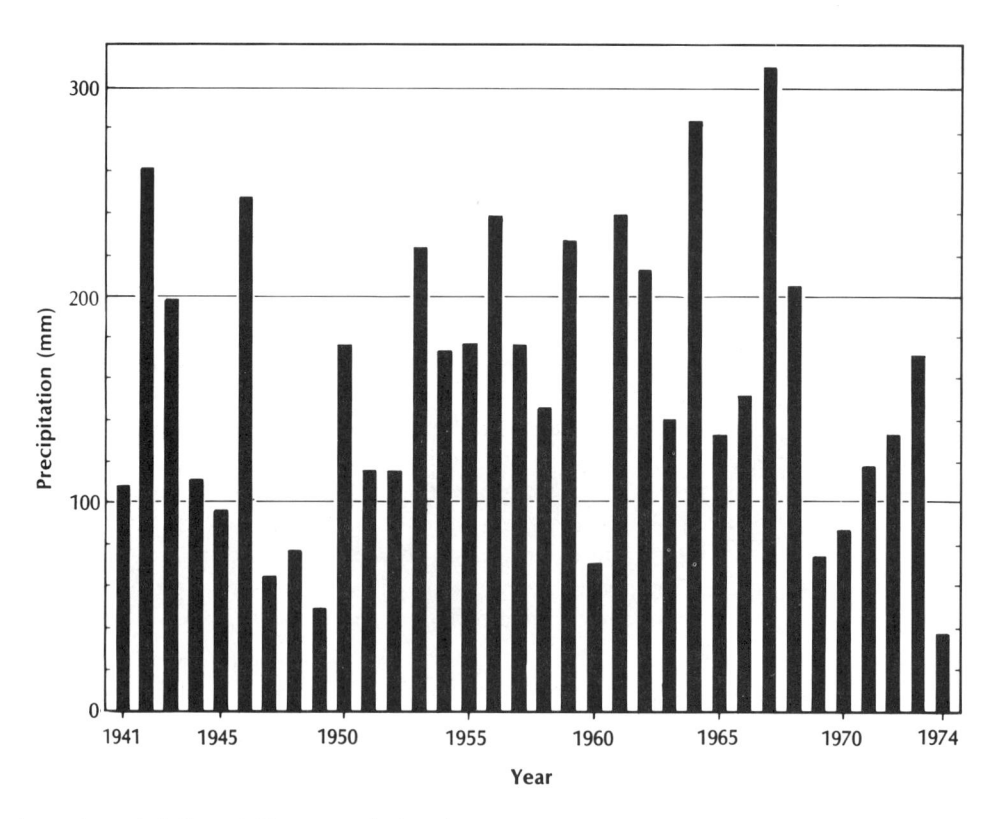

FIGURE 10-6 Annual precipitation at Khartoum, Sudan, from 1941 through 1974.

FIGURE 10-7 Precipitation and carrying capacity. The precipitation over much of the sub-Saharan Africa is subject to marked cycles. Since most of the population engages in subsistence agriculture or herding, precipitation translates into carrying capacity.

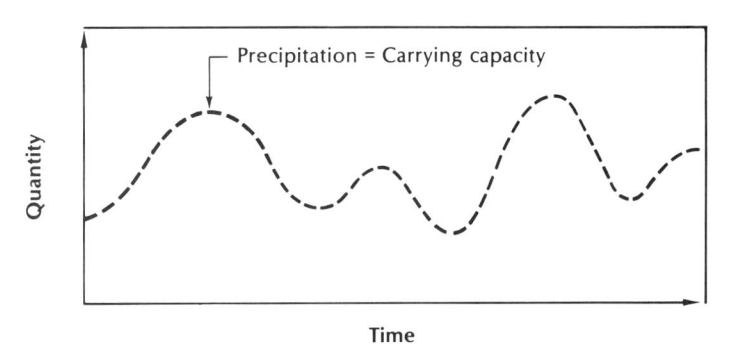

and August (figure 10-8). In the following year precipitation fell in each of the four months from June to September, but only 50 mm was received in total. Both years were disasters for agricultural people, especially for those engaged in crop cultivation, since the two dry months came at critical times in the growing season.

The nature of the ecosystems in much of the Sahelian zone is such that they will not support a large number of people on a subsistence basis. Dry-land cultivation practices are for the most part both primitive and damaging to the land. Cleared areas are farmed for several years, during which the soil gradually deteriorates and loses fertility. When the soil is exhausted, it generally is abandoned on the assumption that it will rejuvenate with time. If the amount of

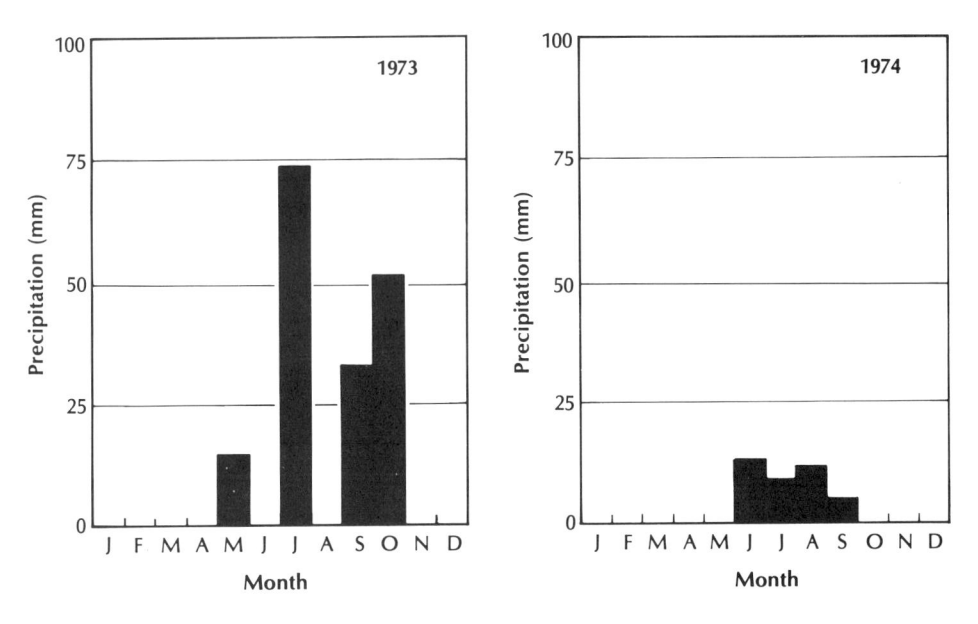

FIGURE 10-8 Seasonal rainfall in Khartoum, Sudan, in 1973 and 1974. In 1973 the total was above average, but there was no measurable rain in the two crucial months of June and August. In 1974 the total was far below normal.

land cultivated in any given year is small enough, and the number of livestock units on the grassland is below the carrying capacity, the ecosystem will suffer little long-range damage.

Desertification takes place when the numbers of people and livestock exceed the capacity of the precipitation to support them. It is not a new problem, but the rate at which it is taking place is unprecedented. Prior to the colonization of Africa there were intermittent periods of environmental degradation. Whenever a dry cycle began, there were overgrazing and overcultivation followed by deterioration of vegetation and soils. Population increased during wet periods and decreased during drought (figure 10–9).

If an area became unproductive the population either died back or moved away. The introduction of colonial administration brought some measures that greatly altered the balance between the population and environment. Intertribal warfare was curtailed and public health measures were introduced, all of which precipitated rapid population growth. In 1980 several of the Sahelian countries had the highest growth rates of any country. Parallel with the increase in human numbers has been an increase in animal populations.

The result of the rapid population growth in the face of fluctuating rainfall is that the demands on the ecosystems become greater than the carrying capacity more frequently and for longer periods of time. As shown in figure 10–10, the demand ultimately exceeds the carrying capacity on a continuing basis.

The result of the increasing pressure on the land is the expansion of the Sahara Desert southward into more humid areas. If this phenomenon is not completely caused by human activity, it certainly is being accelerated by it at a rapid pace. Degeneration into desert usually occurs in scattered patches of bare ground from a few meters to several kilometers across. It is not a desert expansion in the form of an even front of desert surface advancing across the landscape, and hence is not easy to measure. Desertification is taking place at a steady rate in many areas of the world's grasslands and is reported on every continent except Antarctica. It becomes particularly rapid during periods of major drought like the one that struck sub-Saharan Africa in the early 1970s.

When the plant cover is destroyed beyond the minimum required for protection of the soil against erosion, the process becomes irreversible. Less water is retained, the soil dries out, and the temperature increases, speeding the collapse. Bare soil appears with the fine soil particles being removed by either

FIGURE 10–9 Population and carrying capacity in the precolonial era, when population size followed the precipitation cycles. When favorable periods occurred, population expanded; when dry periods followed, the population died back. The shaded areas represent times of environmental degradation.

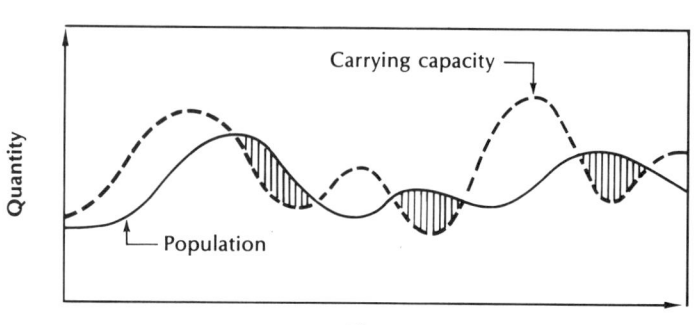

FIGURE 10-10 Population and carrying capacity following colonization, when conditions changed to increase population size greatly. At point A when the carrying capacity began to decline, various forms of aid to the distressed areas kept down the death rate. Thus when precipitation increased again at point B, a much larger carryover population remained from which to continue growth. The result is a population greater than the environment can support even under more favorable circumstances, and desertification is accelerated.

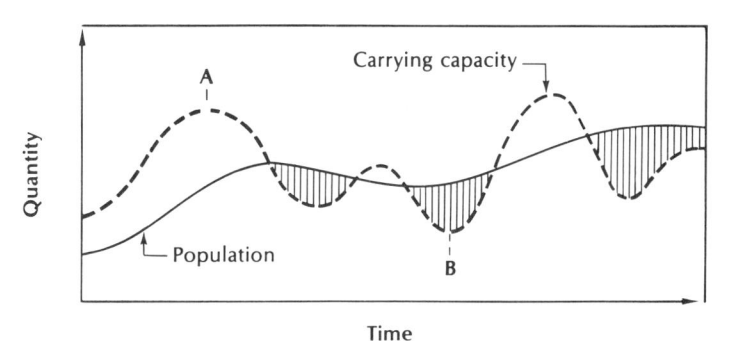

wind or water. This loss of topsoil is substantial and very important. The movement of dust by wind has existed throughout historic times, and hence is by no means a recent phenomenon. The dust picked up from the Sahara and adjacent lands is transported for great distances and in large amounts by the atmosphere. A portion of this dust is transported across the Atlantic Ocean into the Americas. The total drift westward off the African coast may be as much as 60 million tons each year. It is extensive enough to be measurable in the Caribbean and at times, such as in the fall of 1982, in Florida.

The wind erosion is not restricted to dust-sized particles. When large patches of bare ground are formed, the soil is exposed to wind erosion of the larger sand-sized particles, and most of the humus and mineral salts are removed. The surface then is distinguished by denuded deflation pans. During the height of the Sahelian drought satellite images showed that thousands of square kilometers of surface were obscured by blowing sand. The blowing sand and silt further destroys the standing vegetation by stripping the remaining vegetation or by burying it under accumulating dunes.

The only long-range solution to the problem is to keep land use at an intensity level substantially below that which has traditionally been practiced. The levels of animal populations must be kept below the maximum carrying capacity represented by the wetter-than-normal years. This is the only mechanism that will provide immediate relief from the drought hazard and at the same time give the pastures a chance to survive. The determination of these critical levels requires a thorough understanding of the regional environment as well as a close monitoring of the natural changes in precipitation. The present short-run situation is critical. If land use is intensified, desertification and famine are certain. If production isn't increased to keep up with the growing population, famine will also be certain.

11 The Mid-Latitude Climates

GENERAL CHARACTERISTICS
Middle Latitude Circulation
Summer Season
Winter Season
Water Balance

CLIMATIC TYPES
Mid-Latitude Wet Climates
Mid-Latitude Wet and Dry Climates
Summer-Dry Climates
Mid-Latitude Deserts

SUMMARY

APPLIED STUDY 11
Supplemental Irrigation in the Upper Mississippi Valley

REGIONAL CLIMATOLOGY

The climates found in mid-latitudes differ from those of the tropics in two major respects. Whereas in the tropics the variation in radiation and temperature is greater from day to night than it is from season to season, in mid-latitudes the seasonal change is greater than diurnal variation (figure 11-1). In the winter season in mid-latitudes solar radiation is greatly reduced (see chapter 2). Second, in mid-latitudes the primary circulation is quite different from that of the tropics. The subtropical high over the oceans and the polar high act as major source regions for air moving toward the convergence zone centered in the latitudinal range of 50°–60°. These two source areas are very different in character. One is very warm and the other quite cold. While this is sufficient to produce different kinds of air streams in the mid-latitudes, the differences are emphasized by moisture properties of the air masses. Thus, while in the tropical convergence zone the converging air streams are quite similar in temperature and moisture, such is not the case in mid-latitudes.

GENERAL CHARACTERISTICS

Middle Latitude Circulation

The circulation of the middle latitudes is dominated by the upper air westerly flows of the Rossby regime. The equatorward limit of this regime may be considered as the axis of the zone of divergence associated with the semipermanent

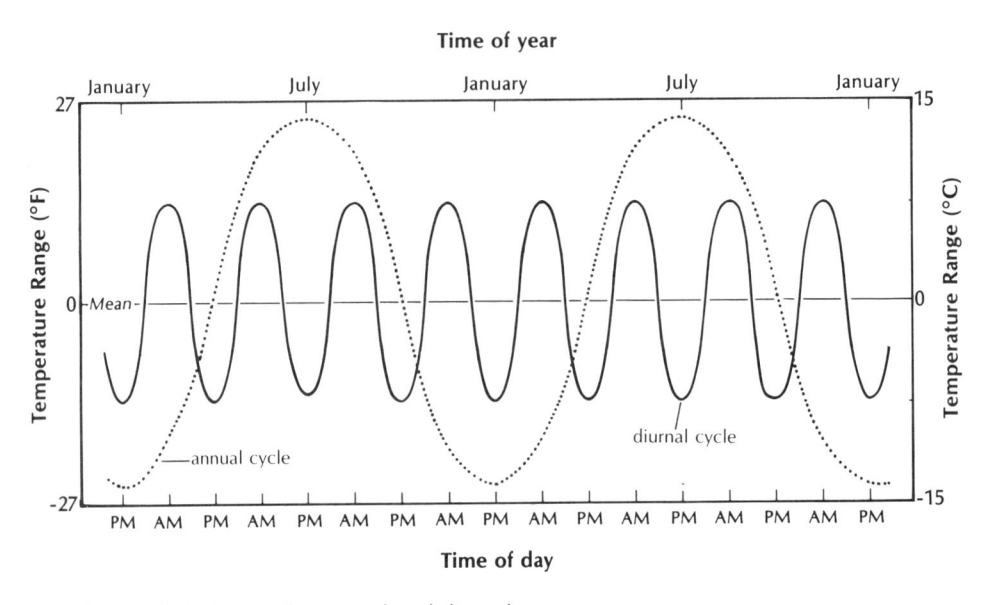

FIGURE 11-1 The relationship between the annual and diurnal energy cycles in mid-latitudes.

high pressure cell that is approximately located beneath the subtropical jet stream. Equatorward of this limit the Hadley circulation of the tropics prevails.

The location of the subtropical jet stream is determined by the temperature gradient that occurs between the equator and the poles. As polar realms become warmer in summer, the gradient is least and the strength of the westerlies diminishes. The jet migrates poleward to attain the average position shown in figure 11-2(a). When a very steep gradient exists, the westerly circulation is strong and the jet migrates closer to the tropics. This is seen in figure 11-2(b), which shows the mean jet stream for the Northern Hemisphere in winter.

It has already been shown that surface circulation patterns are closely related to the flow of the upper air westerlies in middle latitudes. It follows that the migration of the subtropical jet and the weakening of the westerly circulation provide very different conditions from season to season. The movement and relative dominance of air masses will vary, as will the frequency and paths of traveling low-pressure systems. A view of the Northern Hemisphere in figure 11-2(c) and 11-2(d) shows some of the changes that occur in air mass dominance. Note, for example, the extent of cP air masses in winter as compared to summer.

Clearly, the greater difference that is found between the summer and winter provides one of the basic attributes of the middle latitude climates; the general aspects of the prevailing climates can be identified by this division.

Summer Season

The summer season has many of the characteristics of the tropical climates. At the time of the summer solstice radiation intensity at 47° N and S is as high as it is at the equator. Not only is the radiation intensity as great but the duration of radiation is longer than it is at the equator on that date (see figure 11-3). The inequalities of the day and night periods increase with latitude and are coupled with the variation in solar intensity in such a way that the seasonal differences are compounded. During the summer months, solar intensity and duration are high. The result is greater radiation reaching the surface in mid-latitudes than the equatorial zone in summer. During the summer season temperatures will average in the 20° to 25°C range at most stations. Along the equatorward margin, the warm month temperatures may average 26° to 30°C, comparable to the rainy tropics. The warm-month means drop slowly in a poleward direction. Only at latitudes greater than 45° will the summer averages drop below 20°C. July averages differ only slightly from the Gulf Coast to the Canadian border (figure 11-4).

Precipitation generally decreases toward the poles and toward the interior of the continents. The primary reason for this decrease is the increasing distance from the source of mT air. Areas on the west sides of the continents that have their precipitation from mP air normally have less total precipitation than areas receiving precipitation from mT air. The frequency of the thunderstorms decreases rapidly from south to north. Cyclonic precipitation prevails on the polar margins and is more frequent in all sections of the region in the winter. Hurricanes provide another mechanism for inducing precipitation, and some of the heaviest rains along the coastal areas result from hurricanes.

FIGURE 11-2 Upper: The mean jet stream in summer (a) and winter (b). (After J. Namias and P. F. Clapp. From *Introduction to Meteorology*, 3d ed., by S. Petterssen. Used with permission of McGraw-Hill Book Company.) Lower: Principal Northern Hemisphere air mass source regions in (c) summer and (d) winter (From *Introduction to Meteorology*, 3d ed., F. W. Cole, copyright © 1980, John Wiley & Sons. Reprinted by permission of John Wiley & Sons, Inc.)

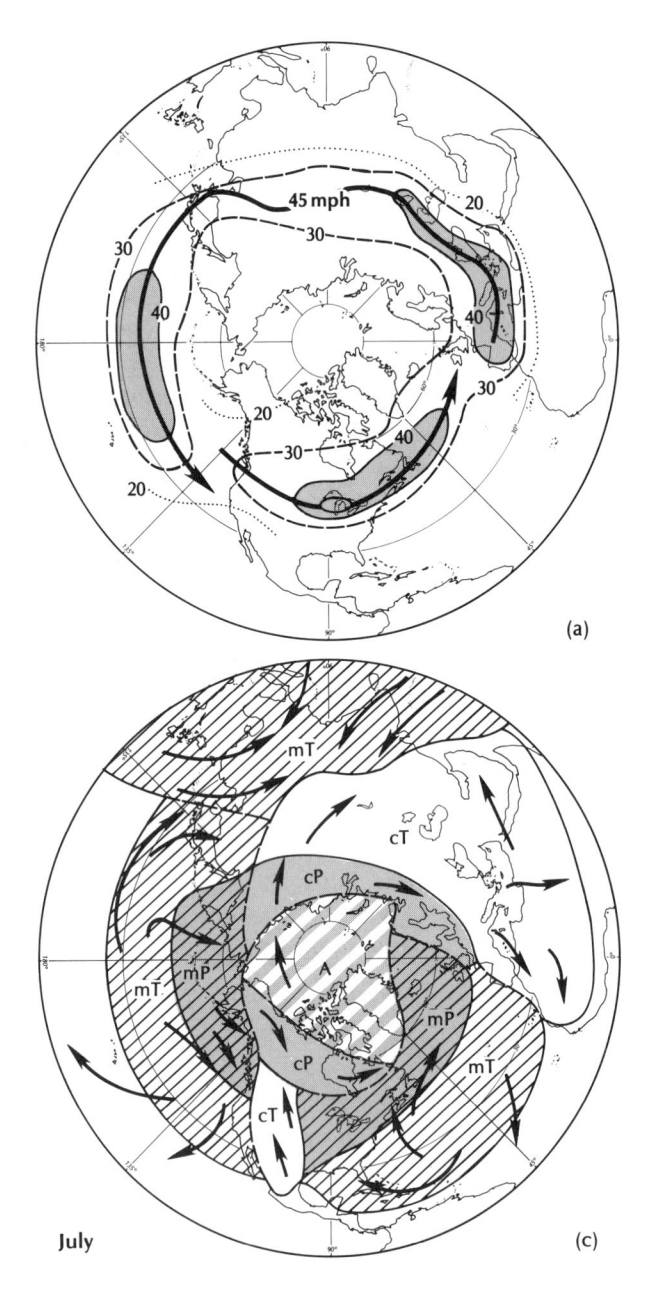

Winter Season

It is the winter season that sets the mid-latitude zone apart from the tropics. All areas in the mid-latitudes experience a range of 47° in the angle of the solar beam between the summer solstice and the winter solstice, with an accompanying difference in radiation intensity at the surface. In the winter season solar intensity is low, and duration is short. Temperatures reflect the change from the summer

(b)

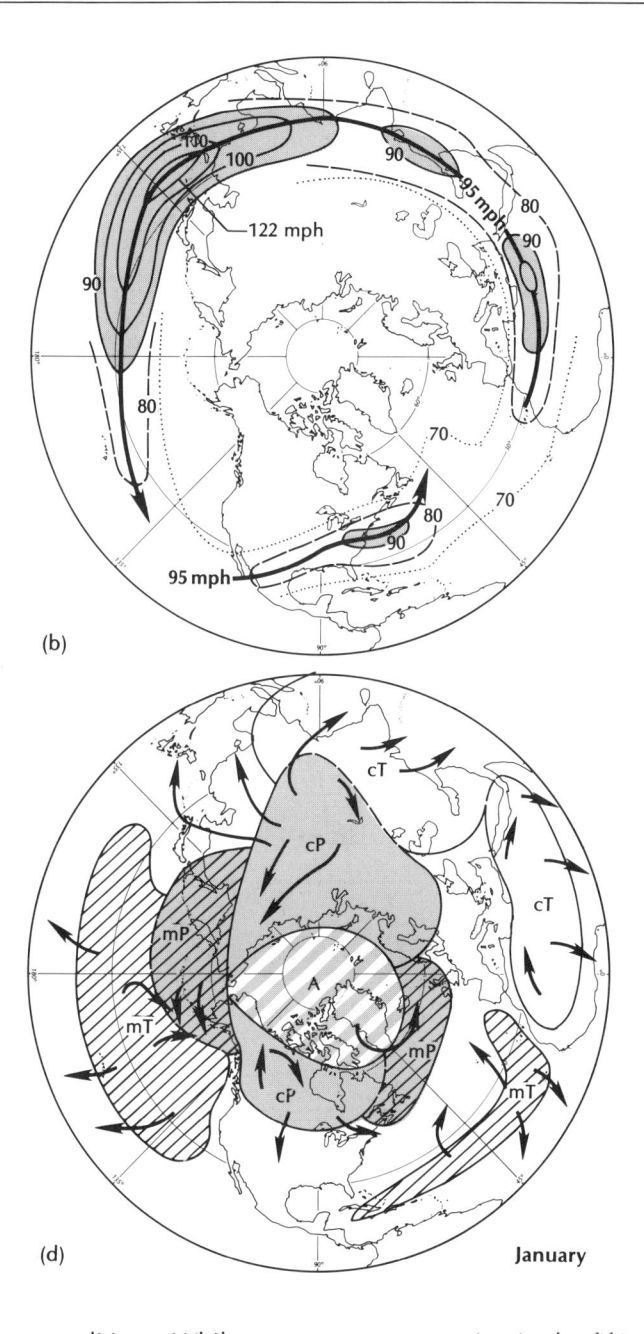

(d) **January**

conditions. While temperatures averaging in the fifties are the rule on the equatorward margins, the low-sun averages drop rapidly to below zero in interior stations in higher latitudes. There is a rapid decrease in the mean winter temperatures with increasing latitude. The annual range increases poleward from 8°C to 40°C at the extreme. The winter extremes have a steep gradient poleward also, but it is not as marked as is the gradient of the average temperatures. Near-zero temperatures have

REGIONAL CLIMATOLOGY

FIGURE 11-3 The length of daylight in summer and winter, and the relative intensity of solar radiation compared to a vertical solar beam.

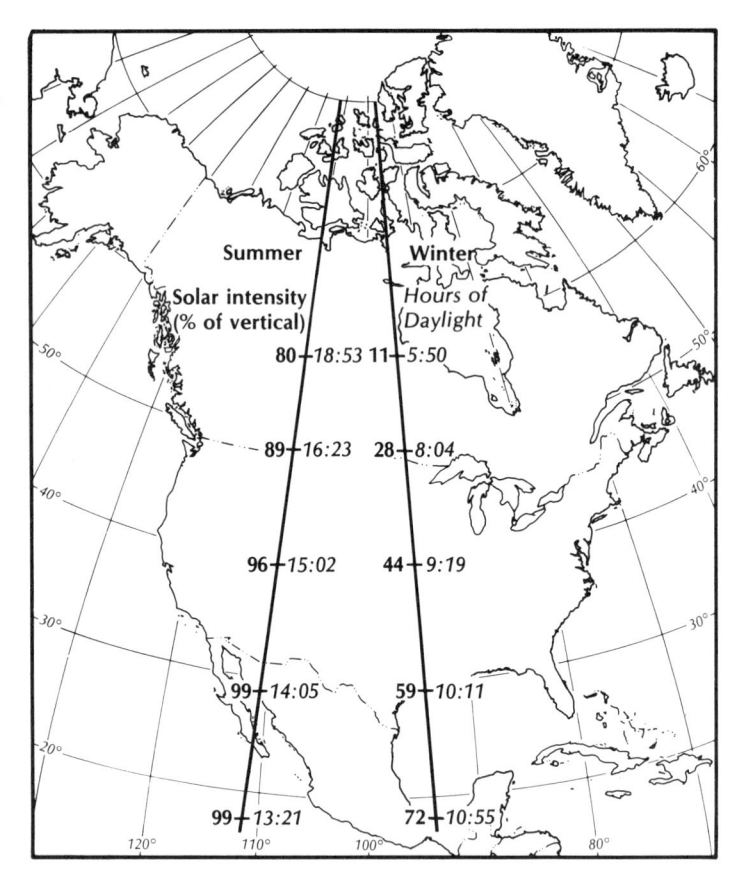

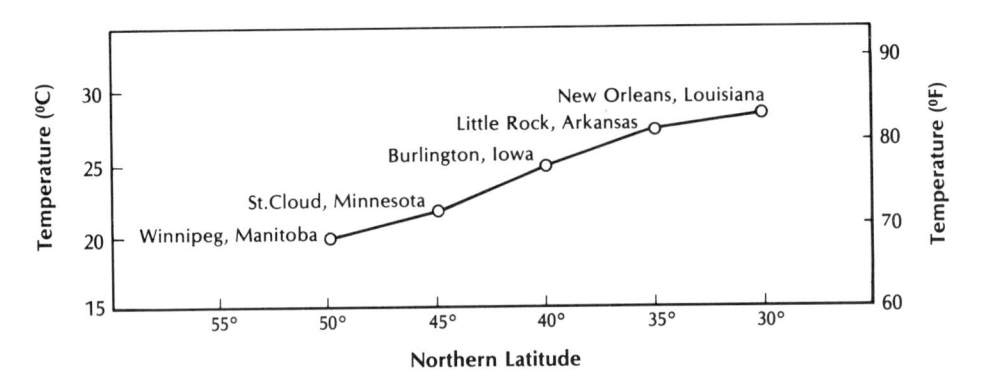

FIGURE 11-4 Mean July temperature along a north-south transect from New Orleans, Louisiana, to Winnipeg, Manitoba (From U.S. Dept. of Commerce, Weather Bureau. World Weather Records. Vol. 1, North America, 1970).

232

been recorded along the equatorward margin, and lows from −31°C to 46°C (−25°F to −50°F) characterize the more northerly locations. Winter diurnal ranges are slightly greater than the summer ranges because the relative and absolute humidity is somewhat lower. The wintertime controls are, in essence, low solar radiation, short days, and moderate atmospheric humidity.

The equatorward migration of the polar front results in the dominance of polar air masses over much of the middle latitudes (see figure 11–2). Periodically, the invasion of very cold arctic air gives rise to *cold waves*. These are defined as any sudden drop in temperatures within a 24-h period such that measures to prevent extreme distress to humans and animals may be necessary. In the United States winter cold waves claim more lives than any other weather-related phenomenon. Statistics compiled by the National Center for Health show an average of 355 deaths per year for the period 1949–1978. While deaths directly attributable to cold take the greatest toll due to weather, evidence indicates that cold indirectly affects the death rate substantially, particularly in areas where cold waves are infrequent. In Florida and Alabama death rates nearly doubled during one severe cold wave in 1965.

Equally cold outbreaks of arctic air have been historically recorded in Europe and Asia. In 763 A.D. a cold wave froze both the Black Sea and the Straits of Dardanelles in Europe. In 1236 the Danube is reported to have frozen solid. In 1468 the winter in Flanders was so severe that wine rations issued to soldiers had to be cut with hatchets and distributed in frozen chunks. In 1691, crazed by the lack of food in the countryside, wolves invaded Vienna, Austria, killing humans for food.

Often high winds accompany the onset of a cold wave. Wind produces cooling, and the effective temperature on flesh is equivalent to a much colder temperature. If, for example, the air temperature is −1°C (30°F) and there is a 60-kph (35-mph) wind blowing, the wind chill is equivalent to a temperature of −20°C (−4°F) in still air. Applied Study 13 provides further details of the wind-chill factor.

Water Balance

In the tropics climates are distinguished on the basis of the seasonal water balance. The same three patterns occur in mid-latitudes. That is, there are regions with a high frequency of precipitation year round, regions with pronounced wet and dry seasons, and desert areas with only occasional precipitation. Thus, while the tropics are subject to one major annual periodicity, that of moisture, the mid-latitudes may be subject to both the seasonality of solar energy and that of moisture (figure 11–5).

The effect of this phenomenon is well displayed in water balances of middle latitude stations (figure 11–6), which indicate a distinct summer maximum of potential evapotranspiration. In some cases, such as in the Mediterranean climates, this maximum is out of phase with seasonal rainfall; as a result, rainfall is more effective, with less lost to the evaporation process. In the continental areas of middle latitudes the summer maximum rainfall coincides with the period of highest rainfall and the time when plants are actively growing. Irrigation is thus an integral part of those areas where agriculture is practiced. In the humid middle latitude areas, precipita-

REGIONAL CLIMATOLOGY

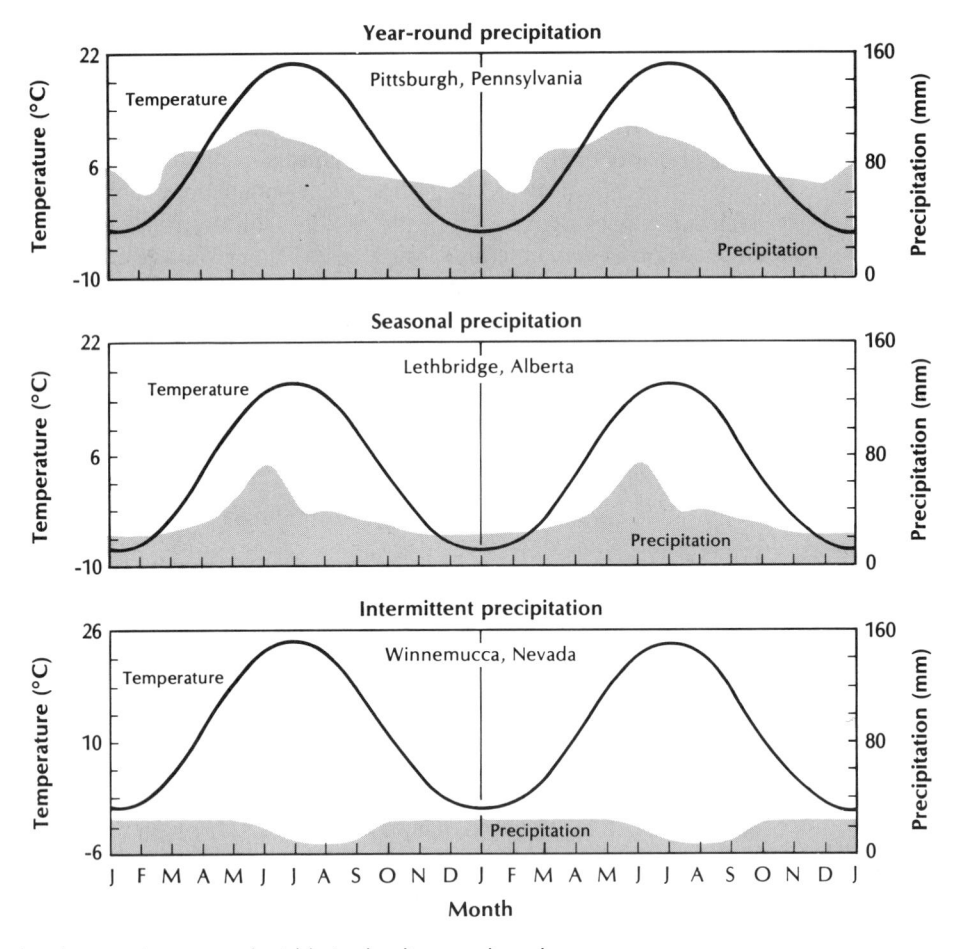

FIGURE 11-5 The three major types of mid-latitude climates, based upon the seasonal pattern of precipitation.

tion may well exceed evapotranspiration for much of the year, apart from droughts, main water shortages occur in summer.

An important aspect of the water balance in middle latitudes is the periodic freezing of streams, lakes, ponds, and soil moisture that takes place where winter temperatures go well below freezing. Most parts of the mid-latitudes experience subfreezing temperatures for varying periods during the winter. When such temperatures occur, soil moisture freezes and, depending on the severity of the cold spell, surface water freezes. This freezing stops the flow of the water runoff. At the same time it decreases the water supply available to plants, producing drought. The winter months tend to be a time of year when moisture accumulates in the environment. Even though winter precipitation is generally less than summer precipitation over the Northern Hemisphere land masses, the demands on the water supply are also greatly reduced. Evapotranspiration withdrawals are reduced in the winter because of lower temperatures and reduced plant growth. Over the land masses it is estimated that 70

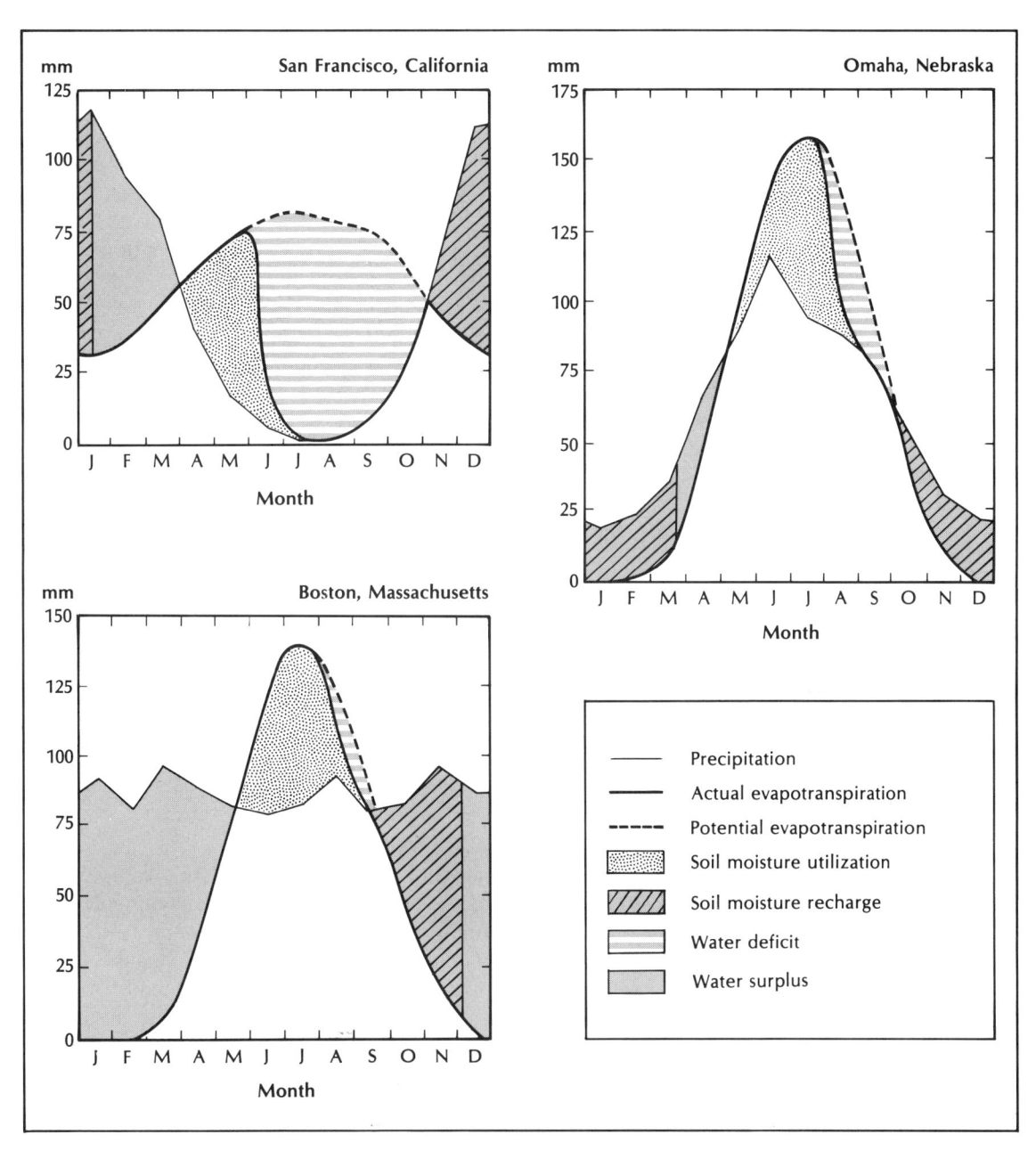

FIGURE 11-6 Selected water balance graphs for North American stations (From Douglas B. Carter, "Basic Data and Water Budget Computation for Selected Cities in North America," No. RS-8 in *The Earth Science Curriculum Project Reference Series*, ed. William H. Matthews, III, 7–9. Published by Prentice-Hall, Inc. Basic data from "Average Climatic Water Balance Data of the Continents," *Publications in Climatology* 17, no. 3 [1964], and from the files of The Laboratory of Climatology, C. W. Thornthwaite Associates, Centerton, N.J.

percent of all precipitation goes directly back to the atmosphere by evaporation. In the winter months, this figure may drop to 40 percent or less, thus allowing moisture to accumulate on and in the soil. Withdrawal of water from circulation by freezing contributes to the annual pattern of stream flow. The United States Geological Survey uses a water year that is different from the calender year in all its calculations. For most of North America, streams are at their low point in September or October. During the summer evapotranspiration is very high and tends to utilize a big share of the water that reaches the surface. Thus in the fall the streams are at low water levels. As temperatures cool, vegetation ceases to transpire as much water, and moisture begins to accumulate. During the winter, moisture accumulates either as soil moisture or as snow. When the spring thaw takes place, streams rise; flooding may occur if there is a large amount of snow present on the land surface. The spring floods due to snow melt are most characteristic of poleward locations and mountain regions but are by no means restricted to them.

CLIMATIC TYPES

Mid-Latitude Wet Climates

Climates with seasonal mT and mP air

Köppen types	Cf, Df
Thornthwaite types	BB', BC'

The mid-latitude humid climates are associated with the subpolar lows over continental areas spread equatorward on the eastern sides of the continents. They tend to stretch nearly across the continents in the Northern Hemisphere in the vicinity of 60° N and S and reach equatorward as far as 25° or 30° along the eastern margins of land masses. This type of climate in North America extends from the Pacific coast of Canada eastward in a crescent to the Atlantic coast. In the United States it includes the area lying east of the Mississippi River and the first tier of states west of the Mississippi. In Eurasia this climate is found from the offshore islands of Great Britain and Ireland eastward into the Soviet Union. It is also found along the Pacific coast of the Orient in latitudes 30° to 60°. In the Southern Hemisphere similar climatic regimes are found in South America, Africa, and Australia. In South America there are two areas, separated geographically. One is in the south of Chile, including the southern end of the central valley; the other portion includes the Pampas of Argentina and Uruguay and extends southward to include part of Patagonia. The area in Africa is very limited, consisting of a small area on the very southeast tip of the continent. Both Australia and New Zealand have sectors with this type of climate. Sample data are given in table 11-1.

Summer conditions are very similar to those found in the rainy tropics. The primary differences are two: cyclonic storms that traverse the region and tempera-

TABLE 11-1 Monthly data for mid-latitude wet stations.

	Jan.	Feb.	Mar.	Apr.	May	June	July	Aug.	Sept.	Oct.	Nov.	Dec.	Yr.
New Orleans, Louisiana 29° 59′ N, 90° 04′ W: 1 m
Temp. (°C)	12	13	16	20	24	27	28	28	26	21	16	13	20
Precip. (mm)	98	101	136	116	111	113	171	136	128	72	85	104	1371
Montreal, Quebec 45° 30′ N, 73° 35′ W: 60 m
Temp. (°C)	−11	−9	−4	5	13	18	21	19	15	8	1	−7	5
Buenos Aires, Argentina 34° 20′ S, 58° 30′ W: 27 m
Temp. (°C)	23	23	21	17	13	9	10	11	13	15	19	22	16
Precip. (mm)	103	82	122	90	79	68	61	68	80	100	90	83	1027
London, England 51° 28′ N, 0° 00′: 5 m
Temp. (°C)	4	4	7	9	12	16	18	17	15	11	7	5	10
Precip. (mm)	54	40	37	38	46	46	56	59	50	57	64	48	595

tures that average from 21° to 26° C with the tropical margins going as high as 29°C, somewhat warmer than the humid tropics.

The summer extremes exceed those of the tropical wet climate. Whereas few stations in the humid tropics have recorded temperatures in excess of 38°C (100°F), most stations in the mid-latitude climate have recorded maximums of 38°C (100°F) or higher. The summer diurnal range is typically 8°–11°C (15°–20°F), again quite comparable with the tropical humid climates. In the three summer months of June, July, and August, the temperatures of the southeastern United States average 1°–2°C (2°–3°F) higher than in Belem, Brazil, and the rainfall is about the same. The nocturnal temperatures are quite often within a few degrees of the diurnal temperatures, as the nights are short. The summer temperatures of the mid-latitude humid climates are controlled mainly by the high solar intensity, long days, and high atmospheric humidity.

The precipitation in these climates is fairly evenly distributed through the year. The annual total is quite variable, depending upon the latitude and continental position. It varies from as little as 510 mm (20 in.) upward to 1780 mm (70 in.) Of U.S. cities Mobile, Alabama, is among the wettest, with a mean annual precipitation of 1.73 m (68 in.). The variability of the annual precipitation is similar to that of the humid tropics, normally less than 20 percent. The seasonal distribution is fairly even, but some areas show a tendency toward either a summer or winter maximum. Frequency of precipitation also varies. Bahia Felix, Chile, averages 325 days a year with rain. There is a 90 percent probability of rain on any and every day of the year. This is not typical, of course, but it does indicate to what extent the precipitation mechanisms can persist.

The precipitation is more varied in form than in the tropics and at times (as described in Applied Study 11) must be supplemented for agricultural production.

During the summer and on the equatorial margins, convectional rainfall is the primary mechanism of precipitation. The southeast of the United States averages 40–60 days per year of thunderstorms. At Tampa Bay, Florida, more than 100 days a year have severe thunderstorms. The frequency of the thunderstorms decreases rapidly from south to north. Cyclonic precipitation prevails on the polar margins and is more frequent in all sections of the region in the winter. Hurricanes provide another mechanism for inducing precipitation, and some of the heaviest rains along coastal areas result from hurricanes.

While the greatest share of the annual precipitation is in the form of rain, snow is a factor to a greater or lesser extent in all places. In the United States the mean snowfall varies from a trace along the Gulf to an average of approximately 1 m (40 in.) through the Great Lakes district. In January of 1977 even Miami, Florida, had snow.

Humidity is generally high. On occasion it will drop substantially but usually for short periods of time. High humidity and high temperatures make the sensible temperatures of summer quite high. As in the tropics, radiation fogs are common on clear nights in summer and fall.

The summer is much like the rainy tropics. Temperatures are high, humidity is high, and convectional showers are common. Summer storms such as tornadoes, thunderstorms, and hurricanes and heat and cold waves bring variety to the summer season. The cold waves are the result of equatorward flow of summer continental polar air from the interior of high-latitude land masses. Though referred to as *cold waves*, they may have temperature above 10°C (50°F), but they represent much cooler air than is normal.

Within the wet climate area there are differences in temperature that have led to subdivisions of the climate. Köppen distinguished between the areas that have mild winters (Cf) and those with cool summers (Cbf). The latter region is found predominantly on the western side of the continents, where summer temperatures are modified by the onshore winds from the oceans.

The mid-latitude wet climates are perhaps best developed on the west sides of the continents from about 45° to 65°. Here the westerlies from the oceans bring frequent mP air masses onshore, resulting in a high percentage of cloud cover, high humidity, and a high frequency of precipitation. The largest area of this type of climate is found in Europe, where it spreads inland a fair distance because there is no north-south mountain range to block the flow of the westerlies from the Atlantic Ocean. In North America the zone extends from northern Oregon to Alaska. In South America it extends from 40° to the tip of the land mass. In both North and South America the extent of the climate is limited by mountain ranges parallel to the coast. Small areas with this type of climate are found in South Africa, Australia, and New Zealand.

The winter season is fairly long but mild, and the summer is cool. The winter maritime air masses are usually at least as warm as the offshore ocean. The winter air is damp and chilling. Fogs are common because of the high humidity. The storm systems often become occluded, bringing more extensive periods of rain and drizzle. For this reason, London, England, is regarded as a foggy, misty, and damp city. London averages 164 days/year with precipitation, which translate into a 45 percent probability of rain on any and every day. On the other hand, the total annual rainfall in London is only 625 mm (25 in.), so that the amount of rain received on each rainy day averages less than 4 mm (0.15 in.). Cloudy weather, fog, and drizzle are indeed typical.

Very cold weather is not common along these coasts. Cold cP air masses develop over land masses, and the westerlies carry them south and east most of the time. Only when an extremely large mass of cold air spreads out so that it expands in a westerly direction do these cold waves affect coast areas. Significantly, winter temperatures are influenced by the warm ocean currents. Figure 11-7 provides an example of the way in which the North Atlantic Drift results in milder winter conditions in northwestern Europe.

In the summer the frequency of precipitation declines somewhat because of the stabilizing effect of the subtropical highs as they move northward. Lowland sites do not receive high total amounts of precipitation, primarily because of the coolness of the air and cyclonic activity that produce the precipitation. Summers are subject to quite cool weather as the result of the influence of the adjacent oceans. The oceans in these latitudes never warm very much, and a cold current flows equatorward along most of these areas. Where highlands exist along these coasts even summer precipitation is frequent, and rain forests of huge trees and dense surface vegetation interspersed with the trees have evolved. The famous redwoods of California and Oregon typify these forests. The Olympic Peninsula of Washington contains a magnificent forest of Douglas fir; these forests extend well into Canada. The green

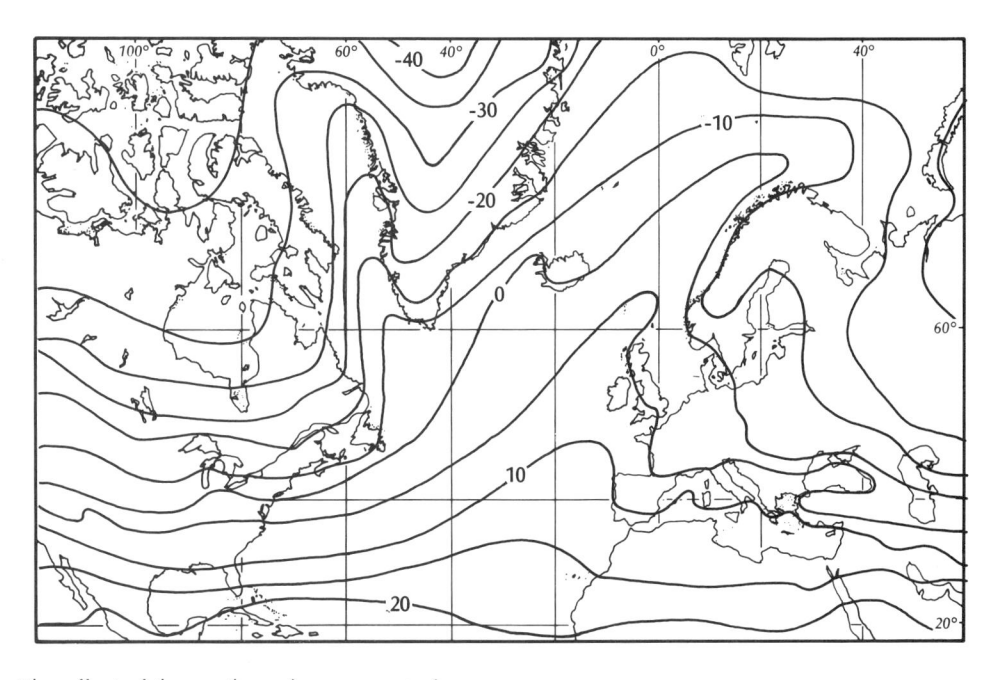

FIGURE 11-7 The effect of the northward movement of warm water in the Gulf Stream. Note the shape of the isotherms (°C).

meadows of Ireland are also a tribute to the frequency of the moist Atlantic air masses.

Mid-Latitude Wet and Dry Climates

Climates with seasonal mT and cP air

Köppen types	Cw,	Dw
Thornthwaite types	BB'w	CB'd
	CB'w	CC'd
	CC'w	DC'd
	DB'w	

As exemplified in table 11-2, the mid-latitude wet and dry climates are characterized by a strong seasonal pattern of both temperature and moisture. The general location of this climate is the interior of the continents in mid-latitudes. This continental location provides the characteristic feature of a large average annual temperature range and a high index of continentality. Figure 11-8 shows the increase in summer temperatures and decrease in those of winter for selected Eurasian stations. The lower diagram shows how continentality varies in Eurasia. The values given in the diagram are based upon the Index of Continentality devised by Gorczynski using the formula $C = 1.7 \times A - 20.4/\sin \phi$. This formula assesses the continentality of a location (C) by using the annual range of temperature (A), while also taking latitude (ϕ) into account. Latitude is given as a sine ratio so that smaller changes that occur in low latitudes can be related to those nearer the poles. As the map illustrates, very high values occur in Asia, a region of maximum continental effect on climate.

Specific areas are the Great Plains of the United States and Canada, the steppes of Eurasia, and small areas of Australia, Africa, and South America. These are continental locations lying where there is seasonal reversal between predomi-

TABLE 11-2 Monthly data for winter-dry stations.

	Jan.	Feb.	Mar.	Apr.	May	June	July	Aug.	Sept.	Oct.	Nov.	Dec.	Yr.
				Denver, Colorado 39° 22' N, 104° 59' W: 1588 m									
Temp. (°C)	0	1	4	9	14	20	24	23	18	12	5	2	11
Precip. (mm)	12	16	27	47	61	32	31	28	23	24	16	10	327
				Bismarck, North Dakota 46° 46' N, 100° 45' W: 507 m									
Temp. (°C)	−13	−10	−5	6	13	18	22	21	14	8	−2	−8	5
Precip. (mm)	11	11	20	31	50	86	56	44	30	22	15	9	385
				Ulan Bator, Mongolia 47° 55' N, 106° 50' E: 1311 m									
Temp. (°C)	−26	−21	−12	−1	6	14	16	14	9	−1	−13	−22	−3
Precip. (mm)	1	2	3	5	10	28	76	51	23	7	4	3	213

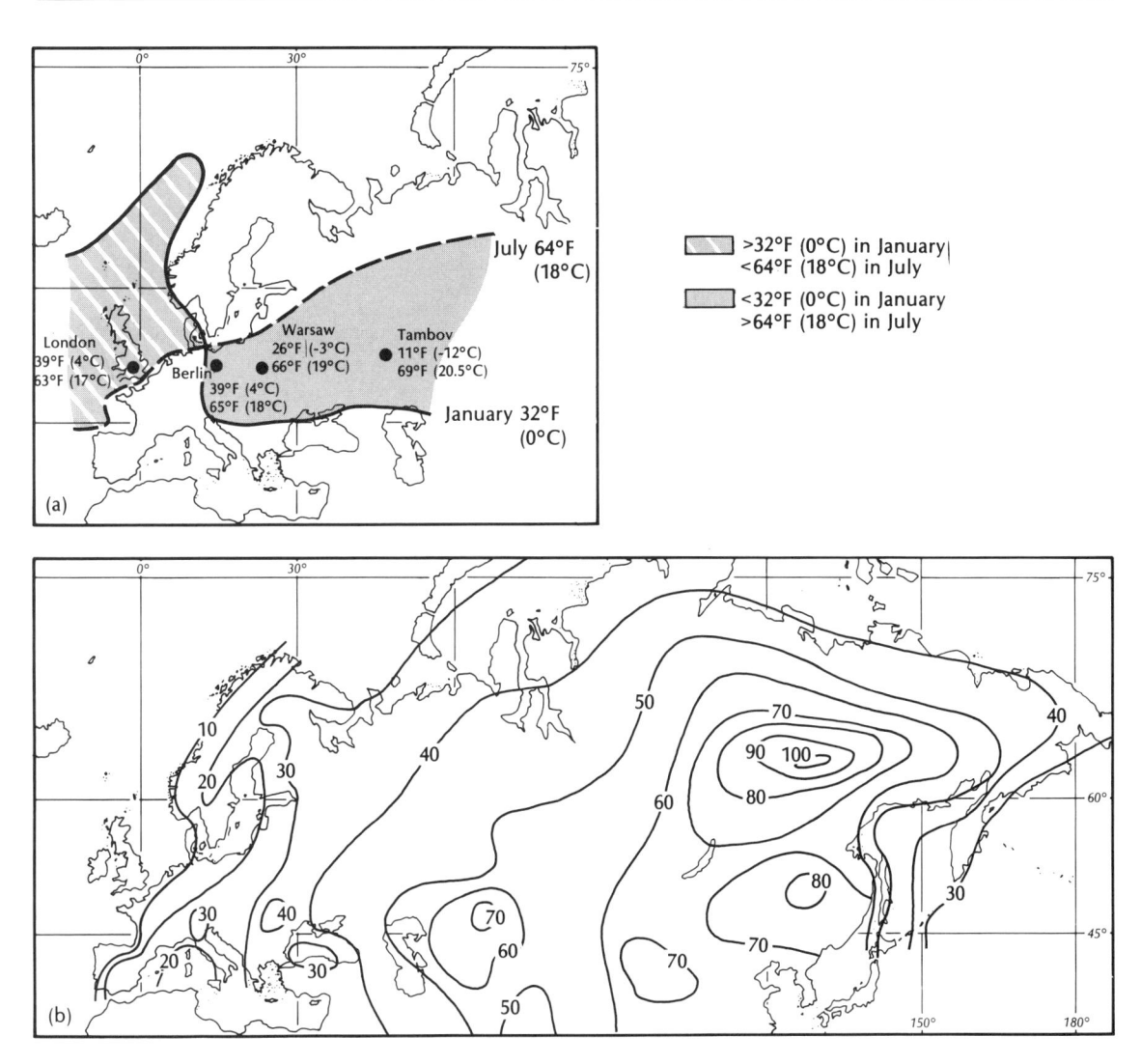

FIGURE 11-8 The influence of continentality: (a) In January, continental areas are colder than neighboring oceans; in summer, the reverse is true; (b) an index of continentality applied to the Eurasian land mass (From K. Boucher, *Global Climates*, [1975, p. 224]. Used by permission of Hodder & Stoughton Ltd.).

nance of atmospheric convergence during the summer (with its high attendant precipitation) and divergence during the winter (with its associated drier atmosphere). These areas also have the most variable climate of any major physical region. They are located in a position to receive maritime tropical air masses in the summer with occasional tropical continental air masses from the adjacent deserts; in the winter they have frequent outbreaks of polar continental air masses with an occasional maritime polar air mass. Each type of air flow brings its particular variety of weather.

A major difference in the response to solar input in this region is the result of lower atmospheric humidity. There are many more clear days, and thus both insolation and earth radiation are at a higher rate. Both summer and winter extremes are greater as the result of more frequent dry air masses. Whereas the extreme highs in the mid-latitude forest regions are predominantly from 38° to 44°C (100°F to 110°F) in the seasonal convergence regions, they will range up to 49°C (120°F) in the warmer sections as they did in the Great Plains in 1980.

The winter is the dry season. This phenomenon favors earth cooling during the long nights with the result that mean winter temperatures are usually a few degrees lower than in the humid climates, and extremes are 3°-5°C (5°-9°F) cooler. It is this climatic region that experiences the lowest temperatures of the Northern Hemisphere. In North America the temperature dropped to −52°C (−62.8°F) at Snag, Yukon, on 3 February 1947; and at Tanana, Alaska, a −51°C was recorded in January 1886. The lowest official temperature of the Northern Hemisphere was −57°C at Verkoyansk, U.S.S.R., and an unofficial −61°C was recorded at Oimekon, U.S.S.R. As a result of less humid conditions and excessive earth radiation, the average annual range is up to 34°C (67°F) in extreme cases and the absolute range up to 100°C (186°F). A winter characteristic of these areas is rapidly changing temperatures. At Browning, Montana, on 23-24 January 1916, the temperature dropped from 6.7°C to −48.9°C (44°F to −45°F)—a total of 55.6°C in 24 hours. Such drastic changes are usually the result of the inflow of air, but adiabatically heated air will cause equally sharp rises in the temperature.

These regions are subject to the effects of a monsoon. During the summer season the excessive terrestrial heating of the Northern Hemisphere causes the subpolar low to expand equatorward, forming a low trough, or cell, in the middle of the continent. The resultant convergence brings moisture to the continent, producing a summer rainy season. During the winter cooling of the land mass, the subpolar trough splits into cells over the oceans, and a ridge of high pressure develops between the polar high and the subtropical high, greatly reducing the probability of precipitation.

Winds are a significant factor in the weather. With the seasonal change in pressure is associated a seasonal change in wind direction. During the high-sun season, onshore or onland winds prevail; during the winter, there is a switch of 90°-180° in prevailing wind direction. The degree of wind shift depends to a large extent on the continental position. The more interior the station, the greater the wind shift. Wind velocities are high and persistent. In the United States, Oklahoma City has the highest average wind velocity of any city, with a mean of 22 kph. Chinook winds are a winter phenomenon that produces extremely rapid changes in temperatures. They occur in the Great Plains from Alberta to Colorado. Spearfish, South Dakota, observed a 2-min rise of 27°C from −20°C to 7°C (−4°F to 45°F) on 22 January 1943.

Atmospheric humidity and precipitation are seasonal in character, and cloud cover varies with humidity. Precipitation decreases with distance from the equator and the sea. Although it is difficult to delimit annual precipitation, most areas receive between 300 and 900 mm (12 and 35 in.). Of the annual total, approximately 75 percent falls during the summer half-year. The annual total is highly variable and apparently subject to long-term cycles as yet unexplained; it has been conjectured that the cycles are associated with sunspot activity. A number of factors are responsible for the great variability of rainfall. Nearly all rainfall east of the Rocky Mountains is produced by mid-latitude cyclones, involving continental polar air from northern Canada and maritime tropical air from the Gulf of Mexico. The Gulf air generally takes a curved path across the eastern United States, moving east of the Great Plains. The dry tropical air masses that move northward across the Great Plains arise in Mexico and contain little moisture.

In the Great Plains the cold fronts are often very steep as they move southeastward, and extremely severe rainstorms develop along them. The unofficial heaviest 12-h rain was recorded at Thrall, Texas, on 9 September 1921, when 810 mm (32 in.) of rain was received. An unofficial rain of 42-min duration totaled 300 mm (12 in.) at Holt, Missouri, on 22 June 1947.

The frequency of hailstorms is somewhat higher here as a response to drier air. An unimportant but interesting fact is that the largest recorded hailstone fell in Potter, Nebraska, 7 July 1928. It measured 137.4 mm (5.4 in.) in diameter.

The winter dry climate is most intensely defined over the Eurasian land mass. Here are found the greatest contrasts between summer and winter weather conditions, with the possible exception of the poles. Much of Asia is subject to seasonal reversal of pressure, wind direction, and precipitation.

Summers are hot and humid with intense summer convectional storms. Along the equatorward fringe as much 70 percent of the annual precipitation occurs in the summer months; toward the interior, this increases to as much as 80 percent. The summer precipitation results from convectional instability, and severe rainstorms are not uncommon. A significant portion of the annual rainfall may come in a single storm. Total annual rainfall may be as much as 1 m along the coastal areas, but this decreases inland to less than 675 mm (25 in.).

In the winter cold, dry, continental air from the Siberian high-pressure zone results in a majority of clear days, with short periods of cyclonic disturbances. In the interior winter temperatures drop below $-34°C$ ($-30°F$). The frost-free growing season tends to be short, less than 100 days in most years, particularly on the northern fringes. Since most of the precipitation occurs in the summer, winter snowfall is not particularly great. Snow tends to remain for long periods, as much as 200 days in the northern interior.

The extreme of this climate is reached in northeastern Siberia. Summers can be warm and winters severely cold. At Oymyakon the range in temperature between the months of January and July is 67°C (115°F), and the range between the highest and lowest temperature recorded is 104°C (187°F). The winter season here is at least 7 months long, and snow cover is over 200 days, often remaining on the ground until May, as short but hot days occur.

Permafrost appears in the coldest parts of the climatic zone. Since often there is little snow to cover the soil, the ground is frozen to a depth of 50 m (150 ft). In the summer the top few centimeters will thaw, leaving a wet soggy soil and extensive swamp.

The Summer-Dry Climates

Climates with seasonal cT and mP air

Köppen types	Cs, Ds
Thornthwaite types	BB's, CB's, DB's

One area with a seasonal rainfall regime differs from the rest, in that the rainy season comes in the winter rather than in the summer season, as illustrated by data in table 11-3. This type of climate is found on the western margins of the continents between 30° and 40°, normally does not spread into the continents very far, and thus is limited in extent. There are three main areas where summer-dry climates are found. The largest is around the Mediterranean Sea and extending eastward into Iran. The second is the west coast of North America from near the Mexico–United States boundary northward into the state of Washington. Along this coast the climate is limited to the area between the Pacific Ocean and the Sierra Nevada and Cascade Mountains. The third major area is across Southern Australia. Smaller areas are found in the Capetown district of South Africa and in central Chile. The climate is often called the *Mediterranean* climate for the large area associated with the Mediterranean Sea, but in total only about 2 percent of the land area of the earth experiences this type of climate.

The seasonal distribution of precipitation makes this climate rather distinct from the rest of the earth's climates. Much of the earth's land area receives most precipitation in the summer or spread fairly evenly over the year. In this climate the winter maximum of precipitation is very pronounced. At Perth, Australia, 85 percent of the annual precipitation occurs in the winter six months. At Istanbul, Turkey, some 70 percent of the yearly total is received in the winter half-year. San Diego, California, receives 90 percent of its annual total from November through April. Santa Monica, California, has recorded only traces of precipitation in the three

TABLE 11-3 Monthly data for summer-dry stations.

	Jan.	Feb.	Mar.	Apr.	May	June	July	Aug.	Sept.	Oct.	Nov.	Dec.	Yr.
Santiago, Chile 33° 27' S, 70° 40' W: 512 m													
Temp. (°C)	19	19	17	13	11	8	8	9	11	13	16	19	14
Precip. (mm)	3	3	5	13	64	84	76	56	30	13	8	5	360
Los Angeles, California 33° 56' N, 118° 23' W: 37 m													
Temp. (°C)	13	14	15	17	18	20	23	23	22	18	17	15	18
Precip. (mm)	78	85	57	30	4	2	T	1	6	10	27	73	373
Rome, Italy 41° 48' N 12° 36' E: 131 m													
Temp. (°C)	8	8	10	13	17	22	24	24	21	16	12	9	15
Precip. (mm)	65	65	56	65	51	30	25	17	66	78	96	97	711

months of June, July, and August; the six months of April through September average only 25 mm out of a total for the year of 381 mm. The occurrence of the precipitation in the wintertime and the relatively low intensity of the cyclonic precipitation permit more efficient plant growth than is the case in the climates where the rainy season is in the summer. Although mean winter temperatures average above freezing, occasional snow falls, even along the equatorward margins. It rarely stays on the ground very long, as daytime temperatures rise high enough to melt the snow fairly rapidly. Total annual precipitation averages less than 750 mm (30 in.) and tends to increase poleward. Table 11–4 provides the annual rainfall for several cities along the west coast of North America. The data show the rapid increase in annual total that takes place in a poleward direction.

A number of climatic and topographic factors have given prominence to these climatic regions as resort areas. In summer the coastal areas are sunny but not as hot as other areas, as the cool ocean waters and the sea breeze offset the intensity of the sunshine. Winter temperatures are mild, with a lot of sunny weather, and short periods of rain. The subsidence associated with the subtropical high tends to keep the sky relatively clear so that it has an unusually bright blue color. The Mediterranean Sea is noted for its brilliant blue color, which is partly due to sky color and partly due to the low organic content of the water.

Like all climates that have a rainy season, these environments have two distinct seasons, each associated with different air masses and different parts of the general circulation. During the summer these areas are under the influence of stable oceanic subtropical highs, giving them essentially tropical desert weather conditions. In the winter months the anticyclonic circulation moves equatorward, allowing the westerlies to bring moisture to the region.

Mid-Latitude Deserts

Climates with seasonal cT and cP air

Köppen types	BWk
Thornthwaite types	Eb', Ec'

The mid-latitude deserts are located basically in the interiors of continents, although they merge with tropical deserts on the west sides of the land masses. Specifically they are found in North America from southern Arizona northward into British Columbia between the Sierra Nevada and Rocky Mountains. In Eurasia they are imbedded in the trans-Eurasia cordillera or lie on the flanks of these mountain masses. All of the central Asian countries contain substantial areas of deserts within their boundaries. The Southern Hemisphere has a much smaller area of desert in

TABLE 11–4 Latitudinal variation in annual precipitation in the summer-dry climates.

San Diego	269 mm
Los Angeles	380 mm
San Francisco	510 mm
Portland	900 mm

mid-latitude locations. The fundamental reason for the diminutive size of the deserts south of the equator is a lack of land area in the latitudes where the deserts would be most extensive.

These deserts have the highest percentage of possible sunshine of any of the mid-latitude climates. Arizona averages 85 percent possible sunshine through the year with an average of 94 percent in June and 76 percent in January. Temperatures reflect the clear skies that dominate the weather. Summer temperatures are the highest of the mid-latitude climates (table 11–5). The summer averages are 5°–8°C (9°–14°F) higher than in the more humid areas, and the extremes of temperature are also higher. During the summer, when the days are long and the sun high in the sky, temperatures will reach above 50°C (122°F). Death Valley in California recorded an official shade temperature of 56.7°C (134°F), which is only 1°C below the highest sea-level temperature ever recorded. Winter temperatures compare with those of the humid climates at the same latitude, primarily because the coldest temperatures anywhere in mid-latitudes are associated with outbreaks of cold arctic air.

This climatic type is the most difficult to place in the pattern of the general circulation. A number of factors seem to be associated with the aridity of these regions. Each is insufficient in itself to produce extreme drought but coupled with other elements results in aridity. Lying between the primary convergence zone of the subpolar low and the subtropical highs, these deserts maintain dry conditions because of topographic barriers separating them from the moisture supply, distance from the moisture supply, and intermittent anticyclonic circulation.

The mid-latitude deserts are not as arid as the tropical deserts. During the winter months the general circulation shifts equatorward far enough to allow occasional passage of weak cyclones associated with the polar front. Precipitation totals

TABLE 11–5 Monthly data for middle latitude deserts

	Jan.	Feb.	Mar.	Apr.	May	June	July	Aug.	Sept.	Oct.	Nov.	Dec.	Yr.
Yuma, Arizona 32° 40′ N, 114° 36′ W: 62 m
Temp. (°C)	13	15	19	22	26	31	35	34	31	25	18	14	24
Precip. (mm)	10	9	6	2	T	T	6	13	10	10	3	8	77
Winnemucca, Nevada 40° 50′ N, 117° 43′ W: 1312 m
Temp. (°C)	−3	0	3	8	12	16	22	20	15	9	2	−1	9
Precip. (mm)	27	24	21	21	24	19	7	4	9	21	20	24	219
Alice Springs, NT, Australia 23° 38′ S, 133° 35′ E: 546 m
Temp. (°C)	28	27	25	20	16	12	12	14	18	23	25	28	21
Precip. (mm)	27	45	18	10	18	15	14	10	6	25	23	39	250
Ashkhabad, U.S.S.R. 39°45′ N, 37° 57′ E: 230 m
Temp. (°C)	2	5	9	16	23	29	31	29	24	16	8	3	16
Precip. (mm)	22	21	44	38	28	6	2	1	3	11	15	19	210

THE MID-LATITUDE CLIMATES

are, of course, very low. Phoenix, Arizona, averages 180 mm (7 in.) of precipitation per year, and Ellensburg, Washington, 230 mm (9 in.). There is little seasonal pattern to precipitation. Phoenix receives a trace of snow and Ellensburg, Washington averages 780 mm (31 in.) of snow each winter season. Moisture efficiency is very low throughout these deserts. In some areas potential evaporation exceeds the precipitation by as much as ten times.

SUMMARY

Mid-latitude climates are characterized by a marked difference in temperature through the year in response to changes in the intensity and duration of insolation. The winter season is accentuated by the incursion of cold air from the polar regions, and it is this winter season that distinguishes these climates from the tropics. The convergence of very different air streams results in the development of actively moving low pressure and frontal systems, which often produce severe storms. These changing atmospheric conditions provide a distinct pattern of summer, fall, winter, and spring seasons. Like the tropics, mid-latitude climates can be subdivided on the basis of the distribution of precipitation through the year. In mid-latitudes the wet-and-dry climates are subdivided into summer-dry and winter-dry regimes. The fact of having the rainy season in the winter in the one case and the summer in the other results in substantially different types of climate.

APPLIED STUDY 11

Supplemental Irrigation in the Upper Mississippi Valley

The success of irrigation is associated with several factors: (1) an abundant supply of fairly good water that can be easily obtained so that cost of water is at a minimum; (2) level terrain that allows for maximum mechanization; (3) well-drained sandy to sandy loam soils; (4) the existence of a highly marketable crop that is extremely responsive to supplemental water additions; and (5) a detailed knowledge of rainfall probabilities during the growing season of the commercial crop.

Supplemental irrigation has increased in the eastern United States in recent years. *Irrigated* land is defined by the United States Census of Agriculture as land watered for agricultural purposes by artificial means. *Supplemental irrigation* is the addition of water (1) in some years and not in others or (2) during only part of each growing season.

Supplemental irrigation differs from the perennial irrigation of the west in two major ways. Normally it is not essential for the production of crops; its greatest benefit is derived from increase in yields or production of higher value crops. Secondly, the individual farmer sustains the investment, in contrast to the government subsidy of the large projects in the west.

The methods used for applying water for supplemental irrigation are essentially the same as those used in the west for perennial irrigation. Sprinkler irrigation and furrow irrigation are the most widely used. *Sprinklers* are overhead systems of pipes that are either motorized or moved by some other means. Center-pivot systems are used, as well as parallel systems. The sprinkler systems are particularly useful where the surface is all uneven and infiltration rates are high. *Furrow systems* involve leveling the ground and allowing water to run between the rows of plants.

A number of factors have favored an increase in the use of supplemental irrigation in the humid eastern states:

> **1.** The recognition that *periodic droughts*, defined as periods of two weeks or more in which soil moisture falls below a level that favors normal crop growth, occur in the eastern states.
> **2.** Rapidly increasing costs of crop production: Capital investment is high, land values have climbed, labor costs have increased, and greater amounts of fertilizers are used. As investments in crops increase, the losses due to poor yields become greater, and expenses of irrigation become more justified.
> **3.** Acreage allotment programs have led to efforts to maximize yields on allotted acreage.
> **4.** The introduction of lightweight aluminum pipe, which makes overhead sprinkler irrigation more feasible.

The benefits to be derived from supplemental irrigation include stabilized and increased crop yields, earlier maturing crops, and a better quality of product, all of which tend to bring a higher gross income from the land.

The Upper Mississippi Valley is defined here as the watershed of the Mississippi River above the confluences of the Missouri and Kaskaskia rivers with the Mississippi, as shown in figure 11-9. The increase in irrigated land is found in three small areas made up of parts of some dozen counties. These three areas—one near Minneapolis and St. Paul, Minnesota; one in central Wisconsin; and one in northwest Indiana—have several attributes in common. Surface materials are all of glacial origin, with both ground moraine and outwash present. Characteristics of slope and relief are similar in both cases, as they are nearly flat surfaces with slopes of less than 2 percent; relief differences are less than 15 m/km^2. The soils are sandy, with a gray to brown A horizon (alfisols).

Irrigation is found most frequently on the outwash, which tends to be better drained. Irish potatoes are naturally well suited to the outwash, except that yields are greatly affected by droughts of periods of five days or longer. When the water supply for potatoes can be controlled, the yields are stabilized and increased 150–400 percent; and a higher quality product, which increases the value of the crop in the fresh food market, is derived.

FIGURE 11-9 Areas of commercial supplement irrigation in the Upper Mississippi Valley.

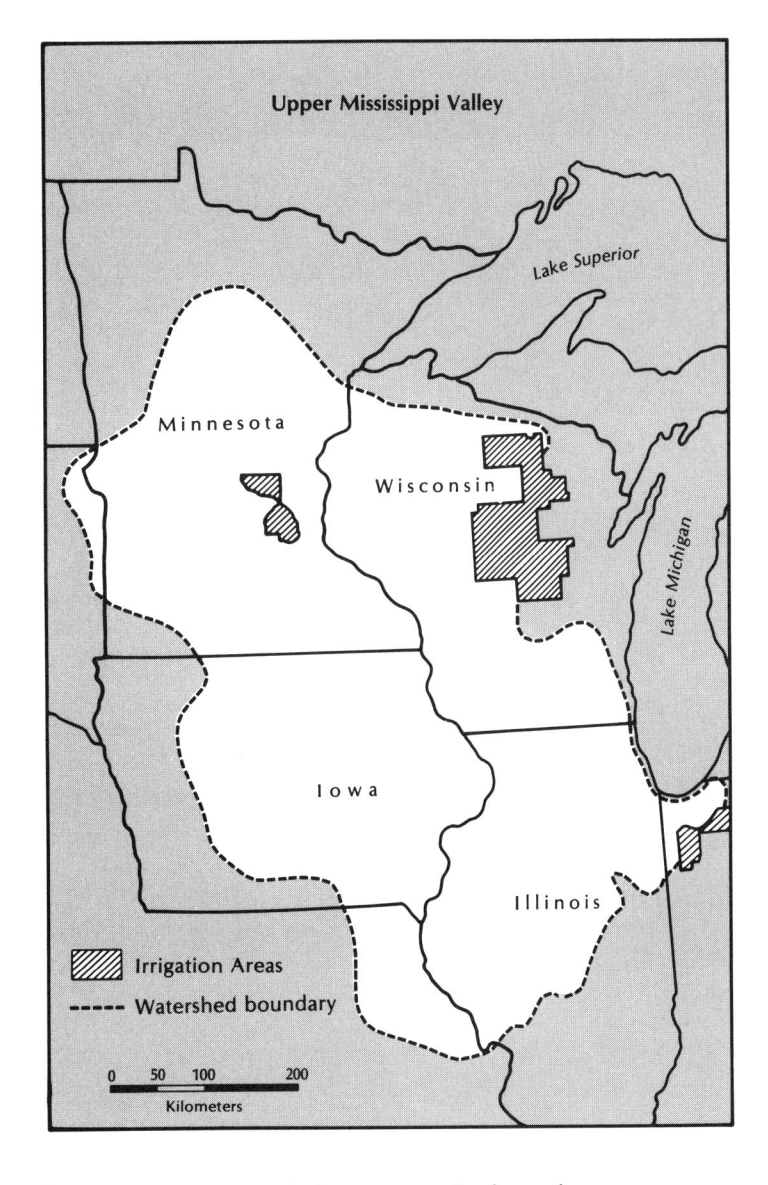

Supplemental irrigation is not needed every year in these three areas because in some years the distribution of rainfall is such that additional water is not advantageous. The amount of land irrigated each year varies but has been increasing steadily. In 1945 there were fewer than 5,000 acres of crop land irrigated in the entire Upper Mississippi Valley; but acreage grew rapidly in the early 1950s and by 1980 had increased twentyfold.

12 Climates of North America

CLIMATE EAST OF THE CONTINENTAL DIVIDE
Summer Weather
Winter Weather

THE CLIMATE OF THE WEST COAST
Summer Weather
Winter Weather
The Effects of the Mountains on Local Climate

SUMMARY

APPLIED STUDY 12
Setting Records—A Case of Probability

REGIONAL CLIMATOLOGY

Chapter 11 outlined the kinds of weather that typify the mid-latitudes and identified the major variations in climate in these regions. In this chapter the weather of North America is examined in greater detail.

Perhaps the most significant aspect of the climate is that the polar front and the westerlies cross the continent. The traveling cyclones that move with the westerlies bring extremely varied weather to much of the continent. Equally significant is the fact that extending north and south, nearly at right angles to the westerlies, is the Rocky Mountain Cordillera. The coastal ranges and interior ranges reach elevations more than one-third the height of the troposphere. These ranges affect the flow of air over the continent to a considerable extent.

The air masses that influence North American weather are listed in table 12-1 along with their seasonal characteristics. The source regions for them are shown in figure 12-1. From the data in the table and the location of the source

FIGURE 12-1 Source regions and paths of air masses affecting the North American continent.

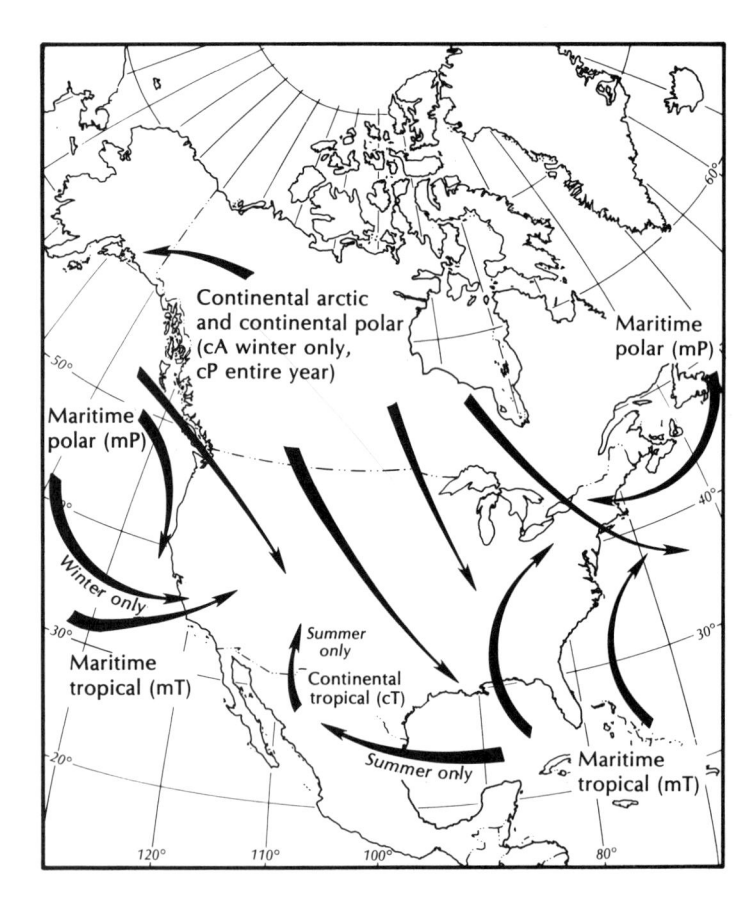

TABLE 12-1 Weather characteristics of North American air masses.

Air Mass	Source Region	Temperature and Moisture Characteristics in Source Region	Stability in Source Region	Associated Weather
cA	Arctic basin and Greenland ice cap	Bitterly cold and very dry in winter	Stable	Cold waves in winter
cP	Interior Canada and Alaska	Very cold and dry in winter Cool and dry in summer	Stable entire year	a. Cold waves in winter b. Modified to cPk in winter over Great Lakes bringing "lake-effect" snow to leeward shores
mP	North Pacific	Mild (cool) and humid entire year	Unstable in winter Stable in summer	a. Low clouds and showers in winter b. Heavy orographic precipitation on windward side of western mountains in winter c. Low stratus and fog along coast in summer; modified to cP inland
mP	Northwestern Atlantic	Cold and humid in winter Cool and humid in summer	Unstable in winter Stable in summer	a. Occasional "northeaster" in winter b. Occasional periods of clear, cool weather in summer
cT	Northern interior Mexico and southwestern U.S. (summer only)	Hot and dry	Unstable	a. Hot, dry, and clear, rarely influencing areas outside source region b. Occasional drought to southern Great Plains
mT	Gulf of Mexico, Caribbean Sea, western Atlantic	Warm and humid entire year	Unstable entire year	a. In winter it usually becomes mTw moving northward and brings occasional widespread precipitation or advection fog b. In summer, hot and humid conditions, frequent cumulus development and showers or thunderstorms
mT	Subtropical Pacific	Warm and humid entire year	Stable entire year	a. In winter it brings fog, drizzle, and occasional moderate precipitation to N.W. Mexico and S.W. United States b. In summer it occasionally reaches western U.S., providing moisture for infrequent conventional thunderstorms

Source: From Frederick K. Lutgens, Edward J. Tarbuck, *The Atmosphere*, 2nd edition, © 1982, p. 203. Reprinted by permission of Prentice-Hall, Inc., Englewood Cliffs, New Jersey.

regions two important attributes emerge: (1) The west coast of the continent is dominated by different types of air masses from different source regions compared to the area east of the continental divide; and (2) there are major seasonal differences in air flow. The western part of the continent is dominated primarily by air streams from the Pacific Ocean, and the continent east of the Rocky Mountains by air streams from the Canadian Arctic and the Atlantic Ocean.

CLIMATE EAST OF THE CONTINENTAL DIVIDE

Weather east of the Rocky Mountains is characterized by day-to-day and month-to-month changes. Shifts in the general circulation and the passage of mid-latitude cyclones bring about most of the changes.

Summer Weather

During the summer season the north-south temperature gradient is at its weakest. The westerlies are reduced in intensity and are shifted northward. Figure 12-2 shows the approximate boundary between the air streams of Pacific origin and those from the Arctic and Atlantic oceans during the month of July. The southern boundary of the continental polar air stretches across Alaska and northern Canada. The boundary between the air from the Pacific and the Atlantic Ocean extends through the Great Basin of the United States and eastward to the vicinity of Hudson Bay.

East of the continental divide, maritime tropical air flows northward over the continent, bringing warm, humid weather conditions. Examples of data for selected stations (New Orleans, Denver, and Bismarck, tables 11-1 and 11-2) have been given. Mean monthly temperature and precipitation data provide a rough indication of seasonal conditions but totally obliterate the variety of weather that takes place in a given season and the differences from one year to the next. Some summers are hotter than normal, others cooler than normal, some rainier, and some drier than normal. In any summer there is a high probability that records of either high or low temperatures will be set. Applied Study 12 provides the rationale for this statement. Figures 12-3 and 12-4 provide additional information on the distribution of precipitation and temperature on the continent.

Heat waves are periods of relatively clear, dry, and unusually hot weather. They develop when the westerlies move unusually far north, permitting the subtropical high to expand and cT air to displace the mT air. Temperatures rise and humidity drops. Heat waves occur in just about every summer someplace in the North American continent. Many severe heat waves have occurred since the colonies were first settled. Major heat waves took place in 1830, 1860, and 1901. The worst period for heat waves was during the drought years of the 1930s. The hottest of those years and the worst on record in terms of fatalities was 1936. Temperatures reached 43°C (109°F) or higher in Nebraska, Louisiana, Minnesota, Wisconsin, Michigan, Indiana, Pennsylvania, West Virginia, New Jersey, and Maryland. Highs of 49°C (120°F) or

FIGURE 12-2 July circulation over North America.

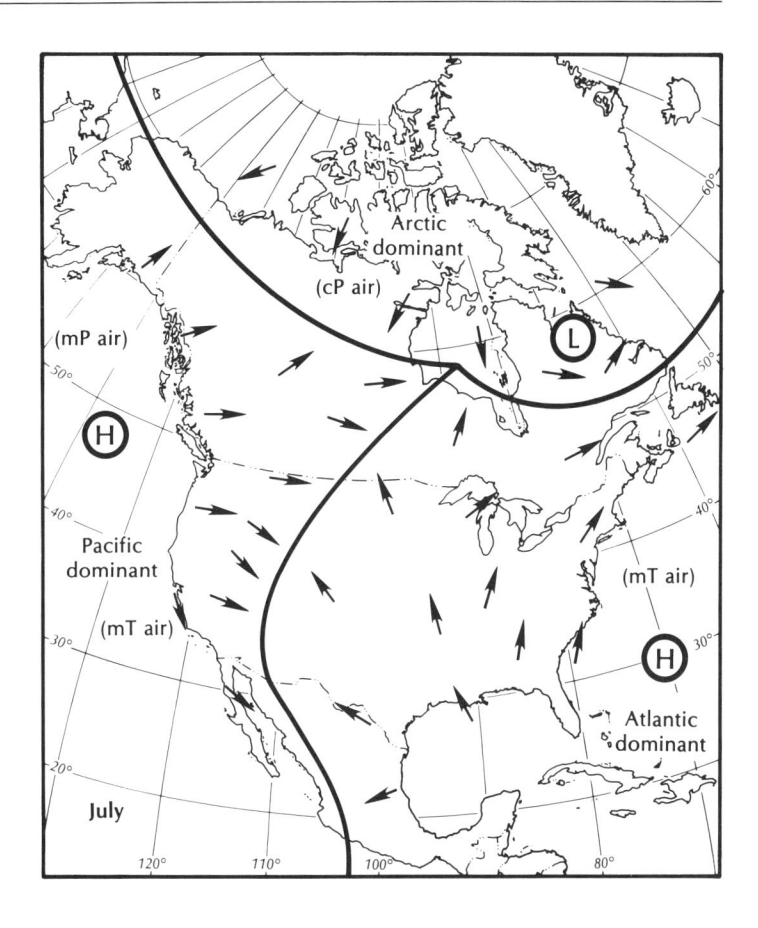

more were reached in the Great Plains states of North Dakota, Kansas, Oklahoma, Arkansas, and Texas. The fatalities approached fifteen thousand in that summer.

A heat wave in the summer of 1980 took at least 1265 lives, a number nearly seven times as many as in the average year. The majority of those who died were the elderly or poor who lived in non–air-conditioned housing. The state most directly affected was Missouri, where over 25 percent of the deaths occurred. The heat wave began in southwest Texas when temperatures went over the 100° mark. By the middle of July most of the central third of the nation was experiencing afternoon temperatures in excess of 38°C (100°F). The hot, dry weather spread over all of the eastern United States until the first week in September. Dallas, Texas, had temperatures of 38°C (100°F) or more every day from 23 June through 3 August. On 13 July several cities reported the highest temperatures ever recorded. They were Augusta, Georgia, 42°C (107°F); Atlanta, Georgia, 41°C (105°F); and Memphis, Tennessee, 42.2°C (108°F). The heat wave caused hundreds of kilometers of concrete pavement to buckle, in many instances exploding violently along seams and cracks. Asphalt roads were softened as the material reached temperatures of more than 66°C (150°F). Trucks moving on the soft surface destroyed many kilometers of road surface.

REGIONAL CLIMATOLOGY

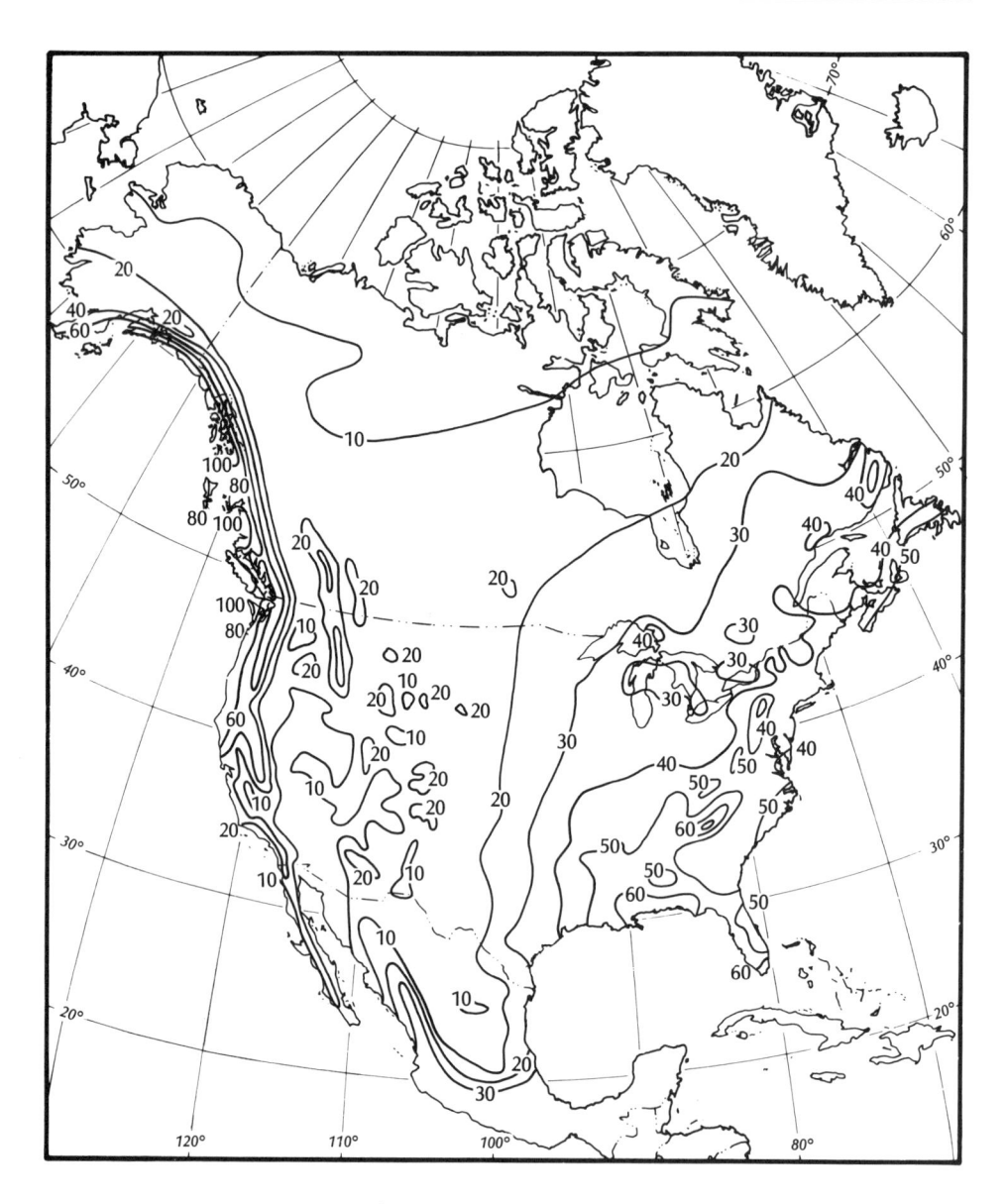

FIGURE 12-3 Distribution of precipitation in North America (From R. G. Barry and R. J. Chorley, *Atmosphere, Weather, and Climate,* 4th edition [1982, p. 225]. Reprinted by permission of Methuen and Co., Ltd. After Brooks and Conner, Kendrew, and Thomas.)

The Great Plains of North America suffer a chronic problem with drought. Evidence indicates that the 40-year period between 1825 and 1865 was one of low rainfall in many parts of North America. It was during this period that the Great Plains were being explored but were not yet extensively settled by immigrants. On the floor of some western lakes the tracks of pioneer wagon trains were ground

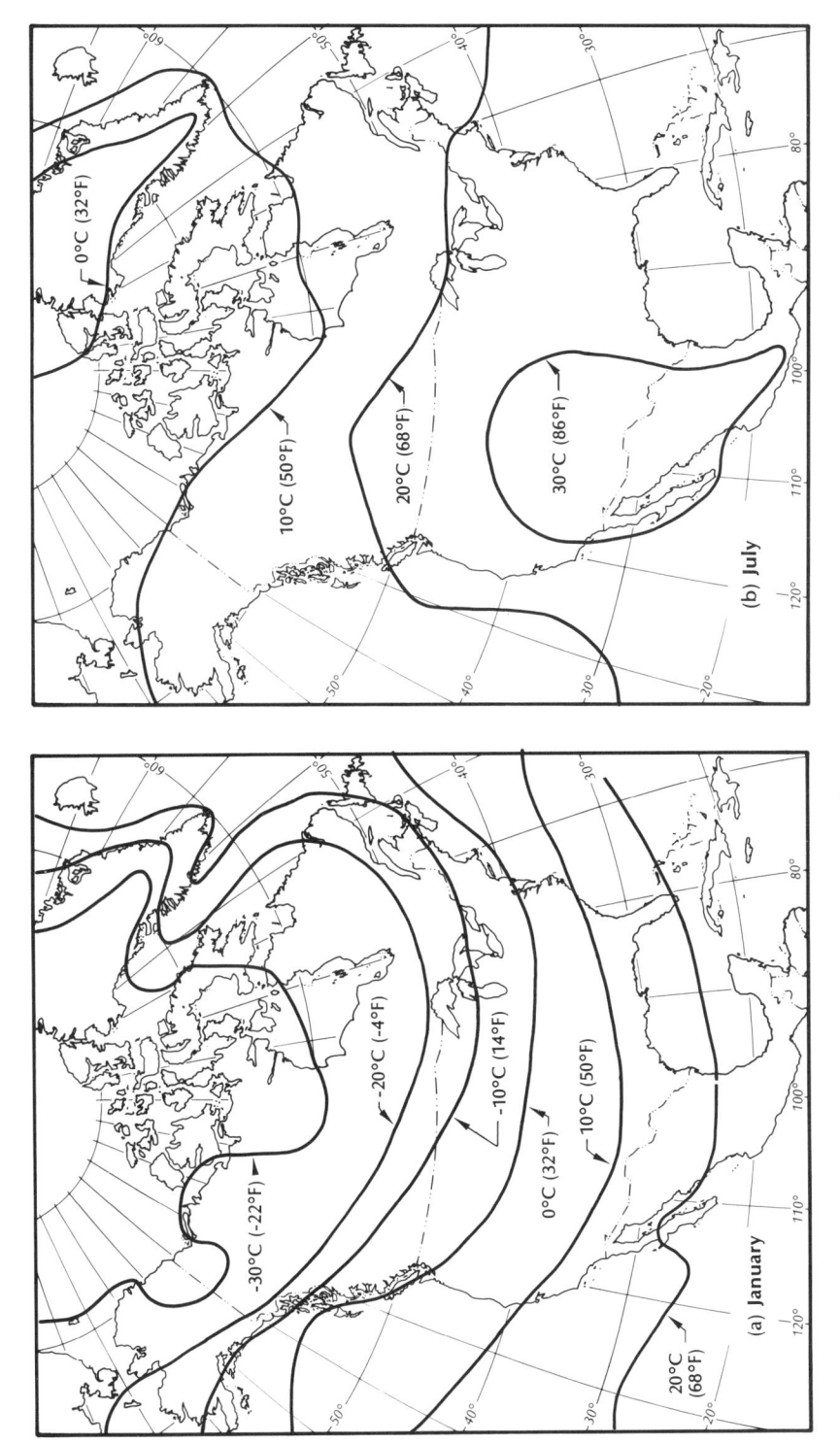

FIGURE 12–4 Mean January and July temperatures over North America. (a) January, (b) July.

into the sediments when the lakes were dry in the 1840s. The lakes refilled and covered the route until the turn of the century. Probably the driest year in the plains states since settlement began was 1860. Kansas, Missouri, Iowa, Minnesota, Wisconsin, and Indiana were all affected. The same area was struck again in 1863–1864 but not quite so severely.

A severe drought peaked in the fall of 1881, spreading over all of the United States and southern Canada east of the Mississippi River system. Many of the wells, cisterns, and springs that failed had never been dry before. Freight trains were seriously delayed by lack of water for steam. The water supply of New York City failed. Again, in the middle of the 1880s, rainfall diminished, culminating in the great drought of 1894 and 1895. This was perhaps the worst drought yet experienced since colonization in terms of intensity and the size of the area affected. Many of the settlers of the Great Plains left their land. Most moved westward, looking for more favorable farming sites. Wetter years returned to the plains and people quickly forgot the drought problem. New settlers moved into the plains states in large numbers, many of whom were driven off by another drought around 1910.

Following World War I, a third wave of farmers moved into the Great Plains on the heels of some wetter than normal years. When the dry years of the 1930s appeared on the scene, a real disaster took place. It was disastrous in terms of the economic system of the Great Plains, the effects on the land, and the social structure of the affected areas. The drought of the 1930s was not of equal intensity over the entire plains or through the decade. It was generally a dry period for most of the United States and southern Canada, with the drought widespread over the northern and central plains and the northern Rockies. The core areas were in Kansas and the Dakotas. The first period was the worst, from mid-1933 to early 1935. It was then that the dust began to blow. Topsoil was removed by the billions of tons, and the dust storms of 1934 and 1935 became evidence to people as far east as the Atlantic coast that there were severe problems in the West. By 1935 an estimated 80 percent of the land in the Great Plains was suffering from erosion, and an estimated one-hundred and fifty-thousand people had moved out of the plains states. The impact of the drought was dramatized in John Steinbeck's great novel, *The Grapes of Wrath*. Two more periods of intense drought occurred in 1936 and 1939–1940. The 1936 period was very intense, but short; the 1939–1940 spell was long, but less intense.

Drought struck the Great Plains again in the 1950s. The flow of some perennial streams decreased to nearly half their normal flow, and salinity tripled. Wind erosion was again a problem, but not so severe as in the thirties. In 1957 an estimated 13,000 km^2 (3.2 million acres) of land had been damaged in the Great Plains, and another 118,000 km^2 (29.2 million acres) was listed as susceptible to blowing.

Winter Weather

During the winter both the Pacific high pressure system and that over the Atlantic Ocean move equatorward, and the influence of the subtropical high over the Atlantic is considerably reduced from the summer. The flow of air from the

Pacific Ocean penetrates farther into the continent, and there is a broad zone of mixing of cP, mT, and mP air through the Great Lakes (figure 12-5).

Most areas in the Northern Hemisphere poleward of 30° experience snow at one time or another, and in North America snow covers the majority of the land for 2-6 months each winter. Higher elevations will sustain snow for much or all of the year. In general, the frequency of snowfall increases with latitude, although there is no direct correlation between the amount of snow that falls and latitude. In the United States the mean annual snowfall varies from a trace along the Gulf of Mexico to an average of 1 m or more in the Great Lakes district (table 12-2). Total annual snowfall does not increase with higher latitudes. As colder weather conditions are found along with lower totals of precipitation, the amount of snowfall declines accordingly. It can snow at all temperatures found in the lower atmosphere, but at temperatures below $-20°C$, the probability of getting much snow is very low because the atmosphere simply will not hold much moisture. The heaviest snowfalls occur when temperatures are from $+4°C$ to $-4°C$. To produce copious snowfall, temperatures must be as warm as possible to have abundant precipitable water, but cold enough for the precipitation to be in the form of snow. It is for this reason that

FIGURE 12-5 January circulation over North America.

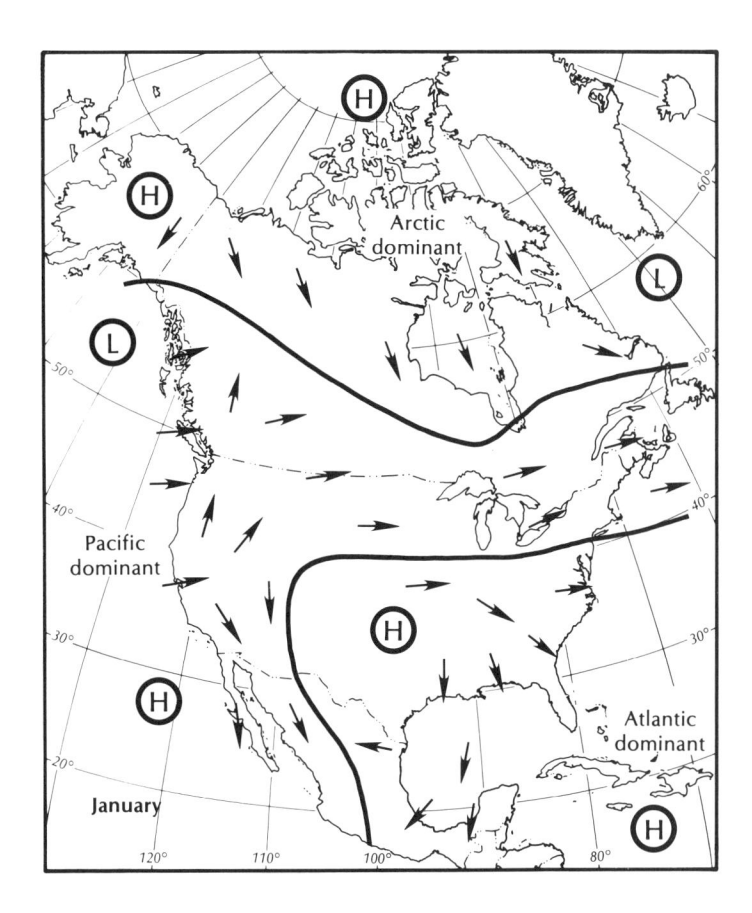

TABLE 12-2 Snowfall records in Great Lakes basin.

Storm Totals		
1 Day		
Bennett Bridge, N.Y.	1.3 m (51 in.)	17 January 1959
Buffalo, N.Y.	1.2 m (48 in.)	10 December 1937
5 Day		
Oswego, N.Y.	2.56 m (101 in.)	27–31 January 1966
Annual Totals		
Old Forge, N.Y.	10.37 m (408 in.)	1976–1977
Steep Hill Falls, Ontario	7.65 m (307 in.)	1939–1940
Herman, Mich.	7.83 m (308 in.)	1975–1976
Tahquamenon Falls, Mich.	8.45 m (333 in.)	1976–1977
Bennett Bridge, N.Y.	8.94 m (352 in.)	1946–1947
Hooker, N.Y.	11.86 m (467 in.)	1976–1977

very heavy snowfalls often occur near the beginning and end of the winter season. In the fall measurable amounts of snowfall begin in late November. It is often jokingly said in the Great Lakes area that a snowstorm can be counted upon during Thanksgiving vacation. In a sense the probability is high for heavy snowfalls, though it is certainly not an annual event. The cP air masses are becoming quite cold and more frequent, and yet warm moist mT air is still flowing northward. One of the earliest heavy snowstorms to move across the United States was 17–18 November 1980. The early storm blocked interstate highways in New Mexico and Texas with drifts up to 1.25 m high. The storm dumped snow in a band all the way from New Mexico to New England. Snow occurring this early in the year will not remain on the ground very long, as soil temperatures are still high. These fall storms have brought disaster to shipping on the Great Lakes. On 8 November 1913 an early fall storm crossed lakes Superior, Michigan, and Huron, sinking nearly a dozen ships and badly damaging more than 50 others. On 10 November 1975 the *Edmund Fitzgerald* went down in an unexpected storm on Lake Superior.

The same kind of phenomenon occurs again in the spring. Warm maritime tropical air begins to move northward, and there are still very cold masses of continental polar air that move south. The mixing can result in heavy snowfalls. Macon, Georgia, received 587 mm (23 in.) of snow from a spring storm in 1973. The great blizzard of 1888 in New England took place on 11–14 March, and in 1978 the major blizzard of the winter in the midwest was on 9–10 March.

Winter storms produce disasters someplace almost every year. The two most deadly winter storms are the blizzard and the cold wave, and it is not uncommon for a cold wave to follow a blizzard. Most winter storms are associated with disturbances along the polar front and the activity resulting from mixing cold dry air (cP) and warm moist air (mT). Many low pressure systems develop or redevelop along the east side of the Rocky Mountains anywhere from New Mexico north into Alberta.

They move eastward, with most traveling over the Great Lakes and out over New England or the St. Lawrence depression every 4–6 days.

The term *blizzard* is believed to have been derived from the German word *blitz*, meaning "lightning." In North America the term came to mean any sudden and violent force or onslaught. During the Civil War it was used to describe a heavy enemy volley of musketry. As near as can be determined, the first use of the term in respect to local snow storms was in the spring of 1870. *The Northern Vindicator*, a newspaper in Estherville, Iowa, used it to describe a severe winter snowstorm accompanied by high winds. The use of the term spread outward from Estherville, and by the severe winter of 1880–1881 it had gained general usage throughout the continent. Today, NOAA defines a blizzard as a storm with winds of at least 56 kph (35 mph) and temperatures below −6.7°C (20°F), combined with enough blowing or falling snow to reduce visibility to less than 0.4 km (0.25 mi). A severe blizzard is one in which wind velocities equal or exceed 72.5 kph (45 mph), temperatures go below −12.2°C (10°F), and visibility is near zero. There have been many great blizzards, depending on the area and decade. One year that produced more than one record breaker was 1888. Two extremely bad blizzards affected large but different areas of the continent. The first took place on 11–13 January and immobilized the Great Plains from Alberta south to Texas. The second was a late storm, striking the eastern seaboard from Maine to Chesapeake Bay on 11–14 March. An average of 1.02 m (40 in.) of snow fell over New England and southeastern New York State. In the space of just a few hours, New York City and other cities in the region were crippled by the storm. Snow drifts in Herald Square in New York reached 9 m in depth. Transportation was crippled. Horse-drawn vehicles couldn't move. The railroads suspended operations because the drifts were so high the trains couldn't get through them. Telegraph communications were out, and mail to and from the city nearly stopped. A food panic threatened to develop in some areas. Fires in New York City became virtual firestorms fed by high winds, and firemen could not respond because they could not get equipment through the snow. The accompanying cold was so intense with the wind chill that the East River froze, and many people crossed back and forth over the ice between Manhattan and Brooklyn. Sparrows froze by the tens of thousands, many of them frozen to tree limbs. Cattle froze to death, some of which remained standing, frozen solid. Human deaths reached 400 from the storm, with 200 deaths in New York City alone.

A region that receives unusually large amounts of snowfall is the lee sides of the Great Lakes, where distinct snow belts associated with the lakes exist (figure 12-6). These snow belts are the result of the winter winds' picking up moisture as they cross the lakes. The lakes remain unfrozen far into the winter, and in some winters none of them freezes over. Whenever the air flowing over the lakes is colder than the water beneath, tremendous amounts of water evaporate into the air. As the air stream moves over the land on the downwind side of the lake it is chilled, and the moisture precipitates out as snow. The greater the difference in the temperature of the air and the water, the greater the evaporation will be and the more likely a lake-effect snowfall will result. Once a lake freezes, the supply of water is largely shut off and the lake-effect snows decline in frequency. The areas frequented by the lake-

FIGURE 12-6 Snow belts of the Great Lakes (From V. Eichenlaub, *The Weather and Climate of the Great Lakes Regions*, [1979, p. 165]. Copyright, 1979, University of Notre Dame Press, Notre Dame, Indiana 46556.).

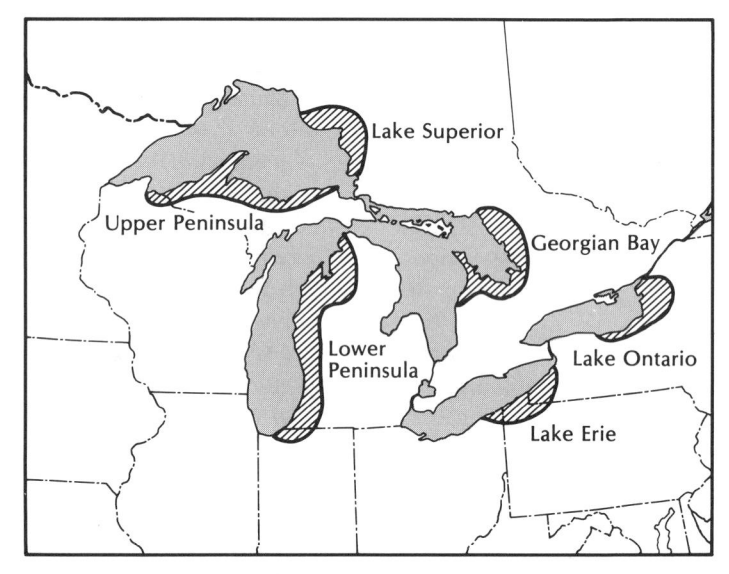

effect storms have substantially higher averages of snowfall than do the surrounding areas. The Keweenaw Peninsula of Upper Michigan receives snowfall averaging over 5 m a year. Several locations have recorded more than 10 m of snow in a given winter (table 12-2). The greatest amount of snowfall officially recorded east of the Rocky Mountains was at Hooker, New York, during the 1976-1977 season, when 11.86 m of snow fell. These snow belts do not extend very far away from the lakes. The maximum is 45-50 km (25-30 mi).

THE CLIMATE OF THE WEST COAST

The weather of western North America is dominated by different air masses from those of eastern North America. Relatively dry subsiding air from the subtropical high and maritime polar air from the Pacific Ocean dominate the weather.

Summer Weather

During the summer when the westerlies are farther north and relatively weak, the intermontane basins and coastal states of California and Oregon are under the influence of the subtropical high and experience clear, dry weather. Summers can be extremely dry. From California to Washington there have been years in which no measurable precipitation fell in each of the months of June, July, and August. Even Vancouver, British Columbia, has experienced a July with no measurable precipitation.

Summer days are long, solar radiation is intense, and temperatures soar. On 18 July 1982, for example, temperatures were over 32°C (90°F) in each of the 17 western states and over 38°C (100°F) in Nevada, Arizona, and New Mexico.

In California, heat waves take one of two forms. The first is the displacement of the subtropical high northwestward so that virtual tropical desert conditions persist for long periods. The second form consists of the hot, dry Santa Ana winds, which flow down out of the high plateaus in the summer. They are also called the *red winds* for the large amounts of dust they carry. They are produced by the development of a high pressure system over the Nevada desert. The air, heated by subsidence and the hot desert surface, is forced outward from the desert. As the air currents descend into the Los Angles basin, they are further heated by compression. Most frequent in September and October, they bring temperatures into the 38°–43°C (100–110°F) range. Often the winds are dust-laden, as in November of 1969, when winds reached velocities of 122 kpm (78 mph) through the passes. During one Santa Ana, the air temperature was 13.9°C and the dew point was −23°C. Such a desiccating wind can raise the fire hazard in mountain forests to an incendiary level. The sudden change in weather and the extreme dryness of the air often make these weather events—and the people who experience them—highly disagreeable.

The fall season often alternates between fire and flood. The natural vegetation is a scrub grassland, which after the long hot summer is explosively dry. Dry thunderstorms at the onset of the rainy season, human carelessness, or arsons will trigger wild fires that destroy thousands of hectares of vegetal cover every year. The first fall rains are often torrential. When the cyclonic storms first move onto the land surface, the land is extremely warm, which increases the instability of the air as it moves onshore. Where fires have raged, the unprotected soil turns into mud; landslides and mud flows destroy much valuable property and leave behind sterile hill slopes. In California nearly every year residential areas are either incinerated, buried in sediment, or simply carried off down the hillsides in a mass of mud, kindling, and plaster board.

Winter Weather

On the west coast of the continent, one of the distinguishing features of the climate is the winter concentration of precipitation. The west coast of the continent in the region from southern California north into Canada falls into this category. Cities in California typically receive an average of 84 percent of the annual precipitation in the winter season, with more than 50 percent of the annual total in the months of December, January, and February. Farther north along the coast, the precipitation increases; but most of the increase comes in the spring and fall and is due to the longer season during which the westerlies flow over the coast. Oregon receives an average of only 73 percent of its annual total in the winter six months. The percentage of precipitation that occurs in the winter decreases poleward; but even Prince Rupert, at 54° N, receives 62 percent of its annual total of 2400 mm in the winter six months. This winter maximum extends inland into the Great Basin of the United States and into the interior valleys of western Canada. While the annual total tends to increase poleward to about 50°, the actual amount received at any given location depends on site characteristics. At some places the totals run fairly high. Monumental, California, receives an average of 3.91 m (154 in.) per year. Heavy rains can produce 24-h totals of 250 mm (10 in.) or more, enough to produce major flooding.

REGIONAL CLIMATOLOGY

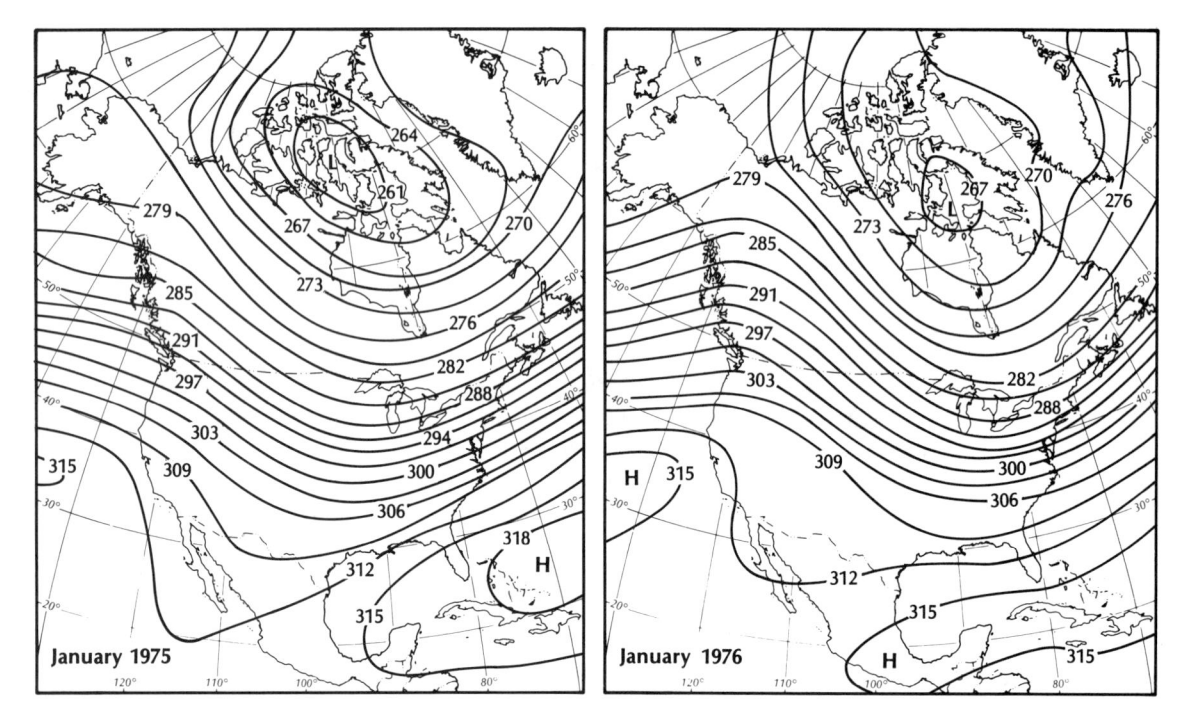

FIGURE 12-7 Mean 700 mb height contours for January 1975-1978 (Source: NOAA).

Since the majority of precipitation falls in the winter, much of it comes as snow. The amount varies from a trace at San Diego, California, to 11.5 m (449 in.) at Tamarack, in Alpine County, California. Generally the heaviest snowfalls occur on the southwestern slopes of the mountains. Like precipitation generally, the amount of snowfall in a given winter varies a great deal from one winter to the next, particularly in the southern mountains. Throughout this western region mountain snow is one of the major sources, if not the major source, of water for irrigation and power.

In North America the areas with the greatest snowfall are the mountain areas of the far west. The mean permanent snowline in mountains in the humid tropics is about 4,700 m (15,400 ft). The snowline increases slightly in a poleward direction to 5,200 m (17,000 ft) near the thirtieth parallel. From there it drops fairly rapidly to near 3,000 m (9,800 ft) at 45° N latitude and about 1400 m (4,500 ft) at 60° N. Actual height of the snowline on a given mountain varies a great deal, depending on orientation to the sun and to the prevailing wind. The snowiest area in North America is apparently at Paradise Ranger Station on Mt. Rainier in the state of Washington. At an elevation of 1,692 m (5,550 ft) and on the Pacific slope, snowfall averages over 15 m a year, and in the 1971-1972 snowfall season, a phenomenal 28.5 m (1,122 in.) of snow fell.

When the winter rains and snows fail, drought comes to the west coast. Such winters occur at intermittent intervals. One such winter was that of 1976-1977. During that winter, a ridge of high pressure extended northward along the west coast of North America, interrupting the usual zonal flow of air across the continent and having drastic effects on the precipitation of the west. Figure 12–7 shows the circulation at the 700-mb level for January in the four years from 1975 to 1978. Between 1975 and 1977, the pattern of upper air circulation had shifted over the continent in such a way that a high pressure ridge extended from California to Alaska. This pattern is shown on the charts by the poleward bend in the height of the 700-mb level over the west coast. This ridge blocked the onshore movement of mT air sufficiently to bring severe drought to California and southern Oregon. In January of 1978 the circulation reverted to a more zonal flow over the continent.

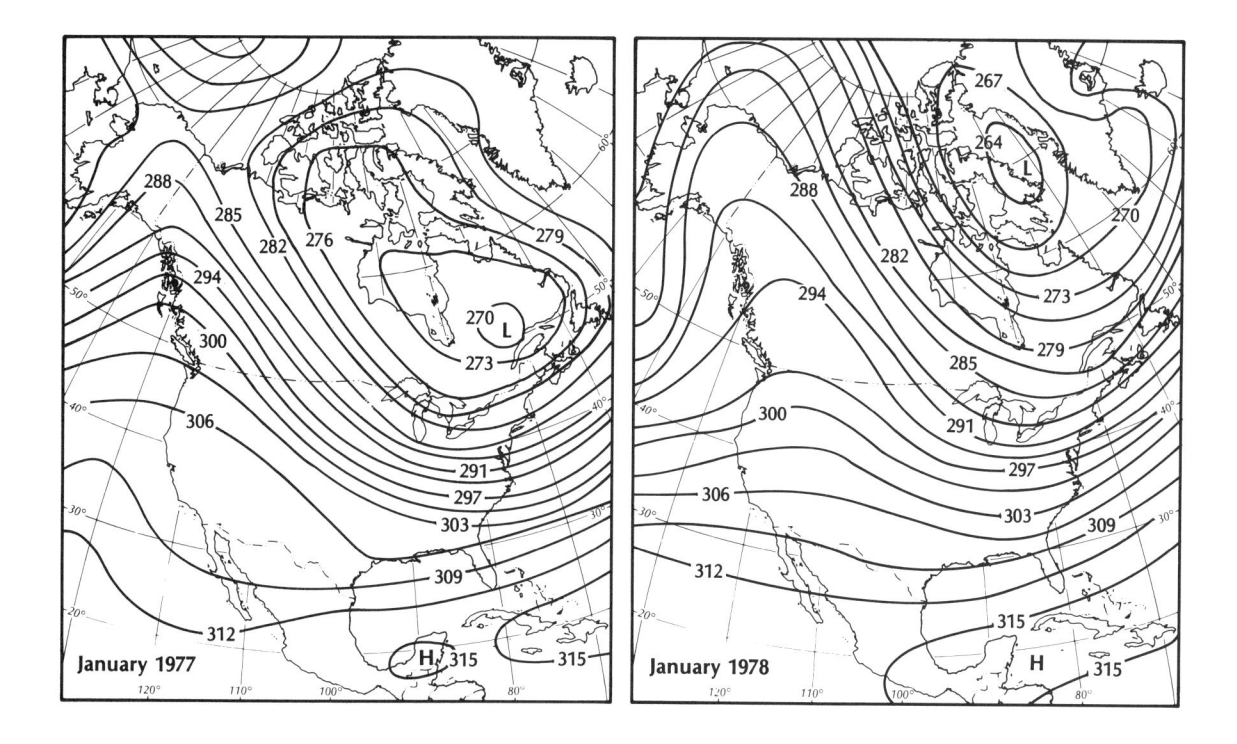

The Effects of the Mountains on Local Climate

The coastal states and provinces have a wide range of local climatic conditions, resulting primarily from the influence of the mountains on temperature and rainfall. The north-south coastline and the westerly winds provide a truly marine climate. This marine climate is restricted to a very narrow zone by the coastal ranges. The climate on the lee sides of the coast ranges is substantially different from that on the windward sides. The effect of the mountains on precipitation is such that some of the wettest and driest areas of North America occur in British Columbia and the Pacific states.

The temperature regime varies within these regions, depending on proximity to the open ocean. Coastal stations have cooler temperatures in the summer and warmer temperatures in the winter than interior stations. The summer cooling in these areas is further enhanced by the cold currents that flow equatorward offshore. San Francisco has a July average of 17°C; Sacramento, farther inland in the Central Valley, has a July average of 25°C. The extremes of temperature at the coastal stations are also moderate for the latitude, reflecting the influence of the onshore flow of air. Fog is a frequent phenomenon along the coasts that is due to the stabilizing effect of the cold current on the moist oceanic air. The fog of the Golden Gate and San Francisco Bay is well known, but the phenomenon is widespread along the west coast of North America as well as in the other areas with this type of climate.

The littoral zone typically has small daily and annual ranges in temperature, and subfreezing weather is infrequent on the coast of California and Oregon. The frost-free growing season in California ranges downward from 365 days along the southern coast. The mildness of the winters on the coast is attested to by the growing of citrus fruit as far north as 40°. The leeward locations are drier and hence have a greater range in temperature. The range between January and July at North Head, Washington, is 8.4C (15°F). Inland a few kilometers at Vancouver, Washington, it is 15.5°C (28°F); and at Kennewick, Washington, it is 23.9° (43°F). The effect of the ocean on west coast temperatures is such that January averages from 45° to 50° N are up to 15°C warmer than on the east coast of the continent. Port Hardy, B.C., averages 2.4°C in January, and September Isles on the St. Lawrence River averages −12.7°C.

West of the coast ranges, and to a large extent west of the Sierra Nevada and Cascade Mountains, severe thunderstorms and tornadoes are infrequent. The primary factor in the development of these storms, the cold front, does not have as steep a gradient as farther inland, since the mP air is relatively warm. Thunderstorms increase away from the coast and with increasing elevation.

SUMMARY

The climate over the North American continent varies from place to place, depending upon the frequency with which mT, mP, cT, or cP air is present. West of the Rocky Mountains the weather is most often associated with either mP air or stable mT air. East of the Rocky Mountains the most frequent air streams are cP and mT. The result of this difference in air mass frequency and the nature of the primary circulation is that west of the mountain ranges most precipitation is from mP air and falls mainly in the winter season. East of the mountains most precipitation is from mT air and either falls in the summer or is spread throughout the year. Seasons change as one type of air stream is exchanged for another. Seasonal changes in the different air streams also add to the variability of weather. The region that experiences the greatest extremes in temperature and precipitation is the winter-dry region of the Great Plains. Here there are no mountains to obstruct the flow of air between the Canadian Arctic and the Gulf of Mexico. In the winter, extremely cold arctic air occasionally flows south as far as Mexico; in the summer, warm air from the Gulf of Mexico or the high plains travels north well into Canada. So changeable is the weather in the Great Plains that an oft-repeated phrase is "If you don't like the weather, just wait for a moment."

APPLIED STUDY 12

Setting Records: A Case of Probability

For those who listen to television weather programs or who read the daily weather accounts in newspapers, finding a report of a new temperature or rainfall record is not a rare event. In fact, using a statistical method known as *probability analysis*, it can be demonstrated that the chances of creating some new record are fairly high.

The *probability* of an event is the degree of certainty of its occurrence over the long run. Probability is denoted by the letter P; if an event is sure to occur (e.g., death), then $P = 1$. For an event that has no chance of ever occurring (e.g., a dog's giving birth to a kitten), the probability is equal to zero. Thus, the probability of an event's occurring ranges from 0 to 1 as the probability increases.

To see how it works, we can use an example of a climatic station that started 10 years ago and is recording January snowfall. In the first January, since no records existed prior to the reading, the probability of a record's being set is 1. In the second January the chances of the existing value's being exceeded or equalled is ½, or 0.5. The third January must now be compared to the two prior years, so the probability of a new record is 1 out of 3 or ⅓, 0.33. By the tenth January, the probability is 0.1. The data can now be used to derive some idea of the chances of breaking a record in any one of the ten Januarys on record. All that is needed is to add the probability derived for each of the individuals. Thus

$$1 + 0.5 + 0.33 + 0.25 + 0.2 + 0.17 + 0.14 + 0.13 + 0.11 + 0.1 = 2.93$$

or approximately 3. This answer tells us that in any three of the ten years it is probable that a new record will be set.

Real world analysis of snowfall data provides some interesting results. Figure 12–8, for example, shows the chances of having a *White Christmas* (defined as a surface cover of snow at least 2.5 cm in depth) in the United States. In this, the probability is expressed as a percentage, where 100 percent is a probability of 1, a certainty.

Clearly, the map is based upon more than ten individual readings; to handle more values it is useful to use simple equations. For example, to find the probability of breaking at least one new record during the period of a year:

$$\text{Probability} = 1 - \left(1 - \frac{2}{N}\right) n$$

In this equation N is the number of years of data available and n the number of events of the element being considered. For example, if a 100-year record existed, then $N = 100$, and if the record mean highest temperature of each month is the event considered, then $n = 12$, each of the calendar months in one year.

REGIONAL CLIMATOLOGY

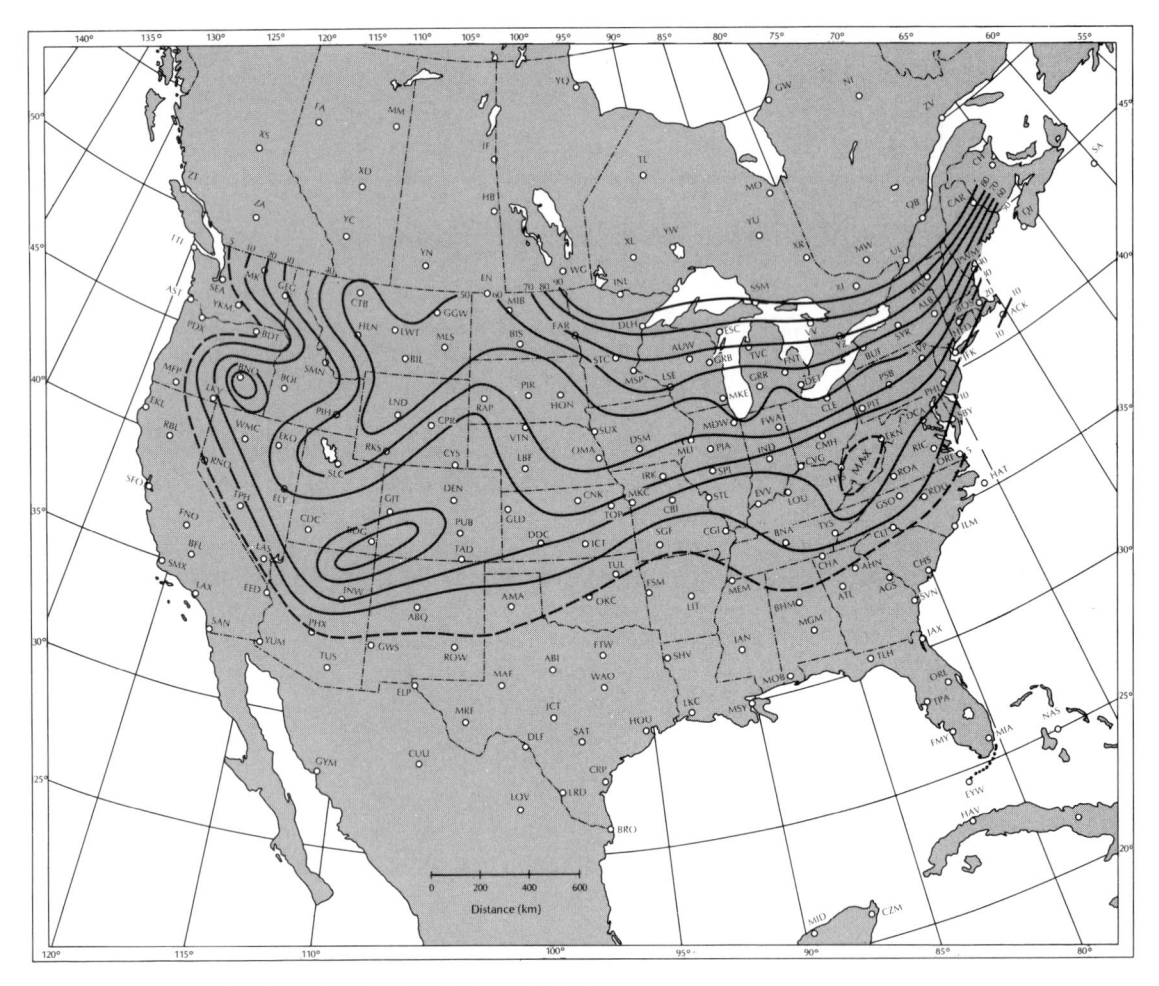

FIGURE 12-8 Probability of a "White Christmas" (2.5 cm or more of snow on the ground) (From "Statistical Probabilities for a White Christmas," U.S. Department of Commerce news release, Washington, D.C., 15 December 1971).

Although at times the numbers may be large, the formula is easy to use. Consider the probability of breaking a mean annual rainfall total in data that have been collected for 100 years. In this case $N = 100$ and $n = 1$. So:

$$\text{Probability} = 1 - \left(1 - \frac{2}{N}\right)n$$

$$= 1 - \left(1 - \frac{2}{100}\right)1$$

$$= 1 - 0.98$$

$$= 0.02$$

This is not a high probability. The probability becomes much greater when n becomes larger, as illustrated in table 12-3, showing the probability of at least one record's (either high or low) occurring in the 100th year of data collection at a station. It can be seen that while the chances of breaking a single annual record, such as snowfall, is slight, the probability of a new record on at least one day of the year is almost a sure thing (P = .9994).

TABLE 12-3 Probability of at least one record extreme (high or low) in the 100th year of observations at a weather station, depending on how many different statistics are examined for a new record.

Number of Statistics	Examples	Probability
1	Annual total snowfall	.0200
2	Annual total snowfall and annual mean temperature	.0396
12	Mean temperature of each calendar month	.2153
36 (3 × 12)	Mean temperature, total precipitation, and greatest daily precipitation for each calendar month	.5168
96 (8 × 12)	Temperature (mean daily maximum, mean daily minimum, extreme maximum, extreme minimum), precipitation (mean monthly, greatest daily, number of days with), and wind (mean speed) for each calendar month	.8562
365	Mean temperature for each calendar day of the year	.9994

13 Polar and Highland Climates

THE ARCTIC BASIN
The Polar Wet and Dry Climate

THE ANTARCTIC BASIN
The Polar Dry Climate
The Polar Wet Climate

HIGHLAND CLIMATES

SUMMARY

APPLIED STUDY 13
Climate and Human Physiology

REGIONAL CLIMATOLOGY

The distinguishing feature of the polar climates is cold weather brought about by low levels of solar radiation. The radiation balance differs from that of the tropics and mid-latitudes in terms of both the diurnal and annual periodicities of solar energy. In the tropics the major periodicity in energy is the diurnal periodicity, in which daily changes in radiant energy and temperature are greater than from season to season. In mid-latitudes there are large fluctuations in both the diurnal and annual periodicity. Generally the more poleward the location, the larger is the seasonal fluctuation in radiation and temperature relative to the diurnal changes. Like the tropics, the polar climates are dominated by a single periodic fluctuation in energy. In the tropics the major fluctuation is diurnal, but in polar areas it is the annual fluctuation that predominates (figure 13-1).

All points on the surface of the earth receive nearly the same number of hours of sunshine during the year, but there is a tremendous difference in the distribution through the year. In the tropics the radiation is distributed fairly evenly throughout the year, with about 12 h of insolation each day. At latitudes higher than the Arctic and Antarctic circles, most of the energy is received in one continuous input during the summer half-year, with the other half-year receiving very little energy. At the time of the summer solstice in the Northern Hemisphere (22 June), the north polar axis is tilted toward the sun by 23.5°. This is the longest day of the year in the Northern Hemisphere, and on this date the sun does not drop below the horizon in the zone from 66.5° N to the North Pole. At Murmansk (69° N) the continual daylight of summer lasts for 70 days. At Spitzbergen (80° N) there are 163 days of continuous light, and at the North Pole there are 189 days of continuous daylight.

In late September, at the time of the fall equinox, the days are everywhere 12 h long. The vertical rays of the sun are over the equator, sunset is occurring at the North Pole, and sunrise at the South Pole. The equinoxes are far more important to

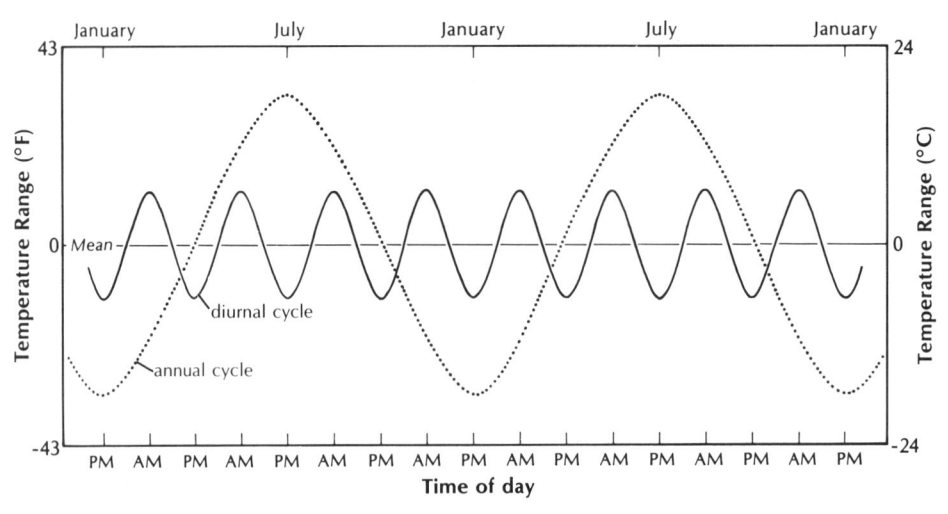

FIGURE 13-1 Relationship between the annual and diurnal energy cycles in polar climates.

the polar regions than are the solstices because they mark major times of change in energy. Basically the equinoxes represent times when solar radiation either becomes significant or drops to near zero.

At the time of the winter solstice the days in the Northern Hemisphere are the shortest. The North Pole is in the midst of a single night 176 days long. At Spitzbergen the night is 150 days long, and at Murmansk continuous darkness lasts for 55 days, from 26 November through 20 January.

During the winter months in the polar regions, direct solar radiation may be absent; but some radiation, including some in the visible range, continues to find its way to the surface. Thus total darkness seldom exists. When the sun is below the horizon, refraction and scattering of the rays bring some sunlight to the surface until the sun is 18° below the horizon (astronomical twilight). Thus at some latitudes there may be several months with no direct sunlight but almost continuous twilight. The stars, moon, and auroras are also sources of light for the poles, so that the darkness is not as intense as might be expected.

Another important element in the polar climates is that the intensity of radiation is never very high for any length of time when compared to mid-latitude or tropical regions. Even on the equatorward margins the sun rarely reaches more than 45° above the horizon. In the summer months, however, although intensity is low, the duration of daylight is quite long, with over 20 h of daylight near the summer solstice. Although summer insolation persists for many hours a day, temperatures do not get very high both because the intensity is so low and because much of the energy is used in melting snow and ice.

While there are very fundamental elements of similarity between the two polar areas, there are also basic differences that influence the climate of each. The most important and obvious difference between the Arctic and the Antarctic is that the Antarctic is a relatively large and high land mass surrounded by water, whereas the Arctic is a sea surrounded by land. The relationship of the land-sea interface is just reversed at the two poles, and this is very significant to the climate of the two areas. The three macroclimates of the polar regions are the polar wet, wet and dry, and dry (figure 13–2). The polar wet and dry climate is found primarily in the Northern Hemisphere, and the other two climates are found mainly in the Southern Hemisphere.

THE ARCTIC BASIN

The Arctic Ocean consists of all the sea surface north of the Arctic Circle, whether covered by ice or open water (figure 13–3). The central part of the Arctic Ocean is frozen most of the year but does have occasional open leads. The central ice pack covers 11.7 million km^2 in winter and melts back to 7.8 million km^2 in the warmest months. The ice drifts from east to west, and, in the process of moving, leads open and close so that some unfrozen areas are always present. The ice ranges in age up to 20 years and in thickness up to 5 m, with the oldest ice being the thickest. In the summer the ice melts away from the edge of North America and Eurasia so that there

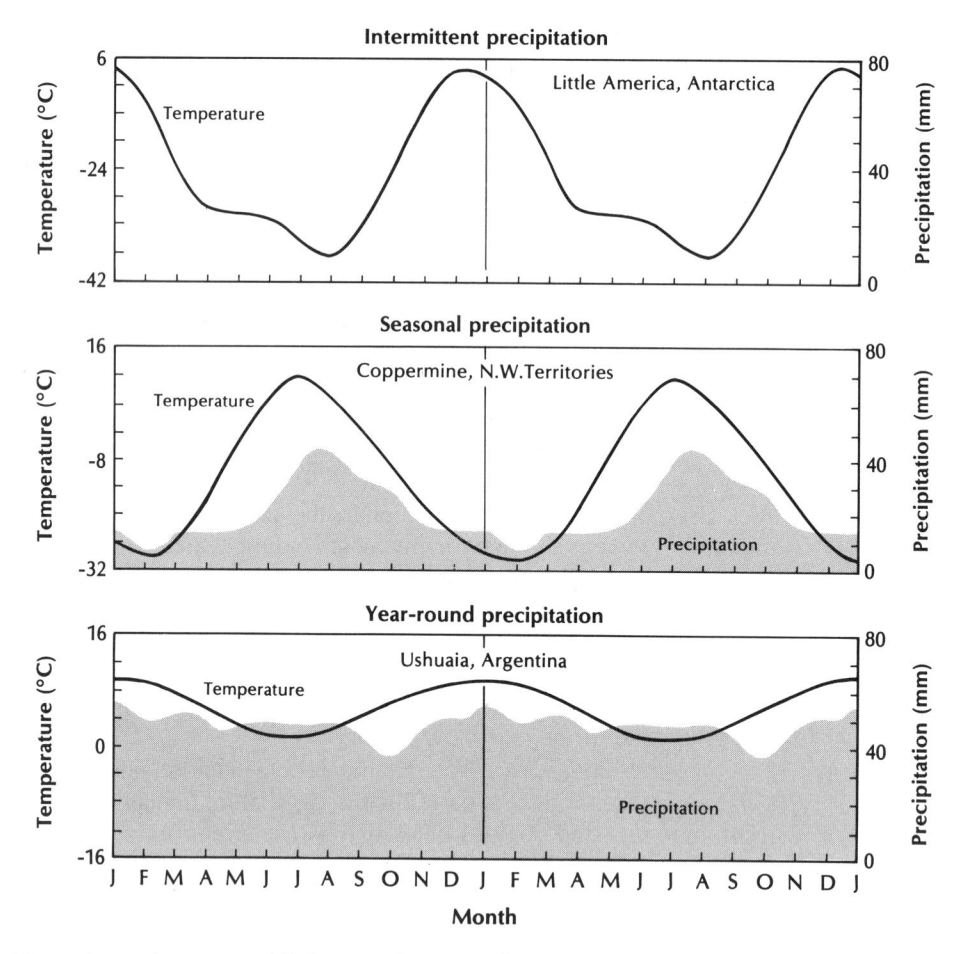

FIGURE 13-2 The polar environments: All three environments have a low solar energy input year-round. The heat from atmospheric water and the sea keeps the average temperature at Ushuaia above freezing.

may be up to 2 months when there is mainly open water with patches of drifting ice. Because of the heat in the Gulf Stream, the Norwegian Sea and Barents Sea remain open all year at latitudes where the sea is frozen in other areas.

The earth-space energy balance in the polar areas is such that the poles are major energy sinks. There is much more energy emitted by the polar regions than is received from solar radiation. In the Arctic Basin radiation reflected and radiated away as long-wave radiation is 60 percent greater than direct solar radiation. For this situation to persist, a supplementary heat source must provide the difference. This energy source is heat carried poleward by air and water currents. It is also this steady flow of energy poleward that maintains Arctic temperatures at a level much higher than insolation would indicate (figure 13-4). For the Arctic Basin the primary source of heat is the inflow from the sea and the atmosphere. This energy is nearly double

FIGURE 13-3 Map of the Arctic Basin.

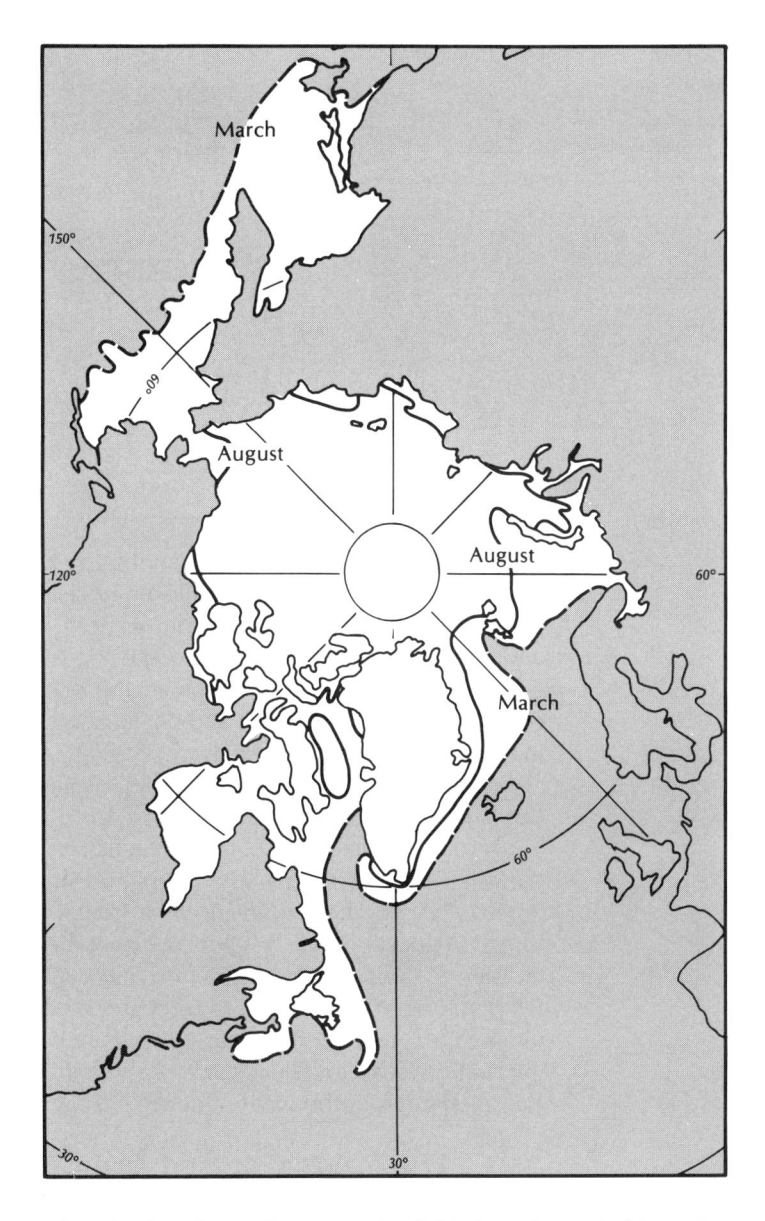

that of solar radiation absorbed at the surface. Much of this heat is carried into the Arctic by warm ocean currents. The water releases heat to the atmosphere by radiation, conduction, and evaporation. The amount of open water compared to ice is of extreme importance, since the heat transferred to the atmosphere from a water surface is more than 100 times that of an ice surface.

Sea water acts differently from fresh water as it cools toward the freezing point. In fresh water the maximum density is reached at 4°C, and fusion takes place at 0°C. In sea water the maximum density is reached just near the fusing point, which is at about −2°C, depending on salinity. One result of this phenomenon is that

FIGURE 13-4 Energy balance of the Arctic Basin.

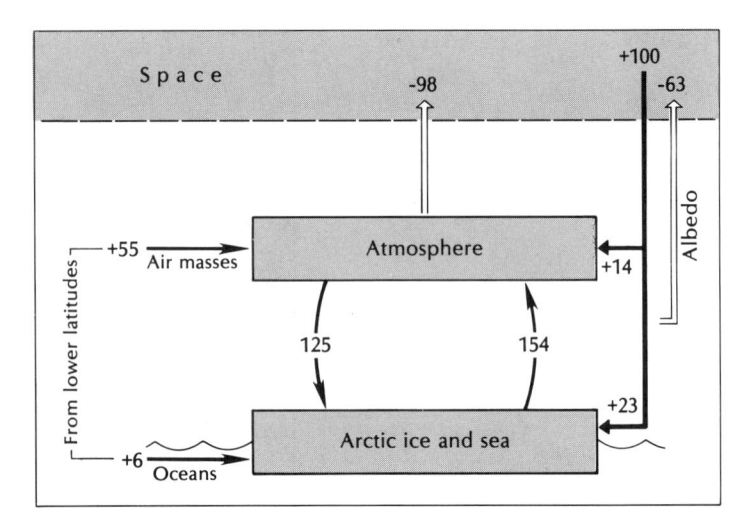

surface water in polar seas may cool up to 6°C more than would otherwise be the case before ice forms. The colder polar sea water is more dense than either fresh water or warmer sea water; as it chills near the freezing point, it sinks to the bottom and spreads out, settling in the deeper parts of the ocean basins. This settling cold water helps promote circulation in the ocean, which in turn helps equalize the energy between the equator and poles. This fact also explains why most of the water in the ocean is so cold. The average temperature of water in the sea is about 3.5°C (38°F). Warm surface water in the tropics and along beaches in mid-latitudes is the exception and not the rule.

The atmosphere over the Arctic Basin never becomes really cold because the water beneath the ice and in open areas does not drop below a temperature of −1.6°C (29°F). The sea, even when frozen, exerts a modifying effect on temperatures. Whenever the air temperature over the ice drops below −1.6°C, there is heat transferred through the ice and from the open water. Over much of the Arctic Basin temperatures over the ice range between −20°C and −40°C, with the lowest temperature yet recorded near the surface being −50°C (−58°F). The influence of the Arctic Ocean as a heat source is such that the coastline represents a very abrupt transition zone in the environment.

The Polar Wet and Dry Climate

Climates with seasonal mP and cP air

Köppen type	ET
Thornthwaite type	E

The polar wet and dry climate, found primarily along the shores of the Arctic Ocean, is characterized by cold winters, cool summers, and a summer rainfall regime. Specific areas experiencing this climate are the North American Arctic coast,

Iceland, Spitzbergen, coastal Greenland, the Arctic coast of Eurasia, and the Southern Hemisphere islands of McQuarie, Kerguelen, and South Georgia. The land areas with this climate occupy only about 5 percent of the land surface of the earth.

A sample of five tundra weather stations in the Northern Hemisphere shows a January average temperature of $-30°C$. Mean annual temperatures are below freezing, but summer temperatures go above freezing at least for a few weeks. The summer period, if measured by above-freezing temperatures, averages 2–3 months but may be as long as 5 months. Mean temperatures in the warmest months seldom go above 5°C. High temperatures range from 21° to 26°C.

Pressure gradients are weak in the tundra, so this is not a stormy region. In fact, it is one of the least stormy areas of the earth's surface. High-pressure systems dominate with cold, still, dry air prevailing, particularly in the winter. The North American Arctic is even less stormy than other areas. Because of the mountainous character of Alaska, the cyclones of the Pacific Ocean are prevented from moving over the Arctic coastline. Some of the storms penetrate, but few compared to other areas. Spring and fall are the seasons with the most active weather, which is true of most high-latitude locations. Winds are light in velocity and variable in direction. The average velocity is less than 15 km/h and less than 10 km/h in the Canadian tundra.

Atmospheric humidity is high during the summer months, averaging 40–80 percent. During the winter it is less, averaging 40–60 percent. Annual precipitation averages less than 250 mm (10 in.) for most stations; more than three-fourths of the annual total falls during the summer half-year. Winter snowfall is only about 50 mm of water because the snow tends to be dry and compact. A deep snow cover is absent over most of the tundra, as winter snow is not excessive and tends to drift easily. Thus, exposed areas are often free of snow, although the sheltered areas along streams may have a deep snow cover. Cloudiness varies through the year in the tundra. In the summer months cloud cover may average 80 percent, but during the winter it drops to around 40 percent.

THE ANTARCTIC BASIN

The Antarctic is not only the coldest area on the earth's surface but the least-known area. Nearly 90 percent of the Antarctic land mass is covered by ice, yet it receives so little precipitation that it classifies as a desert. The ice sheet is comparable to the tropical deserts in amount of precipitation. The extremely cold air is incapable of holding much water vapor. Average annual precipitation probably does not exceed 200 mm any place on the ice cap.

The sheer size of the ice-covered continent dictates that it will have major impact on the climate of the Southern Hemisphere. It is nearly twice the size of the United States and nearly three times greater than the Arctic Basin. The harshness of the environment around this land mass is such that even the existence of the continent was not known until extremely late in history. Throughout much of historic time it was shown on maps as *Terra Australis Incognito* ("unknown southern land"); and it is believed to have been first discovered on 17 November 1820 by Nathaniel Palmer. The first expedition commissioned by the United States government set out in 1838 under Lieutenant Charles Wilks.

The outstanding feature of the Antarctic land mass is, of course, the huge mass of ice surmounting and surrounding the continent. Some 86 percent of the world's ice is associated with this land mass. The *pole of inaccessibility*, or the center of the ice mass, is at about 4000 m (13,200 ft) elevation, and it is 670 km (400 mi) away from the geographic pole. The polar ice cap is not the thickest where it is the coldest. The ice reaches its greatest depth where air temperatures average a little below freezing. The accumulation of ice is quite slow in some areas. Near Byrd Station at a depth of about 300 m (1000 ft) the ice is believed to have formed from snow that fell some 1600 years ago. The rate of movement of the ice seaward is highly variable, in some cases moving seaward as much as a kilometer per year and in some locations becoming buoyant and extending out to sea for great distances. In the Ross Ice Shelf consolidated glacial ice extends almost 800 km out to sea. This floating ice is as much as 270 m (890 ft) thick, with roughly 40 m (130 ft) above water. Icebergs breaking off from this shelf ice are often tabular in shape and may be as much as 8 km across in their longest dimension.

The Polar Dry Climate

Climates dominated by cP air

Köppen type	EF
Thornthwaite type	F

Antarctic temperatures average 16°C (30°F) below those of the Arctic. There are several factors contributing to the colder regime: (1) The area is dominated by a land mass rather than water; (2) the average elevation is substantial, with parts of the ice sheet 3,000 m (10,000 ft) above sea level; and (3) the South Pole is three million miles farther from the sun during its winter season than is the Arctic during its winter. This distance reduces solar radiation 7 percent below that of the Arctic winter. Temperatures are affected by both elevation and proximity to the ocean. Mean annual temperatures range from −19°C (−2°F) to as low as −57°C (−70°F) at Vostok. Minimum temperatures in the winter drop to −57°C along the coast and decrease with elevation and distance from open water. Vostok, a Russian Antarctic station located inland at an elevation of 3950 m (13,000 ft), has recorded the coldest temperature recorded at the surface of the earth, −88.3°C (−127°F). The geographical South Pole is at 3000 m (10,000 ft) elevation; table 13–1 provides illustrative data of the temperature conditions that occur.

In the Antarctic, the coldest temperatures lag far behind the solstice as a result of the long absence of direct solar radiation. At Vostok the coldest days occur after the sun has come above the horizon for a day or two. The sun comes above the horizon there on 22 August, and the coldest temperature yet observed occurred two days later, on 24 August.

TABLE 13-1 Amundsen-Scott Station (South Pole).

Mean annual temperature	20 years (1957–1976)	−49.3°C
Coldest year	1976	−50.0°C
Coldest day of record	22 July 1965	−80.6°C
Coldest month of record	August 1976	−65.1°C
Coldest half-year	1976 (Apr.–Sept.)	−60.7°C
Warmest day of record	12 Jan. 1958	−15.0°C

In summers the temperatures range upward toward the freezing mark but rarely reach it. Over most of the Antarctic it does not get warm enough in summer to melt the surface snow, although there are a few areas that do shed their snow cover.

Diurnal temperature ranges tend to be small, generally less than 5°C (0°F) as a result of the small difference in insolation during the day. The absence of the normal pattern of day and night and the high heat capacity of the snow and ice tend to reduce fluctuations in temperature on a diurnal basis. The extreme nature of the temperature regime of the Antarctic is well shown in the listings given in table 13-2.

The extreme cold of the Antarctic continent influences the climate far from the ice cap itself. Cities that lie on the southern margins of the Southern Hemisphere land mass have average temperatures 3°C cooler than cities at the same latitude in the Northern Hemisphere. Several of the islands in the South Seas support glaciers, including Kerguelan Island and Heard Island. In South America mountain glaciers reach the ocean at latitudes nearer the equator than do those in the Northern Hemisphere.

Winds are a predominant factor in the weather of the polar deserts. Westerly winds prevail at the surface poleward to around 65°, where they give way to low-level easterly winds (figure 13-5) that extend on to about 75°. Cyclonic storms develop over the oceans in the westerlies and move around the Antarctic from west to east. They normally do not penetrate far inland, and they account for much of the precipitation and weather along the coast. Winds of a high enough velocity to move snow and produce blizzard conditions occur at Byrd Station about 65 percent of the time and are of high enough velocity to produce zero visibility about 30 percent of the time. A slight increase in wind velocity brings a substantial increase in blowing snow. The amount of snow moved by the wind varies as the third power of the velocity. It is unfortunate for those working in the Antarctic, but the most accessible locations are also the windiest. Mawson's Base at Commonwealth Bay experiences winds that are above gale force (44 kph, or 28 mph) more than 340 days a year. At Cape Denison the mean wind velocity is 19.3 m/s (43 mph), and during July of 1913 it averaged 24.5 m/s (55 mph).

Strong gravity winds develop and blow off the land masses. These *katabatic winds* are the flow of cold, dense air down a topographic slope due to the force of gravity. They result from the extreme cooling of the air over the ice. A topographic slope of 0.2 percent is about equal to the force of the average pressure gradient over the Antarctic. Where the slopes are steep, the katabatic winds may regularly flow in a direction other than that which would be indicated by the pressure gradient. The strongest winds occur when the pressure gradient and topographic slope are coinci-

TABLE 13-2 World climatic extremes recorded on the Antarctic continent.

	Solar Radiation	
Longest continuous period of direct sunlight	9–12 Dec. 1911	60 h
Highest monthly mean (hours)	Syowa Base (69° S 40° E)	14 h/day
Highest mean monthly intensity	South Pole (Dec.)	955 ly/day
	Temperature	
Lowest mean monthly diurnal range	South Pole (Dec.)	2°C (3.5°F)
Lowest average temperature (warmest month)	Vostok	−32.5°C (−27.6°F)
Lowest mean annual temperature	Cold Pole (78° S 96° E)	−58°C (−72°F)
Lowest mean monthly maximum	Vostok	−66.5°C (−88°F)
Lowest monthly mean	Plateau Station (Aug.)	−71.7°C (−97°F)
Lowest mean monthly minimum	Vostok	−75°C (−103°F)
Lowest daily maximum	Vostok	−83°C (−117°F)
Lowest absolute	Vostok	−88°C (−126.9°F)
	Humidity	
Lowest dewpoint	South Pole	−101°C (−150°F)
	Wind Velocity	
Highest annual mean	Cape Denison	19.3 m/s (43 mph)
Highest monthly mean	Cape Denison (July 1913)	24.5 m/s (55 mph)

dent. The katabatic winds are the prevailing winds over much of the Antarctic, particularly where there is a steep drop from the interior highlands to the coast. Where these winds are fairly constant, they will produce formations of ripples in the snow surface called *sastrugi*. Strong winds and blizzards are often accompanied by rising temperatures because the usual temperature inversion may be broken down, and warmer maritime air will move inland.

Moisture data are even more difficult to obtain for the polar deserts than temperature data. Data indicate that relative and absolute humidity and cloud cover decrease inland. Relative humidity often drops as low as 1 percent. Humidity and cloud cover are also lower in winter as a result of the stronger subsidence and the surface inversion. The cloud decks that appear over the ice caps produce substantially warmer temperatures as they provide a source radiating heat to the ground.

FIGURE 13-5 Average surface winds and January surface pressure over Antarctica (After Schwerdtfeger, Mather, and Miller, From K. Boucher, *Global Climate*, [1975, p. 294]. Used by permission of Hodder and Stoughton, Ltd.).

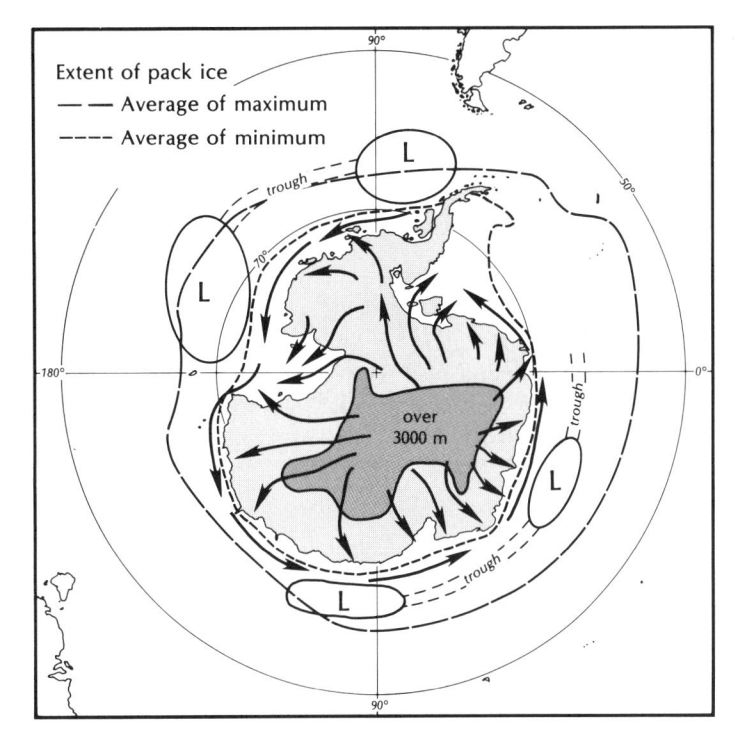

Precipitation may be fairly frequent, but it is necessarily light. Precipitable moisture in the atmosphere is not very great because of the low temperatures and probably doesn't exceed 10 mm (0.4 in.) at any one time. Most of the precipitation occurs as snowfall, with averages of 300–600 mm (12–24 in.) per year. At Halley Bay there is snowfall about 200 days/year. The high winds tend to move the snow in blizzards and as drifting snow. Drifting snow also occurs 200 days/year, and visibility is reduced to less than 1 km on 170 days/year. Some of the precipitation occurs in the form of ice pellets, and some moisture may be deposited directly as condensation from the subsiding air as it reaches the ice surface. Along the coasts, instances of steam fogs drifting inland have occurred when temperatures were as low as −30°C. These steam fogs contribute to the accumulation of moisture.

The Polar Wet Climates

Climates dominated by mP air

Köppen type	EM
Thornthwaite type	AE'

Surrounding the Antarctic is the world ocean, uninterrupted by land masses to a latitude of 40° S except at the tip of South America. Over this ocean the atmosphere is characterized by year-round high humidity, a high frequency of

TABLE 13-3 World climatic extremes recorded in the polar wet climates.

Lowest mean annual diurnal temperature range	Heard Island, Indian Ocean (53°10′ S 74°35′ E)	3.3°C (−0.5–2.8°C) 6°F (31–37°F)
	Macquarie Island Indian Ocean (54°36′ S 185°45′ E)	3.3°C (2.8–6.1°C) 6°F (37–43°F)
Lowest mean annual hours of sunshine	Laurie Island (66°44′ S 44°44′ W)	500 h
Lowest monthly % of possible sunshine	Argentine Island (June) (61°15′ S 64°15′ W)	5%
Highest mean annual relative humidity	Deception Island (63° S 61° W)	90%

precipitation, and low temperatures. The only land areas are a series of islands scattered around the Antarctic continent. Data representative of this climate type are given in table 13-3.

Humidity is very high, averaging over 50 percent all year. Cloud cover is also extensive, averaging 80 percent during the winter and slightly less in the summer. Precipitation frequency is high in all areas, with at least a 25 percent probability of precipitation on any day of the year. The annual total is quite varied, however, as the intensity of precipitation varies with latitude and continentality. The totals vary from 370 mm to 2.92 m (15–117 in.). The annual variability of precipitation is among the lowest to be found anywhere. Snow is a common form of precipitation, but a ground cover of snow is short-lived. Most snows are very wet and likely to last but a very short time.

High humidity and cloudiness, however, tend to have a considerable degree of control over insolation and temperature. The summer averages range up to 10°C (50°F), and winter averages are between −6.7° and 0°C, so the annual range in temperature is not very great. Winter extremes are exceptionally warm for a polar climate. The reason for the mild winter temperatures and low annual range is the marine location of these areas. As the ocean does not freeze, it provides a constant source of heat for the atmosphere. The diurnal ranges are also quite low because of the high humidity. A frost-free season is nonexistent; frost can occur any day of the year, and these areas average over 100 days annually with frost.

In winter there is a much greater frequency of storms with high winds and extreme variability of wind direction than in summer. This is the area that was named the "roaring forties," the "furious fifties," and the "shrieking sixties" by the sailors of the sixteenth century. These names originated as they sailed around Cape Horn in South America.

This climate is distinctive in a number of ways. It is the cloudiest of all the climates, with most of the area experiencing few clear days. No place can equal the South Atlantic zone for the frequency and severity of storms.

HIGHLAND CLIMATES

The tropical, mid-latitude, and polar systems have been distinguished from each other on the basis of periodic aspects of the energy balance, and each of these regimes in turn was subdivided on the basis of periodicity of the moisture balance. *Highlands*, or mountain areas, are found in all of these different climatic regions and hence are subject to the same diurnal and seasonal patterns of solar energy and moisture that prevail in the surrounding lowlands. For example, the coast ranges of California are subject to a diurnal energy pattern, a strong seasonal pattern of energy, and a seasonal moisture regime consisting of a dry summer and wet winter.

The primary aspect of the climate of highland areas is the rapid change that takes place with elevation and orientation of the slopes to the sun or prevailing wind. Radiation, temperature, humidity, precipitation, and atmospheric pressure all change rapidly with height. Radiation intensity increases with height because of a steady decrease in thickness of the atmosphere. La Quiaca, Argentina, for instance, has the highest average annual radiation level of any known surface location (table 13-4). The increase in radiation intensity with height is responsible for the tendency of skiers or hikers in mountain resort areas to sunburn. There is a sharp increase in ultraviolet radiation as the atmosphere thins. Ambient air temperatures decrease with height above sea level at about 1°C /100 m. As the density of the atmosphere decreases and concentration of water vapor and carbon dioxide decrease, the greenhouse effect of the atmosphere drops sharply, and the ground surface heats and cools rapidly. Higher diurnal temperature ranges result. El Alton, Bolivia, at an elevation of 4,081 m (13,468 ft) has a mean annual temperature of 9°C (48°F) and a mean daily range of 13°C (23°F).

The microclimate of a given site in highlands depends also on slope orientation. Orientation with respect to the sun is significant in determining local heating characteristics. A slope facing the sun will have much higher surface temperatures and resultant air temperatures than one facing away from the sun as a result of greater intensity of solar radiation (figure 13-6).

Humidity and precipitation tend to increase with elevation, since the ability to hold moisture is a function of atmospheric temperature and temperature decreases with height. For this reason fog tends to be more prevalent at higher elevations (table 13-5). Moose Peak in Maine averages 1580 h of fog annually, and Mt. Washington, New Hampshire, averages 318 days/year with fog. The seasonal pattern of precipitation is similar to that of surrounding lowlands, but the amount of precipitation changes rapidly with elevation and orientation to the wind. Since highlands provide a mechanical lifting mechanism, precipitation tends to increase up to a point and then to decrease again as moisture is precipitated out. Orientation of slopes with respect to prevailing winds is just as significant as elevation in determining local precipitation characteristics. Windward slopes may receive up to five times the amount of precipitation that is received on the leeward slopes. The rainiest areas found on the earth are where there is orographic lifting of onshore winds. Table 13-6 provides sample data for the effects of elevation on precipitation.

TABLE 13-4 World extremes recorded in highland climates.

Rainfall

Highest monthly mean	Cherrapunji, India (July)	2692 mm (106 in.)
Highest monthly total	Cherrapunji, India (July 1861)	9296 mm (366 in.)
Highest annual mean	Mt. Waialeale (Kauai, Hawaii)	11.68 m (460 in.)
Highest total (4-month period)	Cherrapunji, India (April–July 1861)	18.74 m (737.7 in.)
Highest annual total	Cherrapunji, India (1 Aug. 1860–31 July 1861)	26.47 m (1042 in.)

Snowfall

Highest daily total	Silver Lake, Colo. (14–15 April 1921)	1930 mm (76 in.)
Highest 6 day total	Thompson Pass (26–31 Dec. 1955)	4420 mm (174 in.)
Highest 12 day total	Norden Summit, Calif. (1–12 Feb. 1938)	7722 mm (304 in.)
Highest monthly total	Tamarack, Calif. (Feb. 1911)	9906 mm (390 in.)
Greatest depth on ground	Tamarack, Calif. (9 Mar. 1911)	11.53 m (454 in.)
Highest annual mean	Paradise Ranger Station (Mt. Rainier, Wash.)	14.78 m (582 in.)
Highest annual total	Paradise Ranger Station (Mt. Rainier, Wash.)	25.83 m (1017 in.)

Wind Velocity

Highest 1 hour mean	Mt. Washington, N. H.	77.3 m/s (173 mph)
Highest 24 hour mean	Mt. Washington, N. H.	57.7 m/s (129 mph)

Radiation

Highest mean annual	La Quiaca, Argentina (22° S, 3459 m elev.)	667 ly/day

FIGURE 13-6 Slope orientation and radiation intensity.

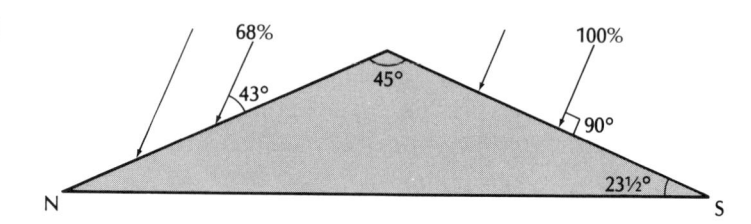

TABLE 13-5 Comparison of days/year with fog at neighboring mountain and valley stations.

Station	Elevation (ft)	Fog Days/Year
Mt. Washington, N.H.	6,280	318
Pinkham Notch	2,000	28
Pikes Peak, Colo.	14,140	119
Colorado Springs	6,072	14
Mt. Weather, Va.	1,725	95
Washington, D.C.	110	11
Zupspitze, Germany	9,715	270
Garmish, Germany	2,300	10
Taunis, Germany	2,627	230
Frankfurt, Germany	360	30

Source: H. E. Landsberg, *Physical Climatology*, p. 140. Gray Printing Co., 2nd ed. 1968.

Since temperature decreases with elevation, a greater proportion of total precipitation occurs as snow in highland areas. The mean snowline for the earth is found at about 4,550 m (15,000 ft) and in the tropics rarely goes above 5,450 m (18,000 ft). The snowline is usually higher on the leeward sides of mountain ranges because there are less snow, more insolation, and Foehn winds that aid in melting. Mountains in Africa, South America, and Asia all exceed this elevation, and snow-capped mountains are found in all latitudes. Extremely heavy snowfalls are common in some highlands. The 25+ m of snow that fell at Paradise Ranger Station on Mt. Rainier in the winter of 1970-1971 serves as an extreme example.

Atmospheric pressure decreases with height at a geometric rate and becomes a significant element of mountain climates. The rapid decrease in pressure has severe physiological effects on humans who are not adapted to it. Flatlanders will begin to feel the effects of the lower pressure somewhere around 2000 m, but it varies from individual to individual. The first phenomenon that most people notice is popping in the ears as pressure is equalized between the inner and outer ear. Other effects include faintness, headaches, nosebleeds, nausea, sleeplessness, and general weakness. These conditions are often temporary, depending on general physical condition. Adaptation to lower pressure usually occurs within a matter of weeks. These responses represent but one component of the influence of high altitude (and

TABLE 13-6 Temperature and precipitation variation with height in Ecuador.

Station	Latitude	Elevation	Mean Temperature	Annual Precipitation
Cotopoxi	0°38' S	3308 m	8°C	1113 mm
Banos	1°23' S	1833 m	16°C	1519 mm
Mera	1°30' S	1063 m	21°C	4770 mm
Santo Dominga de los Colorados	0°15' S	497 m	22°C	3383 mm

high latitude) climate upon human physiology. Applied Study 13 explores this relationship further.

SUMMARY

Cold is the distinguishing characteristic of the polar climates. Both polar regions experience solar radiation of very low intensity, which is the primary factor in producing the cold. The Antarctic is colder than the Arctic because it is an elevated land mass. The Arctic Ocean continually supplies heat to the atmosphere even when covered with ice. For this reason the Northern Hemisphere cold pole is some distance away from the geographic pole. The major continental ice sheets existing on the planet at present are found in Greenland and the Antarctic. Both are related to moderate snowfall and cold temperature. The primary circulation of the atmosphere carries heat into the polar areas to raise the temperature above what it would otherwise be.

Mountain climates are unique in that elevation and orientation to the prevailing winds are the primary variables associated with temperature and precipitation amounts. The seasonal distribution of precipitation and insolation is the same in the mountains as in the surrounding areas of lower elevation.

APPLIED STUDY 13

Climate and Human Physiology

Cold and high-altitude climates pose a distinctive problem to human occupation. One significant aspect of such environments is the stress that they place upon the human body. Humans are warm-blooded animals whose body temperature is usually given as 37°C (98.6°F). To maintain this body temperature requires a balance between heat losses and heat gains. This, in part, is determined by body *metabolism*, which relates the intake of food (for production of energy) to the energy expended by the bodily activities. The heat produced by a resting person is about 50 kcal/h/cm^2 of body surface. This amount of energy is designated the *MET*. Varying activities can be expressed in terms of the MET.

Activity	No. of METs
Sleeping	0.0
Awake, resting	1.0
Standing	1.5
Walking, on level	3.0
Short sprint	10.0

POLAR AND HIGHLAND CLIMATES

Apart from the heat produced by such activities, the body also gains heat through absorption and conduction of energy from the surrounding environment. Losses of energy occur through outward radiation, conduction, and evaporation of moisture from the skin. In normal body temperature, the losses and gains are balanced.

$$M \pm R \pm C - E = 0$$

where M = metabolic heat, R = radiation, C = conduction, E = evaporation.

If more heat is lost than is gained, the equation is no longer balanced; body temperature will decline. If more is gained than lost, body temperature must rise.

The exchanges that occur are influenced by a combination of factors. For example, the loss of body heat is influenced not only by low temperatures but also by air movement. This relationship is expressed by the *wind chill factor*. This chilling effect is the result of the removal of a microlayer of warm air generated by body metabolism. Once the layer is removed, more body heat is required to replace it. As this continues, the loss of the heated layer causes the sensation of cold air at the skin surface. The more rapidly the heat is lost, the colder it feels; and this is a function of wind speed. To express this heat loss in a scientific way, it is possible to generate an equation to measure the heat loss in energy units (e.g., in calories per unit area per unit time). The general public would not find chilling expressed in energy values easily understood so the results have been used to derive equivalent temperatures. These suggest that if the temperature is 25°F and the wind speed is 25 mph, the result is the equivalent to what would occur at a still air temperature of −2°F (figure 13-7). The data in the table

Wind speed (mph)	Actual thermometer reading (°F)											
	50	40	30	20	10	0	-10	-20	-30	-40	-50	-60
	Equivalent temperature (°F)											
	50	40	30	20	10	0	-10	-20	-30	-40	-50	-60
5	48	37	27	16	6	-5	-15	-26	-36	-47	-57	-68
10	40	28	16	4	-9	-21	-33	-46	-58	-70	-83	-95
15	36	22	9	-5	-18	-36	-45	-58	-72	-85	-99	-112
20	32	18	4	-10	-25	-39	-53	-67	-82	-90	-110	-124
25	30	16	0	-15	-29	-44	-59	-74	-88	-104	-118	-133
30	28	13	-2	-18	-33	-48	-63	-79	-94	-109	-125	-140
35	27	11	-4	-20	-35	-49	-67	-82	-98	-113	-129	-145
40	26	10	-6	-21	-37	-53	-69	-85	-100	-116	-132	-148
Wind speeds greater than 40 mph have little additional effect	**Little danger** (for properly clothed person)				**Increasing danger**			**Great danger**				

FIGURE 13-7 Wind-chill equivalent temperatures.

287

suggest the type of clothing that need be worn, given the various conditions. The colder the conditions, the more protection needed. Generally, layered clothing with air spaces between provides the best protection against cold.

Response mechanisms of the body come into effect when an imbalance occurs. Cold conditions cause a reduction of blood flow to body extremities (e.g., feet and hands) so that less heat is lost and body temperature is maintained. This, however, can have a dire effect associated with *frostbite*, the freezing of tissue and cell destruction. Shivering is another response to cold: its function is to increase the metabolic rate of the body. While this occurs, the transfer of blood to the surface layers of the body increases the loss of heat by conduction and radiation. This reaction leads to more shivering, and ultimately the body temperature will fall drastically. Table 13–7 provides more details of these reactions and also provides body reactions to excessive heat.

When the body temperature falls to a level where its mechanisms have no further warming effect, hypothermia occurs. Failure of the thermoregulatory system occurs at about 33°C, death at 25°C. Accounts of suffering from cold are found not only in the writings of Arctic explorers, but also in the accounts of people of more temperate lands who had the misfortune to be stranded in severely cold weather.

TABLE 13-7 Summary of human responses to thermal stress.

To Cold	To Hot
Thermoregulatory responses	
Constriction of skin blood vessels	Dilation of skin blood vessels
Concentration of blood	Dilution of blood
Flexion to reduce exposed body surface	Extension to increase exposed body surface
Increased muscle tone	Decreased muscle tone
Shivering	Sweating
Inclination to increased activity	Inclination to reduced activity
Consequential disturbances	
Increased urine volume	Decreased urine volume; thirst and dehydration
Danger of inadequate blood supply to skin of fingers, toes, and exposed parts, leading to frostbite	Difficulty in maintaining blood supply to brain leading to dizziness, nausea, and heat exhaustion; difficulty in maintaining chloride balance, leading to heat cramps
Increased hunger	Decreased appetite
Failure of regulation	
Falling body temperature	Rising body temperature
Drowsiness	Heat regulation center impaired
Cessation of heartbeat and respiration	Failure of nervous regulation, terminating in cessation of breathing

Altitude creates a special problem for the human body. Since the metabolism of humans depends in part upon the consumption of oxygen, a modification of the amount available has physiologic effects. The gas mixture forming the atmosphere remains in about the same ratio up to 75 km. However, the density of the gases diminishes, and with increasing altitude the partial pressure of oxygen (pO) decreases. At sea level, the partial pressure of oxygen is 212 mb; at a height of only 10 km, it is 55 mb.

The body's need for oxygen does not alter with altitude, with the result that a rapid change from sea level to higher altitudes can cause severe body stress. Even at moderate altitudes, the effects of *hypoxia* (oxygen deficiency) can cause nausea. Above 6 km, the lack of oxygen can seriously damage the brain.

Modern developments allow humans to cope with high elevations—the breathing apparatus of mountain climbers and the pressurized compartments of aircraft are examples—but there still remains the basic physiological problem. In this respect, much has been written about the physiology of people who live their entire lives at high altitudes and how they become acclimatized to the conditions. Long-term acclimatization to reduced pO conditions at high altitudes is exemplified by Andean Indians. It has been noted that they tend to be thickset, of medium height, with large hands and small wrists, disproportionately large chests, and well-developed legs. The squat nature of the body provides the best geometry for reducing heat loss, while the deep chest appears well adapted for obtaining oxygen at high altitudes.

Acclimatization to high altitudes by people native to sea level conditions may last months or even years. In some cases, the acclimatization may never occur. The symptoms associated with acclimatization are numerous and include sleeplessness, lowered threshold of pain, intestinal disturbances, loss of weight, and even psychological and mental disturbances. It should be noted that such ailments are also experienced by those accustomed to high-altitude living when they descend to lower altitudes. During the Spanish control of South America during the sixteenth century, so many Indians died when moved from mountains to sea level that an edict prohibiting forced movement was put into effect. Spaniards themselves also suffered from the effects of living at unaccustomed heights. It has been noted, for example, that for those who lived above an elevation of 10,000 ft, 53 years passed before a healthy child of Spanish parentage was born.

THREE

Climates of the Past and Future

CHAPTER 14: Reconstructing the Past
CHAPTER 15: Causes of Climatic Change
CHAPTER 16: Future Climates

14 Reconstructing the Past

SURFACE FEATURES
Evidence from Ice
Periglacial Evidence
Sediments
Sea Level Changes

PAST LIFE
The Faunal Evidence
The Floral Evidence

EVIDENCE FROM THE HISTORICAL PERIOD
Records of Floods
Records of Drought
Migrations
Contemporary Literature
Evidence of Agriculture and Settlements
The Instrumental Period

THE RECONSTRUCTED CLIMATE

SUMMARY

APPLIED STUDY 14
The Little Ice Age

CLIMATES OF THE PAST AND FUTURE

*U*sing the instruments of today, minor changes or trends in weather are routinely monitored and analyzed. However, the period for which weather instruments have been available is but a tiny fraction of time in earth's history. To understand current climates and to predict future climates, it is absolutely essential that they be considered in the framework of climatic change over geologic time. To achieve this, the climates of the past must be reconstructed. This process reflects much painstaking, detectivelike research because much of the evidence is based upon the relationship of climate to other environmental processes and past life. Prior to presenting an account of how climate has varied, this chapter deals first with the way in which climates are deduced.

SURFACE FEATURES

Evidence from Ice

Study of the processes that modify the earth's surface was the first indication that climates have varied over time, particularly in relation to the existence of ice ages. Perhaps the best-known early researcher in formulating these ideas was the Swiss scientist, Louis Agassiz, whose work began in the early 1800s.

Beginning this research in Switzerland, Agassiz noted that the valleys had a U-shape rather than the typical V-shape of river valleys. Other researchers had previously noted the same phenomenon and commented upon the fact that the U-shaped valley contained a relatively small river, which did not seem to fit the size of the valley. Their explanation was related to the biblical flood so vividly recorded in the Old Testament. It was assumed that the huge valleys must have been eroded by much larger streams when the flood occurred.

Agassiz was also impressed by the presence of large boulders set amid assorted finer sediments that had obviously been transported from an area remote from the location where they stood. These boulders, called *erratics*, aroused much curiosity, and their presence was again attributed to the great biblical flood. Some geologists, however, questioned this catastrophic viewpoint and maintained that the forces that caused the valley terrain and accounted for the huge boulders could only have resulted from the kinds of processes that could be identified at work in the present; that is, they assumed that present processes provided the key to what happened in the past. Accordingly, they believed that the large erratics scattered over many areas of the world had been dumped by ice; basing their ideas upon observed facts, they concluded that the erratics had been deposited by icebergs that floated on an extensive sea that formerly covered large areas of Europe. Using the same reasoning, they decided that the assorted finer materials in which erratics occurred had been derived from melting icebergs and applied the term *drift* to these assorted sediments.

At first, Agassiz was convinced that the iceberg theory best explained the erratics, but in his work he met other scholars who had different ideas. Scientists like Jean de Charpentier and Ignace Venetz had studied the landscape of the Swiss Alps and were convinced that the glaciers they saw had once been much more extensive and were responsible for the valley shapes, the drift, the erratics, and the parallel striations found on hard rock surfaces.

The accumulation of evidence of this type prompted Louis Agassiz to formulate a comprehensive theory of extensive glaciation. He suggested that a great ice sheet had extended from the North Pole to the Mediterranean and that the moraines, striations, erratics, and drift seen in Switzerland had resulted from the action of glaciers. The idea of an ice age was born. Nowadays it seems somewhat astonishing that not until the nineteenth century did the idea of major glaciation covering much of the earth become accepted.

The results of ice erosion and deposition are characteristics of large parts of the Northern Hemisphere. Figure 14-1 shows a typical view of an area that has

FIGURE 14-1 This view of a Swiss Valley illustrates the typical scenery associated with Alpine glaciation.

CLIMATES OF THE PAST AND FUTURE

experienced mountain (Alpine) glaciation. The resulting features—the U-shaped valleys, the hanging troughs, the aretes, and the tarns—are well known and clearly indicative of ice activity. Features associated with continental glaciation differ markedly from those of Alpine glaciation. Depositional features such as the position of terminal moraines and the erratic boulders when studied in detail allow reconstruction of the extent and movement of the ice. The erosional features, glacial striations, ice-gouged lakes, and so on provide similar evidence. Using such findings, it is possible to reconstruct conditions that existed in North America during ice advance. Figure 14-2 provides one interpretation.

While the results of the work of ice are an important interpretive device, glaciers themselves provide a measure of predicting temperature and precipitation conditions. The advance or retreat of glaciers has been used in interpreting climatic change over historic periods. There is evidence that glaciers shrunk about 3000 B.C. and that Alpine snow level was at least 1000 ft higher than today. By 500 B.C., there was a marked readvance followed by a recession. Between the seventeenth and nineteenth centuries, a general resurgence was observed in the Alps and Scandinavia. The twentieth century has tended to be a time of glacier retreat, although at

FIGURE 14-2 The maximum extent of the Pleistocene ice advance in North America. Note the reconstructed areas of climate equatorward of the ice (From *Climate and Man's Environment*, John E. Oliver, Copyright © 1973, John Wiley & Sons. Reprinted by permission of John Wiley & Sons, Inc.).

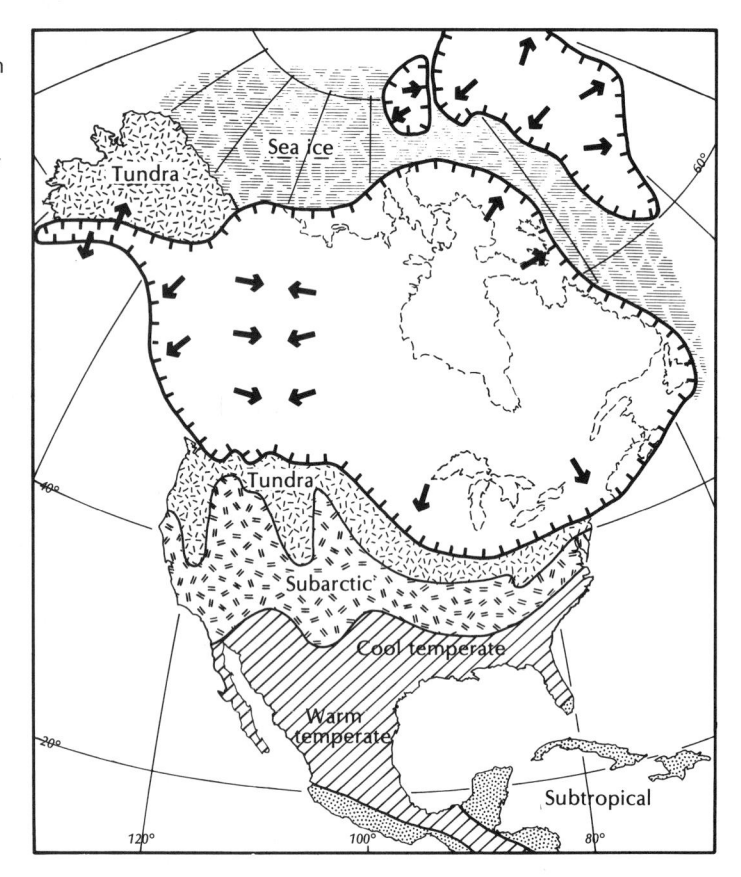

present we seem to be in an in-between stage, with some glaciers growing, others disappearing.

Periglacial Evidence

It is not, however, features associated with the ice itself that supply all geomorphic evidence regarding ice ages. Areas not directly affected by the great ice sheets experienced a climate quite different from today's. Some areas that are now quite dry experienced much wetter climates as a result of modified circulation patterns. Many inland basins, for example, have been occupied by large lakes as a result of the higher precipitation. Such *pluvial* lakes, so named because they resulted from increased precipitation in earlier times, have been widely identified.

The western part of the United States, particularly in the Great Basin area, shows fine examples of the extent of such lakes. In this area it is estimated that glacial lakes Bonneville and Lahontan were enormous. At its maximum, Lake Bonneville occupied some 50,000 km^2, an area approximately the size of Lake Michigan. Evidence of the extent of this great lake is found in the present Bonneville salt flats and in the *strand lines*, the shore areas indicative of the former level of the lake. Such lakes would necessarily modify the drainage systems, and the formation of a lake and its eventual overflow might lead to a totally new drainage direction.

Sediments

Sediments deposited during geologic time offer evidence of the climatic environment in which they were formed. Sedimentary beds themselves provide information: if a sedimentary rock is identified as having the same physical and chemical characteristics as one now being formed under, for example, arid conditions, then the sedimentary rock can be assumed to have formed in a similar way. Proof of this is offered by *evaporite* (salt) deposits.

Salt deposits can be formed when, on a long-term basis, evaporation exceeds precipitation where water is available from other sources. Water will evaporate to leave salts, formerly in solution, as deposited sedimentary rock. In interpreting the climate responsible for the great geologic evaporite beds of Texas, Germany, and the U.S.S.R., it is fortunate that similar deposits are being formed at the present time.

Sea Level Changes

Changes in the level of the oceans can occur either when the volume of water or the volume of the ocean basins decreases or increases. Many factors can cause either of these to occur; but most of them, such as the accumulation of sediments on the ocean floor or the extrusion of igneous rocks into the oceans, take an exceedingly long period of time. Rapid changes result mostly from the alternating accumulation and melting of ice. It has been suggested that if the present Antarctic ice sheet were to melt, there would be a worldwide increase in water sufficient to cause a sea level rise of 60 m (200 ft). The melting of ice from a large area would, however, be accompanied by an isostatic readjustment of the land; downwarping would occur in the oceans because of the weight of water added. Even allowing for this, melting of the Antarctic ice would still cause a rise of 40 m (135 ft), sufficient to flood most of the major ports of the world.

The rise and fall of sea level during the Pleistocene is an important guide to the periodic glacial and interglacial events. Submergence and emergence of coastal areas and marine terraces point to the amount of water tied up as ice; using appropriate dating methods, such evidence provides a guide to glacial advance and retreat.

PAST LIFE

The study of past life on earth provides much information about the conditions that once existed. Both plant and animal fossils have been widely used in the reconstruction of past ecologic conditions, and a great deal can be learned about the climates under which they lived.

The Faunal Evidence

Invertebrate fossils Fossils without backbones are widely used in establishing the geologic sequence of rock, and they have proved valuable for reconstruction of past climates. However, caution should be exercised because it is possible to draw biased conclusions from incomplete evidence. This restriction may apply particularly to the invertebrates because many fossil species lived in fresh or salt water; they were not directly affected by the atmospheric climate because the sea, lake, or river in which they lived acted as a buffer to direct exposure.

Isotopic temperatures While the physiology of fossil animals in relation to modern species is used to determine paleoclimates, much information has recently been gained from the chemistry of invertebrate organisms. Of particular note in this respect is the use of isotopes.

Oxygen has three isotopes, O^{16}, O^{17}, and O^{18}, with O^{16} (oxygen 16, the lightest) comprising 99.7 percent of all oxygen molecules. In 1947, H. C. Urey presented a paper on the rates at which the different isotopes go off when water evaporates. The ensuing discussion led Urey and later Emiliani to use the observations for determining water temperatures in which fossil organisms lived. The method depends basically upon the relative abundance of O^{18} in the organism; the amount depends upon the ratio of O^{18} to O^{16} in the water in which the animal lived, a function of the temperature of the water. Using a modern shell that had grown under known temperature conditions, isotope ratios could be established and standards acquired. An early experiment using a *belemnite* (a cigar-shaped fossil with concentric growth rings resembling a tree) gave significant results.

The study of isotopes has played a highly significant role in research completed by CLIMAP (Climate: Long-range Investigation, Mapping and Prediction).

CLIMAP investigates ocean-atmosphere changes over the past million years to obtain fundamental knowledge of what produces climatic variations and how the changes might be predicted. The keys to acquisition of CLIMAP data are cores taken from the layers of mud that cover ocean floors.

The layers that exist contain billions of microscopic skeletal remains of plankton. When the creatures die, their remains fall to the ocean depths to be incorporated in the mud layers. Since the tiny organisms are adapted to temperature, each species lives within a certain ocean climate; if that climate changes, they drift away, to be replaced by plankton better adapted to the new conditions. Thus the fossils found in the mud layers provide a record of temperature changes in the oceans—hence in the atmosphere above the ocean. To obtain a sample of the undisturbed layers, cores up to 30-m long are taken from the ocean floor. Analysis of the cores is a lengthy task, since 20–50 species and up to 500 individuals exist in each few centimeters of core.

Dating the time at which the organism was deposited is achieved through carbon 14 analysis. The basis of this technique is that skeletons contain both ordinary carbon and a minute trace of the isotope carbon 14. The proportion of carbon 14 to carbon 12 remains fixed while the organism is alive. After it dies, the carbon 14 begins to decay; by knowing the ratio of carbon 12 to carbon 14, one can determine the age of the shell.

Oxygen isotopes are also used in analyzing the cores. As already noted, the heavier oxygen 18 is mixed with ordinary oxygen 16. Oxygen 16 is found in greater proportion in snow; if a large amount of oxygen 16 is trapped in huge ice sheets, the amount of oxygen 18 in the oceans will rise. This process will be reflected in the fossil remains in the core. Thus when the ratio of oxygen 16 to oxygen 18 is lower, it represents time of ice advance—an ice age. Using this and other information, CLIMAP researchers produced a map (see figure 14–3) of the climates of earth as it might have been in August 18,000 B. P. (*B. P.* is the abbreviation for Before Present, where the year 1950 is used as the base.)

Vertebrate animals These fossils provide important clues to past climates. Much can be interpreted from their fossil distribution and their physiology as related to environment. The great changes in vertebrate life over geologic time result in quite different interpretive methods. (Refer to table 1–4 for an outline of geologic time and the geologic ages named in the following account.)

The Devonian of the Lower Paleozoic is sometimes referred to as the "Age of Fishes" because they were one of the dominant life forms of that time. Many of the fish were air-breathing, similar to the modern lungfish. Since the conditions under which modern lungfish live are known, by analogy it might tentatively be suggested that the ancient lungfish lived under similar conditions—namely, a warm, wet/dry seasonal climate. Amphibian development, with fossil remains over widely spaced geographic areas, would tend to confirm a widespread warm climate. Such warm conditions probably continued through the Carboniferous. Cold-blooded amphibians and reptiles, unless specially adapted, require warm air temperature to maintain their body temperatures. Again, the cosmopolitan distribution points to widespread warmth for much of the period.

CLIMATES OF THE PAST AND FUTURE

FIGURE 14-3 Computer depiction of world surface features 18,000 years ago (upper) and today (Source: EDS [NOAA]).

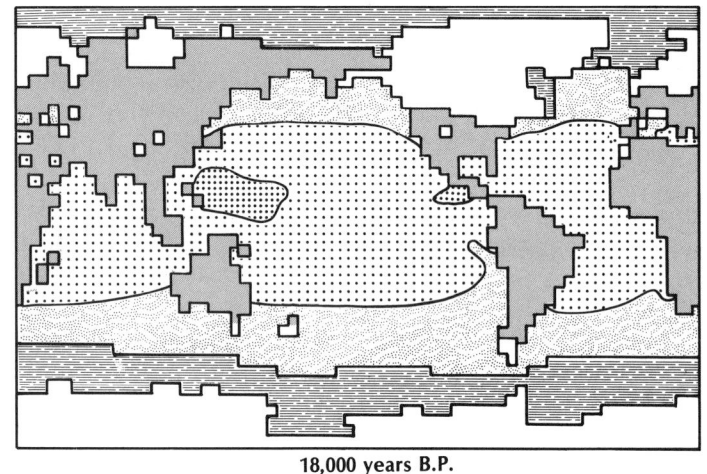

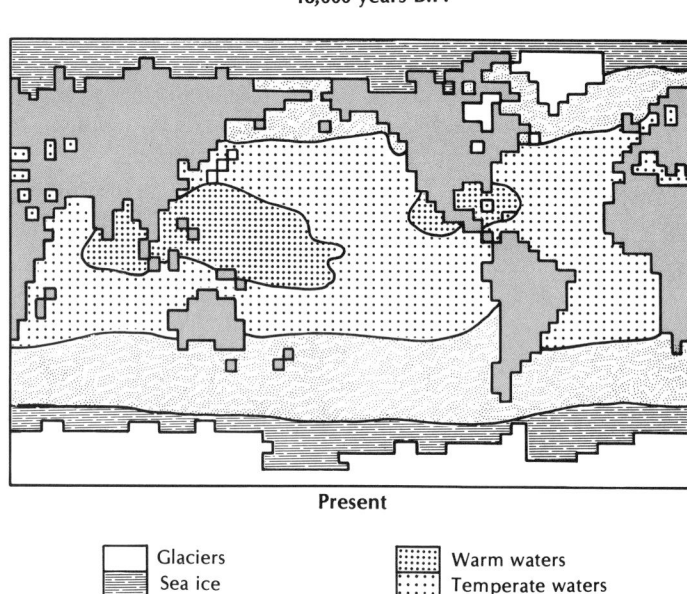

18,000 years B.P.

Present

Glaciers
Sea ice
Land
Warm waters
Temperate waters
Cool waters

The Permian, representing the rapid emergence of reptiles with dinosaurs preeminent, is an interesting period for the use of vertebrates to establish past climates. The reptilian method of reproduction—the amniote egg—allowed reptiles to live away from water sources and become much more independent of land-water distribution than the amphibians. This is not to say, however, that such animals could exist without water. The wide distribution of reptiles in the Permian would seem to indicate that water was available on the land and that conditions were similar over broad areas of the globe.

The extinction of the dinosaurs in Cretaceous time has promoted much discussion; a widely held view is that progressive and increasing world aridity might have been the cause of their extinction. There are many reasons to question such an interpretation, not the least of which is that while other animals became extinct (for example, belemnites and ammonites), related reptiles (notably crocodilians) did not. Note, too, that many plants, which are usually more responsive to changes of climatic regime, did not pass from the face of the earth. There are a number of other theories to explain the demise of the dinosaurs. One, for example, invokes temperature rather than moisture change. It has been suggested that changing temperatures might have influenced sperm production in the dinosaurs; because they could not reproduce, they became extinct.

While the physiology of vertebrates is probably the most widely used method for interpreting the ecologic conditions under which they lived, important evidence is also offered by the way in which they are fossilized. Some are found, for example, "right side up," in a standing position. To explain their death, it might be assumed that they became bogged down in a swamp environment; by relating such evidence to surrounding deposits and other indicator fossils, it becomes possible to reconstruct the environments in which they lived. The fact that many fossil remains are found close together indicates death of animals through a catastrophe. Obviously, the catastrophe can take various forms—freezing or drought, for example—but by correlation to other past climatic indicators, its nature might be deduced.

The Floral Evidence

Plant distribution provides an important guide to the distribution of climate at the present time; the same is true of paleoclimates. Identification of vegetation patterns and their changes over time is widely used to interpret past climates. Often the evidence is used in relation to other environmental features. For example, it has already been noted that the extent of mountain glaciers varies. A change will be reflected in mountain vegetation, particularly the elevation of the tree line on the mountain. This feature can be traced over time and has been used to evaluate climatic trends in selected areas.

The physiology of plants, like that of animals, again provides much information. The development of drip leaves in plants is indicative of their existence under very moist conditions; the fossil remains of plants with thick, fleshy leaves are probably indicative of arid or semiarid climates.

The interpretation of the climate of the Carboniferous, a time of prolific vegetation that gave rise to great thicknesses of coal measures, provides examples of this use of plant physiology. Many of the fossil plants in the Carboniferous appear to be related to the horsetail and the club mosses, both representative of a marsh or swamp environment. Such an interpretation is endorsed by fossils that suggest plants with layered roots such as those found in modern bog plants and by many minor structures that appear to indicate that some of the plants actually floated on water. Trees lacked a development of growth rings, indicative of a climate without marked seasonal differences; the dominance of trees over herbaceous plants would further indicate a swamp environment. In all, the representative vegetation suggests a warm, moist climate that favored a luxurious, if wet, plant cover.

Similar evidence has been used to help reconstruct the climates of late Paleozoic and Mesozoic rocks. More recent deposits can be interpreted by pollen analysis; for much more recent plant distributions, tree ring analysis is used.

CLIMATES OF THE PAST AND FUTURE

Pollen analysis The study of pollen grains, or spores, is termed *palynology*. Its success depends upon the fact that many plants produce pollen grains in great numbers (e.g., a single green sorrel may produce 393 million grains; a single plant of rye, 21 million grains) and that pollen is widely distributed in the area in which plants are found. Most important, the outer wall of the pollen grain is one of the most durable organic substances known. Even when heated to high temperatures or treated with acid, it is not visibly changed. This is important, for pollens possess morphological characteristics that allow identification of groups above the species level.

For pollen to be of value in interpreting the past distribution of vegetation and hence inferring the climate that occurred, it is necessary to obtain a layered sequence of the pollen. As shown in figure 14-4, this often occurs in ancient lakes or peat bogs where seasonal pollen deposits would be covered by sediments. Cores taken would show a sequenced pattern.

Pollens from the cores are identified and a frequency distribution of plant types derived. Thus, a high proportion of spruce pollen in the lower core might give way to oak pollen at higher levels. This variation would indicate that a vegetation change had occurred over time and that the difference could be related to a passage from cool to warmer climatic conditions. Even a relatively crude classification of pollen type (for example, those from trees compared with those from other plants) could provide a rough guide to changing climatic conditions. The change from pollen associated with the nontree climates of the cold tundra to tree climates might indicate an amelioration of climatic conditions. In-depth statistical counts obviously provide a good deal more detail. Much work in this has been completed in Scandinavia, where the first palynological stratigraphy was devised.

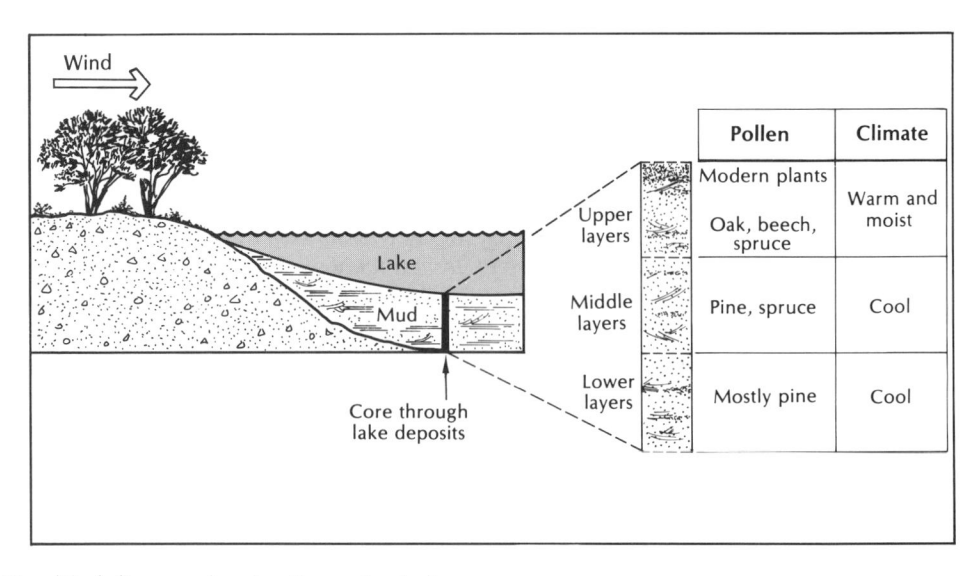

FIGURE 14-4 Simplified diagram showing the method of reconstructing past climates using pollen analysis (From J. E. Oliver [1981, p. 223]. Used by permission of V. H. Winston and Sons.).

302

Despite the important progress using this method, it does have shortcomings. A vegetation cover only attains maturity after a fairly lengthy period of time, and it is quite feasible that the vegetation established through pollen analysis represents a successional stage that is not totally representative of the prevailing climate. In some areas, the vegetation cover is mixed; and it becomes difficult to establish any dominant type that can be related to climate. It has also been pointed out that from Neolithic times, people have interfered extensively with the forest cover, and human-induced changes might well give misleading results.

Dendrochronology Tree ring study was pioneered by A. E. Douglas and his colleagues at the University of Arizona. Initial studies were used in an attempt to relate seasonal growth of trees to sunspot cycles, and a great deal of significant work was completed. In the quest for ancient living trees, the *Pinus aristata* was found to be 4000 years old. Analysis of such ancient trees permitted reconstruction of climates of the American southwest during various settlement periods.

Tree ring analysis depends upon the fact that growth rings record significant events that happened during the life history of the tree. Growth rings are formed in the xylem wood of the trees. Early in the season, the xylem cells are smaller and darker. The abrupt change from light- to dark-colored rings delineates the annual increments of growth. Study of these rings, their size and variations provides information about the varying environmental conditions to which the tree was subjected and provides a field of investigation known as *dendrochronology*. The method is, of course, most valuable in determining conditions that existed in a relatively recent part of geologic history and is widely used in archeological research.

EVIDENCE FROM THE HISTORICAL PERIOD

Many researchers have used historical records to establish climatic changes that have occurred during the brief existence of the human species on earth, and their findings have allowed fairly detailed reconstruction of climates over the past 6000 years. The type of evidence used is highly variable; the following merely provides some guidelines to the methods.

Records of Floods

In relation to the significance of floods, it has been noted that a deluge such as that described in the Book of Genesis occurs in the legends and folklore of almost every ancient people. Such evidence cannot, of course, be used to substantiate flooding on a world scale; but many writers have related the common theme to great changes in sea levels that occurred in former times. More scientific evidence of climates can be derived from Egypt, where careful records of the Nile floods have been kept for many thousands of years. By relating the occurrence and height of floods, it is possible to reconstruct the pluvial conditions in the source areas of the Nile.

A recent study of historic Nile floods for the period between 640 A.D. and 1921 showed that very low floods occurred in the years 930–1070 and 1180–1350.

The highest flow periods were 1070–1180 and 1350–1470. By comparing these flow periods with lake levels in Lake Chad, a high correlation between the Nile discharge and rainfall in equatorial East Africa was revealed. Interestingly, the time of low discharge in the Nile showed an apparent correlation to cold climates in Europe.

Records of Drought

The periodic effects of severe drought have frequently been used in reconstruction of past climates. As an example, the growth and abandonment of large dwellings in the southwest United States have been attributed to drought. The abandoned dwellings of Chaco Canyon and Mesa Verde show that they supported quite dense, highly active communities. By 1300 A.D., large settlements had become deserted. Reasons other than drought have been suggested to account for the abandonment, but tree ring analysis shows that between 1276 and 1299 A.D., practically no rain fell in the areas. Droughts such as this have been used to explain mass migration to less arid areas.

Migrations

The mass migration of peoples from drier to wetter areas has certainly occurred over history. The interpretation of the causes for this is problematic. The historian Arnold Toynbee, for example, suggested that the outpourings of Arabs under Mohammed were climatically induced, as was that of the Mongols under Genghis Khan. This view has been countered by many writers who suggest that it is a simplistic explanation.

Careful research has, however, indicated that some pre-Columbian Indians of the Great Plains did migrate as a result of precipitation variations. A study by archeologists and climatologists showed, for example, that Indians of the Mill Creek culture of Iowa deserted what appeared to be a thriving community about 1200 A.D. Through careful investigation of the evidence left by the culture, Bryson and Baerreis were able not only to reconstruct the fact that the precipitation had declined but to construct a climatic map of the conditions that had prevailed.

Contemporary Literature

Medieval chronicles contain many references to prevailing weather conditions. Unfortunately, although not surprisingly, these pertain to exceptional weather events rather than day-to-day conditions. These events include such unusual phenomena as the freezing of the Tiber in the ninth century and the formation of ice on the river Nile. While these sources do not supply a continuous record, it is possible to obtain an overall view of the usual conditions by assessing the number of times given events are recorded. The freezing of the river Thames provides one such example. Between 800 and 1500 A.D., only one or two freezings per century are recorded. In the sixteenth century, four freezings are known; in the seventeenth, there were eight; while six are known to have occurred in the eighteenth century. One can only suppose that a progressive cooling caused the frequency to increase.

Such chronicles also provide insight into climate through inference of other factors. William of Malmesbury, writing about the Gloucester region of England in 1125, stated

> [the area] exhibits a greater number of vineyards than any other county in England, yielding abundant crops of superior quality. . . . they may also bear comparison with the growths of France.

Since the fifteenth century there has been no wine industry in England, and it might be supposed that this is related to a worsening of the climate since those times.

Contemporary literature has aided in reconstruction of one of the most complete sequences of climate over the last thousand years. The data for Iceland (figure 14-5) are essentially based upon the following sources:

1864 to present	Actual meteorological instruments
1781–1845	A reconstruction of weather conditions as derived from the relative severity and frequency of drift ice in the vicinity of Iceland
1591–1780	Historical records combined with incomplete drift ice data
900–1590	Information from Icelandic sagas indicating times of severe weather and related famines

Such a complete record as this can be used as a base guide to the climate of the entire North Atlantic area.

Icelandic Temperatures

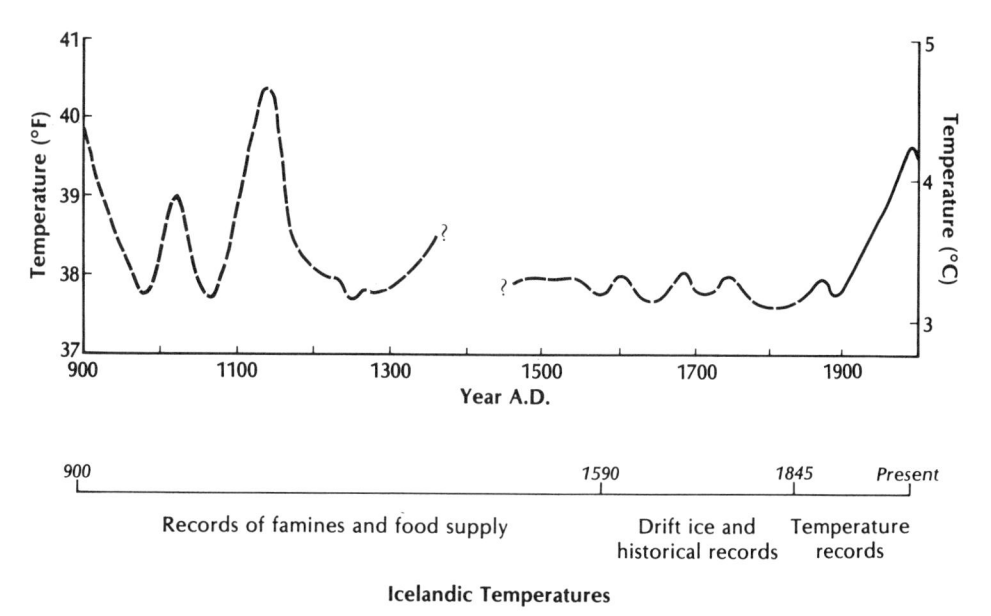

FIGURE 14-5 Among the most complete sequences of temperature over the last 1000 years are those reconstructed for Iceland.

CLIMATES OF THE PAST AND FUTURE

Evidence of Agriculture and Settlements

Apart from documented evidence such as that described above there is considerable physical evidence of agricultural activity, of a given type, in areas where it can no longer be practiced. Perhaps the best-known type of evidence in this respect is the existence of irrigation systems that are no longer in use; it has been pointed out, however, that cultural rather than physical differences could account for their falling into disrepair. Evidence of agriculture, together with other indicators, has been used to reconstruct the climates that existed at the time Greenland was settled by the Vikings. Under the leadership of Eric the Red, the Vikings passed from Iceland, which they had settled in the ninth century, to Greenland. While an icy

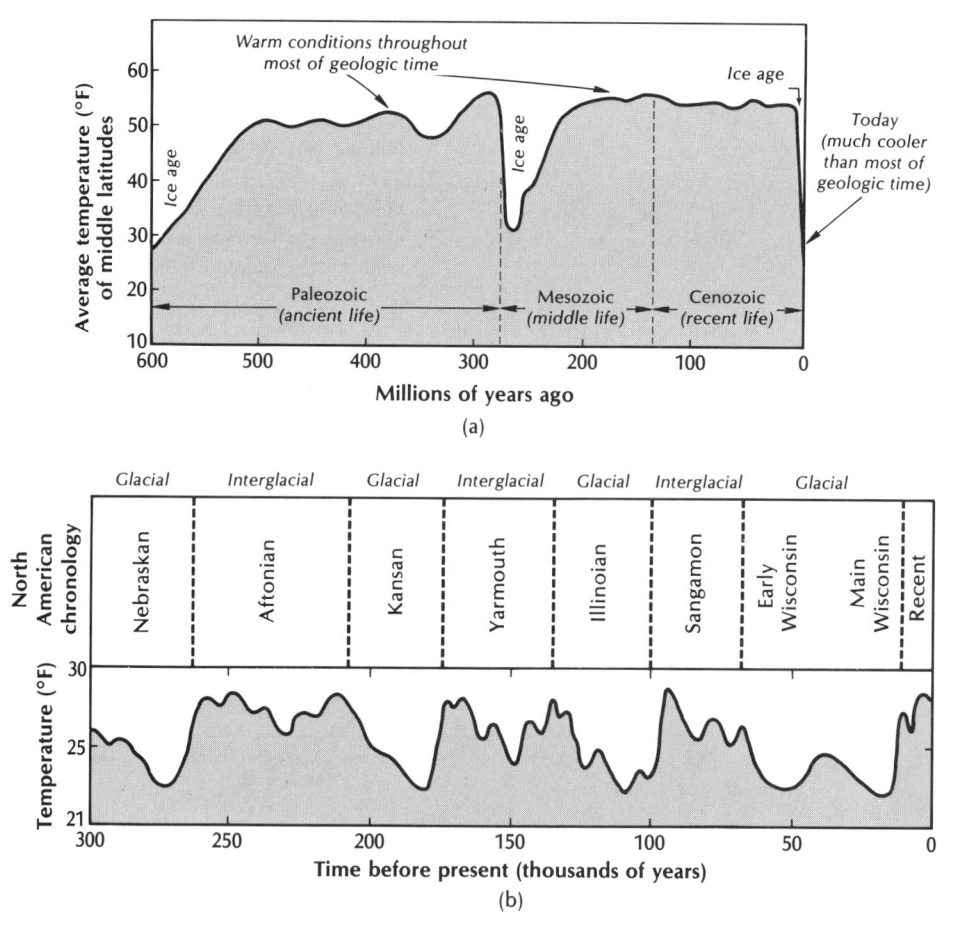

FIGURE 14-6 Reconstructed temperatures for various time intervals: (a) The last 600 million years; note that the most recent cool period—the Pleistocene—is barely distinguishable at this scale. (b) Enlargement of the graphical scale shows that the Pleistocene consisted of both glacial and interglacial times. Names apply to North American glaciation.

land, it supported sufficient vegetation (dwarf willow, birch, bush berries, pasture land) for settlement. Two colonies were established, and farming was permitted. The outposts thrived, and regular communications were established with Iceland. By 1250, Greenland was practically isolated from outside contact, with extensive drift ice preventing the passage of ships. By 1516 the settlements had practically been forgotten, and in 1540 a voyager reported seeing signs of the settlements, but no signs of habitation. The settlers had perished.

Whether this was due to the deteriorating climate or to invasions of other peoples is not known, although a Danish archeological expedition to the sites in 1921 found evidence that deteriorating climate must have played a role in the demise of the people. Graves were found in permafrost that had formed since the time of burial. Tree roots entangled in the coffins indicated that the graves were originally in unfrozen ground and that the permafrost had moved progressively higher. That food supplies had been insufficient was shown by examination of the skeletons found; most were deformed or dwarfed, and evidence of rickets and malnutrition was clear. All the evidence points to a climate that became progressively cooler, leading eventually to the isolation and extinction of the settlers.

The Instrumental Period

Accurate measurement of climate variables could not occur until instruments were devised to measure them. Although some indication of prevailing weather conditions was accumulated by Greeks and Egyptians (e.g., Hippocrates' *Airs, Waters, and Places*—400 B.C., Aristotle's *Meterologica*—350 B.C., Ptolemy's diaries of local weather—first century A.D.) precise observations really began with the invention of the thermometer by Galileo in 1593 and the barometer by Torricelli in 1643. Some weather data, particularly in Europe, were collected as early as 1649. As late as 1800, however, there were few weather stations, with 12 in Europe and 5 in the United States. The great impetus for collection and use of weather data came with the invention of the telegraph in the 1830s; since that time, more and more data have become available (see chapter 1).

THE RECONSTRUCTED CLIMATE

Synthesis of the evidence derived from the methods outlined has allowed the essential details of climatic change over millions of years to be reconstructed. In figure 14–6(a), the generalized temperature that prevailed for the last 500 million years is shown. It is seen that average world temperatures have been appreciably higher than those of today. These warm temperatures have been interrupted, at periods of about 250 million years, by dramatic declines in temperature, declines that produced the ice ages. We are living in one of these cooler periods of time, a cooling that began at the onset of the most recent ice age, the Pleistocene.

The Pleistocene, figure 14–6(b), was not merely a time when ice advanced and retreated but consisted of a number of ice advances (*glacial* times) separated one from another by periods of relative warming (*interglacial* times). The graph shows four basic ice advances that have long been recognized; recent work suggests this is a simplification of what actually occurred. The interglacials themselves experienced fluctuating temperatures, with times of relative mild conditions punctuated by cooler conditions. At its maximum extent, the ice cover of the earth was three times greater than that of today. At the present time ice sheets cover 3 percent of the earth's surface: at the maximum glacial advance, they occupied 9 percent (table 14-1).

The climate since the retreat of Pleistocene ice has varied and may be represented by a series of graphs derived from various types of evidence. Figure 14-7 provides a generalized curve for the past 15,000 years based upon evidence derived from a variety of methods. Widespread deglaciation began abruptly about 14,000 years ago, and most of the ice had melted by 10,000 years ago.

Subsequent to the ice melt, temperatures of the Northern Hemisphere rose slowly until by about 2000 B.C. they attained their highest temperatures since the onset of the ice age. This period is called the *climatic optimum* or *thermal maximum*.

TABLE 14-1 Characteristics of existing ice sheets and of the maximum quaternary ice cover.

	Area (10^{12} m²)
Existing glaciers	
Greenland	1.80
Spitsbergen and Iceland	0.07
Canadian Archipelago	0.15
North America	0.08
Europe and Asia	0.17
South America	0.03
Antarctica	12.59
Total area	14.99 (3% of earth's surface)
Total ice volume*	2.5×10^7 km³
Equivalent sea-level change	70 m
Maximum quaternary glaciation	
Greenland	2.30
Spitsbergen and Iceland	0.44
Alaska	1.03
Cordillera	1.58
Laurentide	13.39
Scandinavia	6.67
Europe	0.09
Asia	3.95
South America	0.87
Antarctica	13.81
Other	0.04
Total area	44.17 (9% of earth's surface)
Total ice volume*	7.5×10^7 km³
Equivalent sea-level change	210 m

Source: After Flint (1971).
*Based on the present ice thickness of 1700 m in Greenland and Antarctica

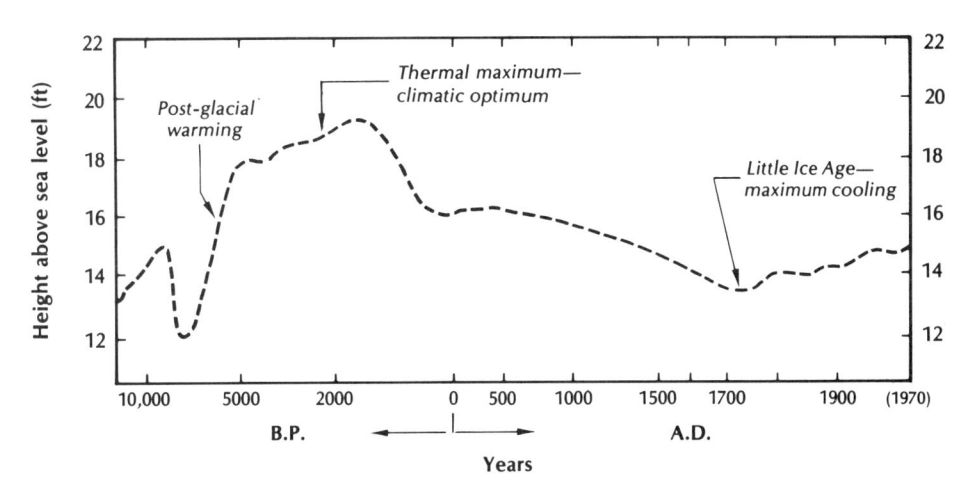

FIGURE 14-7 A reconstruction of temperature conditions over the last 10,000 years. This depiction is obtained from varying heights of the snow lines above sea level in Northern Europe.

Thereafter, temperatures declined, reaching a minimum in the Little Ice Age. This cooling is well displayed in the series of graphs in figure 14-8. Each of the graphs illustrates a climatic record derived from different evidence. Essentially they show (1) the record of oxygen isotopes in ice cores from the Greenland ice cap; (2) the Icelandic record already described; (3) 50-year means of observed and estimated temperatures in central England.

While there is not a perfect match between these reconstructions, they do clearly show periods of relative warmth and cooling. Between 1100 and 1400, a period of somewhat milder conditions prevailed—but this was followed by a fluctuation that eventually led to the Little Ice Age (Applied Study 14).

A view of temperature variations over the past century can be obtained from meteorological data. Figure 14-9 provides a reconstruction of annual mean temperatures of the Northern Hemisphere. The index suggests that a warming trend began in the late 1800s and continued up to about 1940. Since that time the mean temperatures of the Northern Hemisphere have declined.

The data in the figure, as averages, mask their geographic variability. Largest changes in temperature have occurred in polar regions of the Northern Hemisphere, with the tropics experiencing minimal change.

Of much interest to climatologists is the dynamic explanation of the atmospheric conditions that occur during periods of different climates of the past. To be able to explain why different events actually took place requires an understanding of circulation patterns. Obviously, given the incomplete data base, this is a very difficult task; however, much research is continuing in the field, and models of circulation have been constructed. Examples of this type of research are presented in chapter 16.

FIGURE 14-8 A comparison of conditions using three interpretive methods. Shaded parts of each graph indicate warmer times.

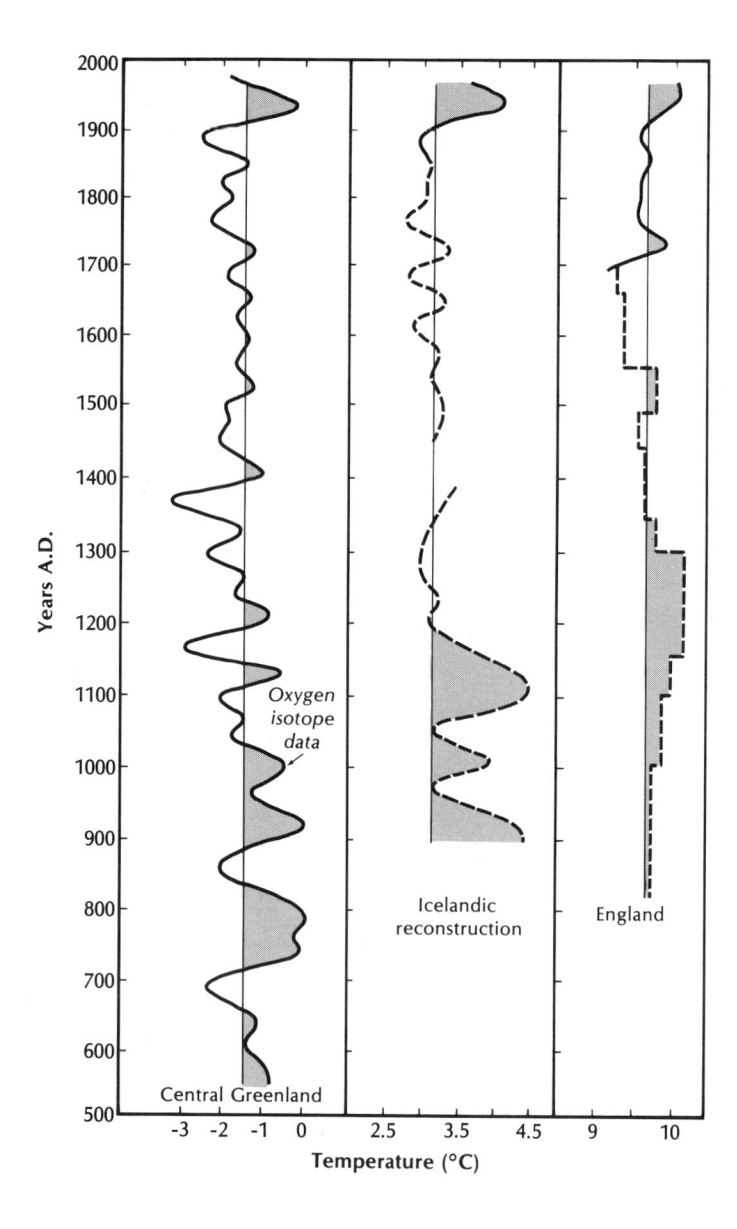

FIGURE 14-9 Recorded changes of annual mean temperature of the Northern Hemisphere (Source: EDS [NOAA], based upon data by Budyko and Asakura).

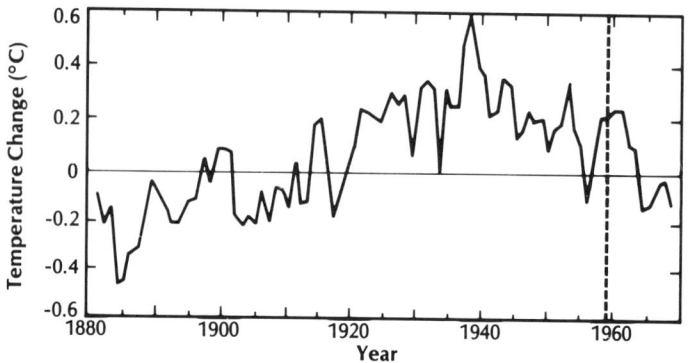

SUMMARY

In obtaining data for reconstructing climates of the past, the impact of climate on other environmental features is used. By relating the role of climate to fossil life, sediments, and surface erosion and deposition, it is possible to deduce the climate that prevailed. Modern research, however, has added important new methods to the reconstruction, with isotope analysis being of major significance.

The actual reconstructed climate shows that throughout much of geologic time, the earth experienced warmer conditions than exist today. These warm conditions have been periodically interrupted by cooling periods. The current climate on earth is representative of a cool time following an ice age.

APPLIED STUDY 14

The Little Ice Age

Since the retreat of the ice cover associated with the last great continental glaciation, climates of the earth have fluctuated. Periods of relative warmth have been followed by cooling, to be again followed by a warmer time. Each of the identified warm and cool periods has been named, with perhaps the best known being the Little Ice Age. The precise meaning of this term changed over time but it is now used exclusively to describe a cool period that began in the fifteenth century and lasted until the middle part of the nineteenth century. While precise dates do vary, we can consider the Little Ice Age as extending from 1430 to 1850.

There is both physical and historic evidence to support the cooling. Significant physical evidence comes from the advance and retreat of glaciers in Europe. During the period, glaciers enlarged to the extent that some Alpine villages were overwhelmed by ice. The photographs in figure 14-10 provide visual evidence of how severe the threat was. The figure on the top is from an eighteenth century engraving of the scenery near Chamonix in the French Alps. The second is a photograph taken from the same vantage point 120 years later. The change is remarkable.

History is replete with vivid descriptions of the climate that occurred during this time. It marked the end of the Norse settlements in Greenland that had begun in the tenth century. In fact, in 1492 the Pope complained that none of his bishops had visited the Greenland outpost for 80 years because of ice in the northern seas. He was not aware of the fact that the settlements were gone. In Europe itself, rivers froze and grape vines were killed by frost. In North America, the soldiers of the American Revolution suffered in the cold weather, although at times the ice was used as a resource. British troops, for example, were able to use a frozen river to slide their guns across the ice from Manhattan to Staten Island.

CLIMATES OF THE PAST AND FUTURE

FIGURE 14–10 Changes in mass of a mountain glacier near Chamonix, France (Source: EDS [NOAA]).

It is from North America that there comes a well-known account of life during the last years of the Little Ice Age—a description of the year 1816, known as "the year without a summer." The year began with excessively low temperatures across much of the eastern seaboard. But as spring came, it seemed to be normal—cool but not excessively so. In May, however, the temperatures plunged; in Indiana there was snow or sleet for 17 days, which killed off seedlings before they had a chance to grow. The cold weather continued in June, when snow again fell to devastate any remaining budding crops totally. It is reported that no crops grew north of a line between the Ohio and Potomac rivers, and scanty returns were obtained south of this line. In the pioneer areas of Indiana and Illinois, the lack of crops meant that fishing and hunting had to supply the food. Reports suggest that raccoons, groundhogs, and the easily

FIGURE 14-11 *The Hunters in the Snow* by Pieter Brueghel the Elder (Kunsthistorisches Museum, Vienna).

trapped passenger pigeons were a major source of food. The settlers also collected many edible wild plants, which proved hardier than cultivated crops.

During the period of the Little Ice Age, many famous historical events took place. Colonial expansion from the European base, the Black Death, the French and American revolutions, and the Napoleonic Wars are a few examples. That such events should occur during a cold time has led to many speculations about the role of climate in these events. In many cases, the relationship has been examined deterministically, suggesting that climate may have been the causal element underlying the event. In no way can this be proved. From a climatic standpoint, it is perhaps more meaningful to suggest that the prevailing climates may have contributed, in a few instances, the trigger required for social upheaval. A case in point is the French Revolution in 1789. The basis for the revolution had been laid long before the devastating effect of climate on wheat yields in northern France in 1788. From the bad harvest that resulted came the bread riots that preceded the revolution. Climate did play a role, but it was not the causative factor.

CLIMATES OF THE PAST AND FUTURE

The Little Ice Age provides the historical climatologist a most interesting period to study, and many different observations have been made. The image of the period shown by contemporary artists is one different from that of today. Paintings of scenery and activities in the Low Countries show winter scenes in which ice and snow are central to the theme. One famous painting by Pieter Brueghel the Elder, shown in figure 14-11 (page 313), provides an excellent example of this image.

15 Causes of Climatic Change

THE NATURAL CHANGES
Earth-Sun Relationships
Solar Output Variation
Atmospheric Modification
Distribution of Continents
Variation in the Oceans
Other Theories

CLIMATIC CHANGE: THE HUMAN IMPACT
The Role of Surface Changes
Atmospheric Changes

THE CLIMATE OF CITIES
The Modified Processes
The Observed Results

SUMMARY

APPLIED STUDY 15
Acid Rainfall: A Change in Quality

From all of the evidence that is available, it is clear that climate has changed over time. Given this fact, it is natural that the possible cause of such change has been the basis of much study. The question of what causes climates to change has not been completely answered. This, however, is not because of a lack of theories. In the 100 years or so since the magnitude of climatic change was realized, it has been estimated that for every year that has passed, a new theory has been postulated. Not all of these have been satisfactory, for some have failed to account for the two basic ingredients of a viable explanation. They must (1) explain the onset of ice ages through geologic time, that is, account for long periods of warmth with cooler interruptions; and (2) account for the warming and cooling periods that occur within an ice age. Not all theories can do this, especially in quantitative terms.

The search for the explanation of climatic change has become increasingly important in recent years. Human activities have resulted in a modification of both the atmosphere and the earth's surface to such a degree that the changes are now an integral part of explaining climatic variations. Given this fact, this chapter deals with climatic change in two parts. First, what may be termed the natural causes are considered. This discussion includes theories that explain changes that occurred long before the human habitation of the earth. Second, the role of people in inadvertently modifying climate is examined. Obviously, the two causes are not independent, for natural changes are still taking place while people are modifying the system. However, once the two components are dealt with, they can be combined in an attempt to predict what will happen in the future (see chapter 16).

THE NATURAL CHANGES

The basic reason for climatic changes on earth is essentially very simple. Change must be related to the flows of energy into and out of the system and the ways in which energy is budgeted within the earth-ocean-atmosphere system. Unfortunately, the explanation of the flows and interchanges is very complex and requires an examination of all parts of the system. Because of this complexity, it is convenient to divide theories of climatic changes into a number of categories. These include the following factors:

- ☐ Variations in earth-sun relationships
- ☐ Variations in energy output by the sun
- ☐ Atmospheric variations modifying the flow of energy
- ☐ Changes in the position of continental land masses
- ☐ Variations in heat stored in the oceans

Each of these is considered separately.

Earth-Sun Relationships

Variations in the earth's motion around the sun explain diurnal and seasonal differences in the amount of solar energy arriving at the surface (see chapter 2). However, the angle of the earth's axis and the distance of the earth from the sun are not constant values as they vary over time. The actual orbital variations that occur are discussed in the following sections.

The obliquity of the ecliptic This term refers to the angle of the axis in relation to the plane in which the earth revolves around the sun. At the present time the angle is 66.5°, which gives an obliquity angle of 23.5°. This angle is not constant; on a cycle of a period of about 41,000 years, the angle varies some 1.5° about a mean of 23.1° (figure 15-1[a]). The effects of changing obliquity are illustrated in figure 15-1(b). The top diagram shows the present position, and the lower some hypothetical cases. An obliquity of 0° would lead to equal lengths of day and night over the globe to result in a lack of seasonal changes, which would cause well-defined climatic zonation. Another extreme is an angle of obliquity of 54°. Such an angle would produce great extremes in the lengths of summer and winter days and nights. For example, at the December solstice position shown in the diagram, much of the

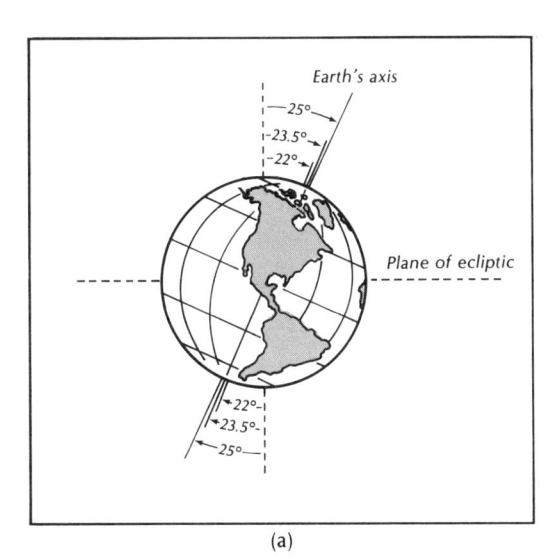

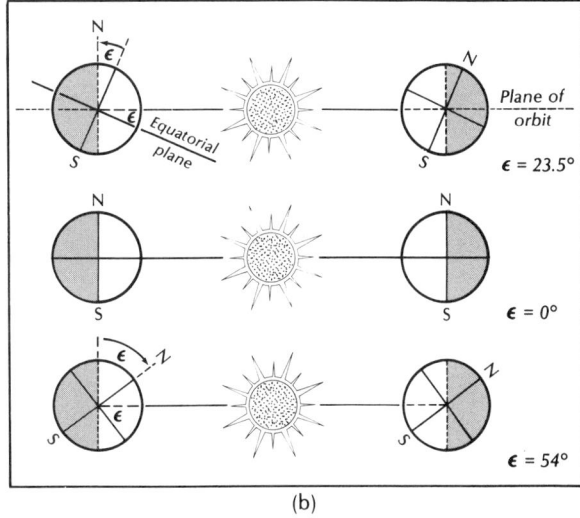

FIGURE 15-1 Obliquity of the ecliptic (a) The amount of tilt of the earth's axis changes between 22° and 25° over a 41,000 year period. The present value is 23.5° (From *Weatherwise*, 35 (3) June 1982, p. 112, a publication of the Helen Dwight Reid Educational Foundation); (b) In the three cases shown here (present angle, 0°, and 54°), the earth's climate would change appreciably (From *Climate and Man's Environment*, John E. Oliver, Copyright © 1973, John Wiley & Sons. Reprinted by permission of John Wiley & Sons, Inc.).

Northern Hemisphere would have 24 h of darkness. Extreme temperature differences would occur from summer to winter. Although the actual changes in the angle of obliquity are not as large as these examples, they are sufficient to cause distinctive changes in the zonation of earth's climates.

Earth's orbital eccentricity The earth moves around the sun in an elliptical orbit; the eccentricity of the orbit is derived by comparing the path to that of a true circle (figure 15-2[a]). Currently the orbit is relatively close to a circle: its eccentricity, measured by the method shown in figure 15-2(b), is 0.017. Over the past million years, this value has changed from almost circular ($e = 0.001$) to an extreme value of $e = 0.054$. This change influences the amount of solar radiation intercepted by the earth and also modifies the dates at which the solstices and equinoxes occur. This factor is used to derive the precession of the equinoxes.

Precession of the equinoxes Because of varying earth motions, the days on which the earth reaches its closest and most distant point from the sun (*perihelion* and *aphelion*, respectively) change over time (figure 15-3). At present, perihelion occurs during the Northern Hemisphere winter; in 10,500 years the date of perihelion will pass to the Northern Hemisphere summer season. The present situation will be reversed, and lowest input of solar radiation will correspond to winter in the Northern Hemisphere, the land hemisphere, in which extreme cold temperatures are attained.

FIGURE 15-2 Changes in earth's orbit around the sun: (a) Over time, the earth's orbit changes from elliptical to almost circular (From *Weatherwise*, 35 (3), June 1982, p. 113, a publication of the Helen Dwight Reid Educational Foundation); (b) Eccentricity of the orbit. Two foci are assumed: the sun and the point F. C is the center of the distance from the earth's position at aphelion and perihelion. Eccentricity is measured by dividing the distance *l* by the distance *a* ($e = l/a$) (From *Climate and Man's Environment*, John E. Oliver, Copyright © 1973, John Wiley & Sons. Reprinted by permission of John Wiley & Sons, Inc.).

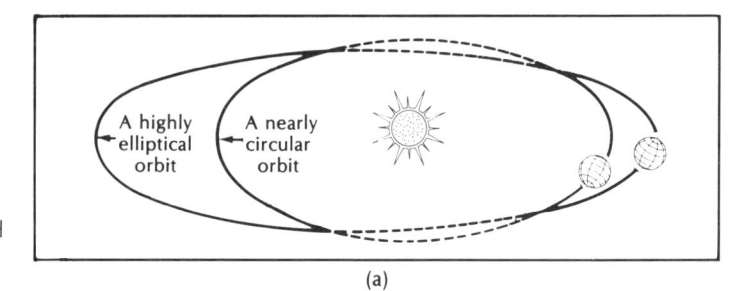

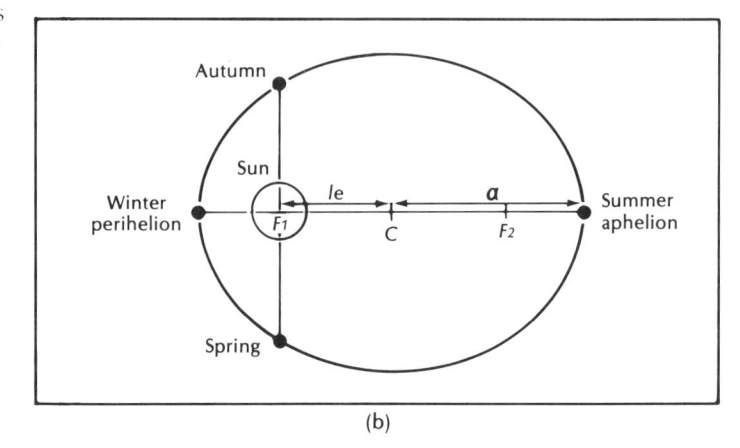

FIGURE 15-3 Precession of the equinoxes occurs as the dates of solstices and equinoxes move along the orbit.

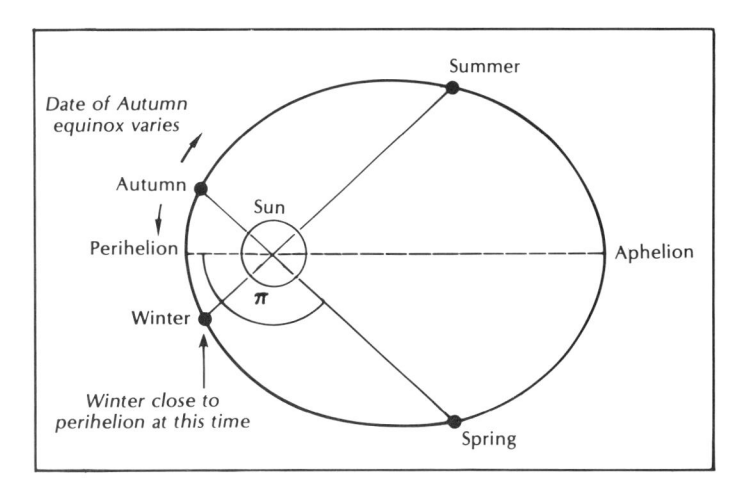

The dates at which these events occur can be calculated. It is possible to reconstruct the times at which the various cycles of change reinforce one another to produce very high or very low radiation values within a hemisphere. In fact, long before the advent of high-speed computers, the Yugoslavian scientist, Milutin Milankovitch, actually derived values going back thousands of years. Although his derived values provide the basis for understanding the relationships, more recent workers have constructed models of the cycles and their relative weightings in influencing climate. Currently, many climatologists think that these orbital variations are at the base of climatic change and have postulated future conditions from the research findings. Using the orbital variation theory, it is suggested that a long-term cooling trend, which began some 6,000 years ago, will continue.

Solar Output Variation

In most theories of climatic change, it is assumed that the output of energy from the sun is constant or nearly so. However, if it is considered that a fluctuation of less than 10 percent in output from the sun could explain all of the climatic changes that have occurred on earth, then it is natural that theories of climatic change have used changes in the sun as their basis.

There are two main approaches that have been considered. The first visualizes an actual change in the radiating temperature of the sun over long time periods. The second considers shorter times and deals with periodic phenomena, specifically sunspots.

One model that deals with the longer-term changes assumes that the sun contains substances other than hydrogen and helium. Diffusion of these substances from the core of the sun toward the surface would produce a barrier to the flow of energy from the core to the surface. This barrier would have the effect of ultimately reducing the amount of energy passing from the sun's surface and producing an ice age on earth. Although this is an attractive model, there is no physical evidence that it could occur; it cannot be considered, at least at the present, a viable explanation. There is, however, much evidence that sunspots occur, and they have been intensively examined.

To the naked eye, the visible portion of the sun, the *photosphere*, appears a bright disk. Magnification of this surface reveals dark, circular indentations that are called *sunspots*. Sunspots are complex electromagnetic phenomena that represent a cool region on the photosphere; with a temperature of some 4600°K they are appreciably cooler than the 6000°K of the photosphere. Observation of cyclical fluctuation of sunspots has led to much speculation on their role in weather and climate on earth.

Records of sunspots have been kept for many years; because they are relatively easy to observe, they are used as an indicator of solar activity. Their frequency has been found to follow a number of cycles, the best known being the 11-year cycle. In this, the sunspots increase to a maximum with a hundred or so visible at a given time and then diminish until only a few, or none at all, are seen. Other cycles that have been recognized include the 80-year Gleissberg cycle and one with a cycle frequency of 205 years.

Just how sunspot cycles influence weather and climate is a matter of controversy. Some researchers claim to have found correlations between sunspot activity and weather on earth. Others find no relationships. Even the physical mechanism by which they influence the atmosphere is not well defined, for the wavelengths of emission have different effects on different parts of the atmosphere. A further complication is that sunspot activity coincides with an active *solar wind* (a stream of ionized gases that escape the gravitational pull of the sun), and it is difficult to separate the two effects.

Recent research suggests that the shorter sunspot cycles are incidental to a much larger scale of variability. It has been found that in the period 1645–1700 very few sunspots occurred. This minimum, called the *Maunder minimum*, corresponds to the Little Ice Age on earth. In contrast, maximum sunspot activity took place between 1100 and 1250, a distinctively warm time in the Northern Hemisphere. Many climatologists now think that this longer cycle plays a significant role in the explanation of the variability of temperature since the end of the Ice Age. While this is the case, much research remains to be completed before any definitive answer is attained.

Atmospheric Modification

The amount of energy available at the earth's surface depends upon the extent to which the energy is modified as it flows through the atmosphere. Similarly, the amount of energy retained through the greenhouse effect is also a function of the atmosphere. Because of these factors, it is quite easy to theorize causes of climatic change resulting from variations in the transmissivity and absorptivity of the atmosphere.

Of much interest in this respect is the role of volcanic activity in modifying climate. The eruption in 1980 of Mount St. Helens, in the state of Washington, brought to public knowledge the possibility of weather and climate changes resulting from the addition of dust particles to the air. However, the possible role of volcanoes in modifying climate goes back much further. It was noted, for example, that the

1883 eruption of Krakatoa in the East Indies resulted in a 3-year reduction in the amount of solar radiation measured at Montpellier Observatory in France. Such did not occur in the case of Mount St. Helens, and subsequent research has shown that it has had minimal climatic effects. The causal link between volcanic activity and climate relates to the injection of large quantities of dust and fine particles into the stratosphere.

Large eruptions such as Krakatoa, Katmai (Alaska, 1912), and Tambora (Indonesia, 1815) can be shown to have a short-term effect on energy flows. Mechanisms for cleansing the atmosphere lead to particle fallout from the air. Herein lies a major problem of the volcanic dust theory of climatic change. To account for major glaciations, it is necessary to assume a long-term residence of volcanic dust in the atmosphere. One hypothesis is that during active times of earth-building forces, continued volcanic activity over long time periods would have an extended effect; once glaciation was initiated, the role of volcanic dust would be secondary to changing surface albedos. Unfortunately, study of deep-sea cores seldom shows layers of volcanic ash beneath deposits indicative of a cooling of climate.

Other changes in atmospheric composition that influence climate include modifications of ozone, water vapor content, carbon dioxide, and oxides of nitrogen. While these do change naturally over time, the major modifications are, at present, a result of human activities. It has already been noted that the ozone layer is put in jeopardy by addition of chlorofluoromethanes; the relative impact of the other changes is discussed later in this chapter.

Distribution of Continents

Modern geophysical research has provided a unifying theory to the old idea of continental drift, the concept of *plate tectonics*. There is now little doubt that the present positions of the world land masses are but a transitory location in the long-term evolution of the continents and the oceans. Given this, one set of theories of climatic change deals with the relative location of continental land masses in relation to the position of the poles and the equator.

When the continental distribution during the last two great earth-cooling periods is reconstructed, an interesting pattern results. Reconstructed maps for the Permo-Carboniferous glaciation (250 million years ago) and those of the most recent Pleistocene glaciation show that in both cases there is a concentration of land masses in the polar realms. The presence of large land masses in polar areas is much more conducive to glacial formation because land masses lack the heat storage and transfer mechanisms of oceans. It has been suggested that a primary requirement for the formation of great ice caps is the polar location of continents. Of course, there have been times when such a location occurred but no glaciation resulted. It seems that the relative location of the continents may well be an important factor in ice age occurrences, but actual ice formation must rest upon other causal factors.

Intimately related to the idea of moving continents, although often treated as a separate theory of climatic change, is the role of mountain building and continental uplift. As an explanation of this theory, one has only to think of the formation of permanent snow on Mount Kilimanjaro, a mountain located astride the equator. With increasing height of land masses, the potential for ice formation is greatly increased.

Geologists have long noted the relationship between times of extensive mountain building periods and ice ages. For example, both the Permo-Carboniferous and recent ice ages were preceded by extensive mountain building periods. There is, however, a lag time of millions of years between the mountain building and the onset of the glaciation. To this criticism must be added the fact that some mountain building periods—for example, the Caledonian period in Europe 370–450 million years ago—did not give rise to ice ages.

Despite these criticisms, it is generally agreed that mountain building, or at least the presence of high mountains, certainly contributes toward an optimum condition for ice formation. The idea that mountain building can influence climatic change is reinforced by modern research that shows that mountain ranges do influence upper-air circulation patterns by changing upper-air vorticity. *Vorticity* is a measure of the amount of circulation characteristics of the air; it is found that on the leeward side of the mountains, a cyclonic circulation—*positive vorticity*—is induced in air flow.

Variation in the Oceans

Modern research in climatology is paying increasing attention to the role of the oceans in the climatic system. Variations in sea-surface temperatures have been linked to changing circulation patterns and weather anomalies. In terms of climatic change, the oceans have received attention, providing in some cases the basis of entire theories of climatic change. Some of the ways in which the oceans influence the prevailing climates on earth include the following:

1. As a factor in the relative elevation of land: a drop in sea level would increase the heights of the continents and enlarge land masses in area.

2. As a heat storage mechanism the oceans are less variable than the continents, and changes in the relative temperature of oceanic waters would influence world climates. Variations in the oceans could occur because of changes in salinity, evaporation rates, and the relative penetration of solar energy.

3. As a mobile medium, ocean water plays a significant role in the redistribution of energy over the earth's surface. Ocean currents transport large amounts of heat, and any changes in their relative extent would have extended results. One example of this role is illustrated in figure 15-4, which shows paleogeographic maps of oceanic circulation during the warm mid-Tertiary and the circulation that existed in the cold conditions of the late Pleistocene. The essential difference is that a zonal circulation of water occurred because of the equatorial passage of the Tertiary. This closed by Pleistocene times, and the whole circulation pattern was modified to encourage ice formation.

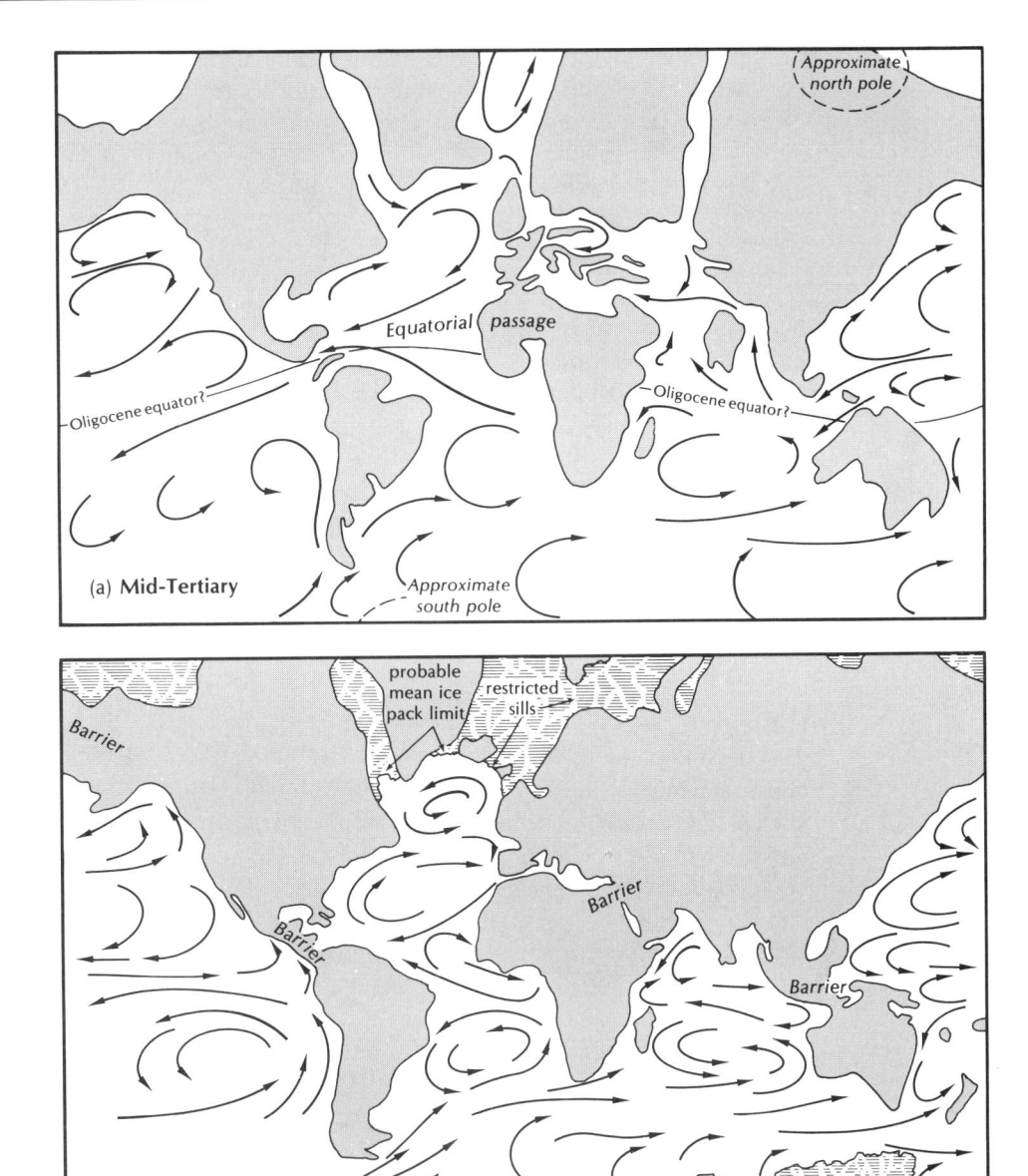

FIGURE 15-4 World oceanic circulation in (a) the middle Tertiary and (b) the late Pleistocene. Note how the world-encircling equatorial passage was closed by Pleistocene times, whereas a limited north-south exchange occurred when ice occupied large areas during the Pleistocene (From *The Encyclopedia of Atmospheric Sciences and Astrogeology*, Edited by Rhodes W. Fairbridge, Figures 2 and 3—page 464. Copyright © 1967 by Dowden, Hutchinson & Ross, Inc. Stroudsburg, Pa. Reprinted by permission of the publisher.).

Although the oceans are highly significant in the nature of climate that occurs over the earth's surface, many researchers think that their role in climatic change may be a secondary effect. That is, the oceans react to other changes, such as a modified atmospheric circulation; they do not create them. It has been suggested that the atmosphere leads and the ocean follows.

Other Theories

The theories presented above are a partial representation of those that have been suggested. Other researchers have introduced ideas ranging from the possible influence of the periodic passage of the earth through an interstellar "dust" cloud to variations in atmospheric water vapor caused by both natural and human activities. Despite all of these ideas, there is no single theory that can account for all of the observed events; it is evident that earth's climates result from a spectrum of causal elements. This is well shown in figure 15–5, which presents in diagrammatic form the components of systems that could be modified to cause a significant change in the climate of the earth.

CLIMATIC CHANGE: THE HUMAN IMPACT

In the short period, in terms of geologic time, that people have inhabited the earth, they have brought about massive changes in the environment. These changes have had a significant impact upon the earth's climate. Many modifications have occurred, but the two most influential types are changes that modify the earth's surface and the addition of carbon dioxide to the atmosphere. These are covered in some detail in the following pages. Discussion of this component of the human impact on climate is followed by a study of what must be considered the most closely investigated modified system, the urban environment. The climate of cities provides a case study of the multitude of climatic efforts that result from the total change of a natural environment.

The Role of Surface Changes

Humans have been altering the environment since they first controlled fire, domesticated animals, and originated agriculture. Modifications began in an early epoch, when the hunters and gatherers used fire to make hunting easier and to drive game during the hunt. Records from early explorers of Africa refer to massive fires, which probably represented the annual burning over of grazing areas south of the Sahara. In their visits to the Americas, European explorers noted that Native Americans used fire to improve hunting grounds and catch game.

The result of these activities was the deforestation of large areas of the world. In the tropical realm, it may be that the savanna grasslands are a response to deforestation by fire; in temperate regions, grasslands in North America and eastern Europe, prairie and steppe, may at least be a partial response to burning of woodlands that once existed.

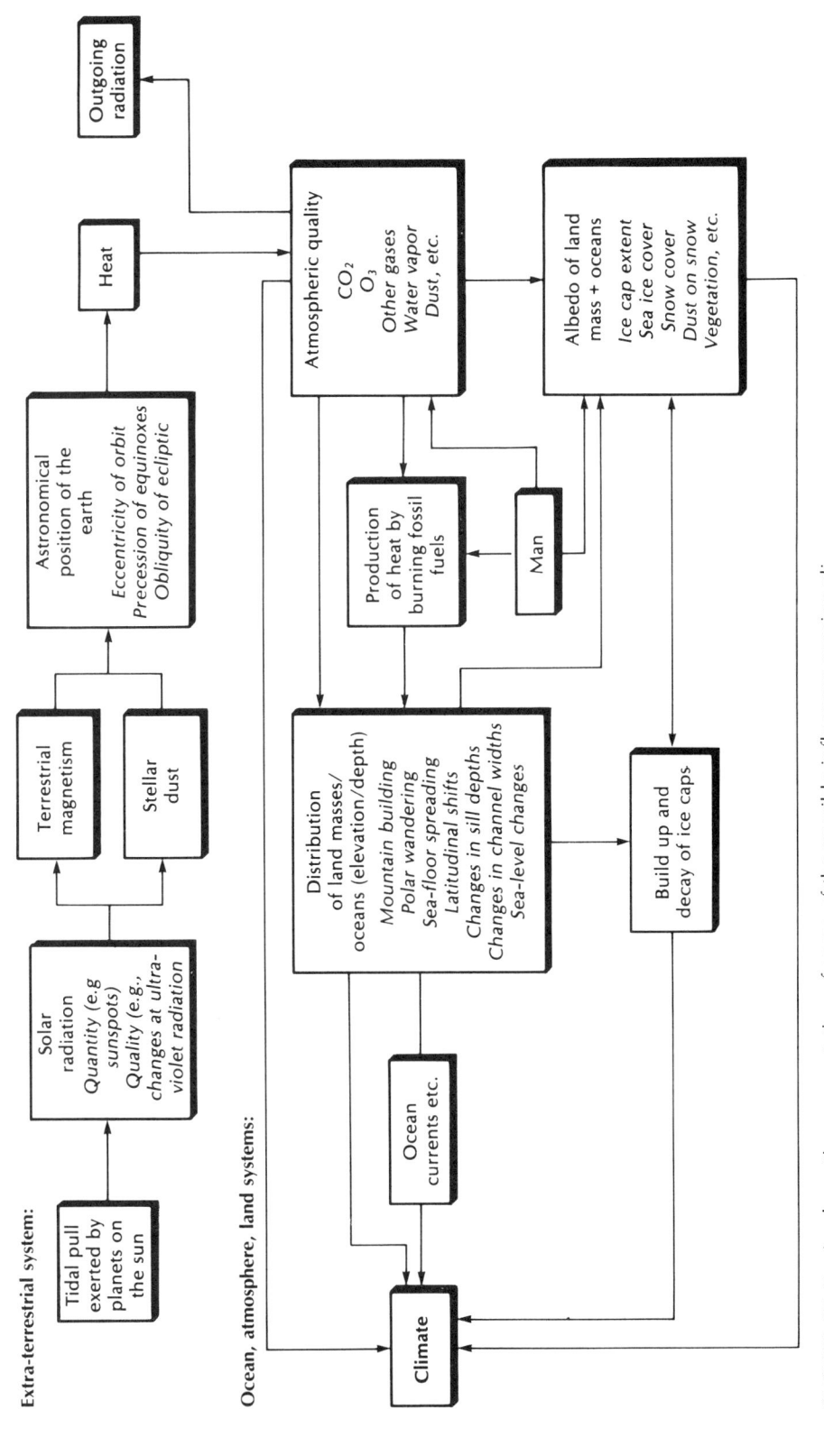

FIGURE 15-5 A schematic representation of some of the possible influences causing climatic change (From A. S. Goudie, *Environmental Change* [Oxford, England: Oxford University Press, 1977], p. 203.).

With the development of agriculture, deforestation became even more extensive. Once-extensive forests in China, the Mediterranean basin, western and central Europe, and North America were cleared for farming. The extent of the change is illustrated by the fact that 50 percent of central Europe was converted from forest to farmland over the last 1000 years. But deforestation was not the only extensive alteration. The misuse of marginal lands led to overuse and the eventual destruction of vegetation that once existed. This process resulted in the creation of a desertlike environment and began the process of desertification. Desertification has occurred in India, Africa, and South America.

The advent of a technological society, such as that in which we now live, created further changes. Destruction of the environment in the quest for raw materials, creation of artificial lakes, generation of energy, expansion of farmlands, urbanization, and other processes have significantly changed the face of the earth.

The result of the cumulative changes is their modification on the energy interchange that occurs at the earth's surface. It was pointed out in chapter 2 that the surface climate that occurs is a function of the energy that arrives at the surface and the way it is utilized. Of considerable importance in this respect is the amount of energy that is reflected from the surface, energy that does not enter the heat balance of the system. The amount of energy reflected depends upon the albedo of the surface, and changes in surface cover over time have appreciably altered values. Table 15-1 provides examples of how much albedo may have changed as a result of the human impacts outlined above. It will be noted that urbanization (estimated at 0.2 percent of the earth's surface) is not included in the above list. It is omitted because its role in surface modification is quite variable, and it may lead to a lowering of albedo.

The eventual result of the long-term changes listed is a reduction of surface temperatures. The worldwide temperature decrease resulting from the change has been estimated at about 1°K; this value is, of course, open to question because albedo changes lead to other modifications (e.g., cloud cover and dust) that also play a role in determining the earth's average temperature. Despite this, it is clear that surface change has had a significant impact on earth's climate as a whole and most certainly on the local areas where changes are the greatest.

TABLE 15-1 Changes in land use and albedo.

Land Type Changed	Earth's Surface (%)	Albedo
Savanna to desert	1.8	0.16 to 0.35
Temperate forest to field, grassland	1.6	0.12 to 0.15
Tropical forest to field, savanna	1.4	0.07 to 0.16
Salinization, field to salt flat	0.1	0.1 to 0.25

Atmospheric Changes

The pollution of the atmosphere through human activity has many effects (see Applied Study 15). In terms of climatic change, however, the most significant is the addition of carbon dioxide to the atmosphere through the burning of fossil fuels. The significance of carbon dioxide in the world climate system is its role in the greenhouse effect. Carbon dioxide acts to retard the flow of terrestrial radiation back to space and thereby increases the temperature of the troposphere.

Throughout earth's history there have been changes in the carbon dioxide content of the atmosphere because of natural variation. Within the earth-atmosphere-ocean system there are sources of the gas (volcanic activity, fire) and sinks (vegetation and the oceans) that remove it from the atmosphere. Given sufficient time, an equilibrium is reached between sources and sinks (figure 15-6). Prior to the Industrial Revolution it is thought that a near-equilibrium situation prevailed. Since humankind began using the carbon locked in coal, petroleum, and natural gas for combustion purposes, the source products have far outweighed the amount entering sinks. As a result, there has been an increase in the carbon dioxide content of the atmosphere.

Figure 15-7 shows the long-term increase in carbon dioxide in the atmosphere and some projections for the future. In the middle of the nineteenth century, carbon dioxide content of the atmosphere was estimated at 290 parts per million (ppm). By the middle part of the twentieth century, it was measured as 310 ppm. Such significant changes are certain to have an impact on world temperatures. Although it is very difficult to separate the role of a single variable in creating global temperatures, a number of authors have estimated the increase that has occurred. Table 15-2 lists one estimate of the impact of the increase in carbon dioxide. The

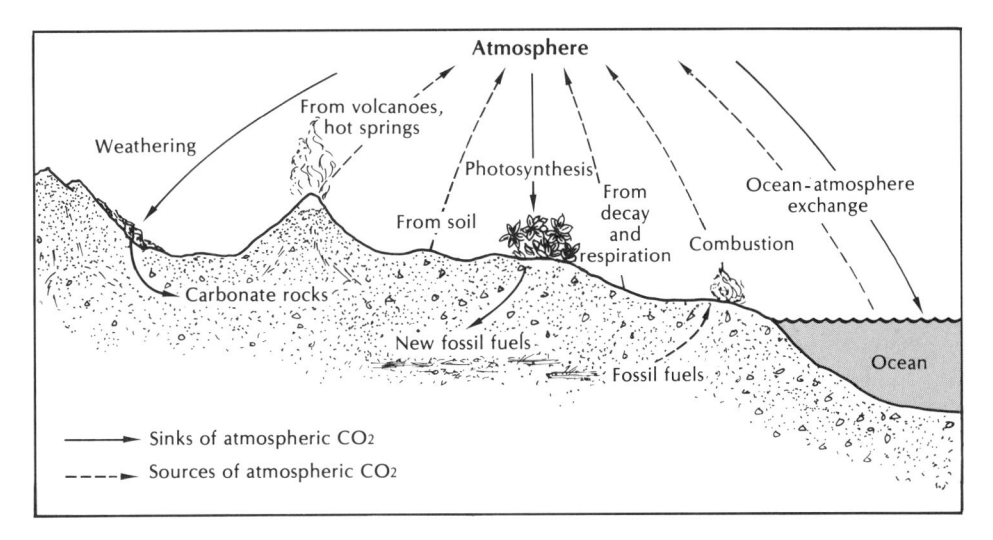

FIGURE 15-6 The carbon cycle, showing sources and sinks of carbon dioxide.

FIGURE 15-7 World production of carbon dioxide from 1860 (From *Weatherwise* 33 (6), December 1980, p. 254, a publication of the Helen Dwight Reid Educational Foundation.)

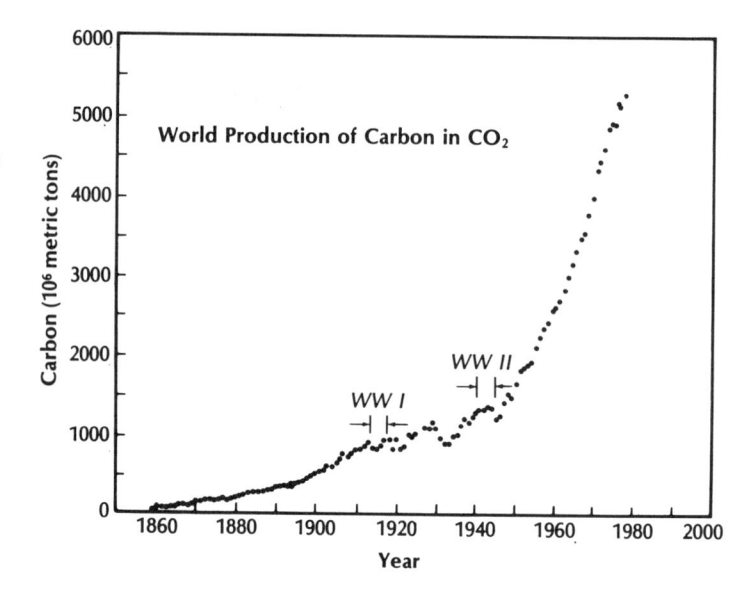

table also provides future estimates and suggests that the role of a modified greenhouse effect will lead to a marked warming of the earth's climate. Various scenarios have resulted from this prediction, and some are discussed in the next chapter.

Much discussion has also centered on the role of particulates in climatic change. It would seem logical to assume that dust and industrial particles would interfere with the flow of energy to the surface and thus lead to a cooling of the atmosphere. In fact, some climatologists advocate this point of view as a major factor in climatic modification. On the other hand, if the particles are not highly reflec-

TABLE 15-2 Reconstruction and prediction of atmospheric CO_2 content and global temperature increase.

Year	CO_2 in Atmosphere (ppm*)	Global Temperature Increase (°C)
1900	295	0.02
1910	297	.04
1920	299	.07
1930	302	.09
1940	305	.11
1950	309	.15
1960	314	.21
1970	322	.29
1980	335	.42
1990	351	.58
2000	373	.80
2010	403	1.10

*parts per million

Source: After Broecker (1975). The original work provides sources and limitations of data presented.

tive—for example, those with a dark color—then their role in reflecting solar energy will be minimal and result in only a slight change in the albedo of a region. Such particles, however, may absorb solar radiation and thereby increase the amount of energy available, leading to a rise in temperature. This effect could be heightened if the particles absorbed terrestrial radiation, especially wavelengths that normally pass through atmospheric windows to space. This process would lead to an enhanced greenhouse effect and a corresponding rise in temperature.

Clearly, there is uncertainty about the way in which pollutants in the atmosphere influence local global temperatures. Not only is there the question of their impact upon the energy flows but questions are also raised about the residence time in the atmosphere of the particles. Unlike volcanic dust, which is often injected into the stratosphere to remain for a number of years, industrially generated dust is largely confined to the troposphere. With the cleansing mechanisms at this level, the residence time may be measured in days.

The many changes in climate that occur because of human activities are epitomized in the most modified of all human environments, the city. So many changes occur in the urban environment that it can be considered as a case study of the human ability to modify climate inadvertently. As such, the urban climate requires special consideration.

THE CLIMATE OF CITIES

The constructed environment of a city creates a totally different climatic realm from that which occurred prior to its founding. Consider the following:

- ☐ Concrete, asphalt, and glass replace natural vegetation.
- ☐ Structures of vertical extent replace a largely horizontal interface.
- ☐ Large amounts of energy are imported and combusted.
- ☐ Combustion of fossil fuels creates pollution.

These and related factors modify the climatic process in the urban environment.

The Modified Processes

Figure 15-8 provides a graphic summary of the modified flows of energy in a city. Of prime importance is the energy characteristics of the asphalt/concrete/glass environment as compared to that of the vegetation covered surfaces of rural areas. For the most part, the city surface has a lower albedo than nonurban areas and greater heat conduction and more heat storage. During the day the city surfaces absorb heat more readily, and after sunset, they become a radiating source that raises night temperatures.

The energy flows are further modified by the geometry of the city buildings (figure 15-9[a]). Walls, roofs, and streets present a much more varied surface to solar radiation than undeveloped areas. Even when the sun is low in the sky, a time when

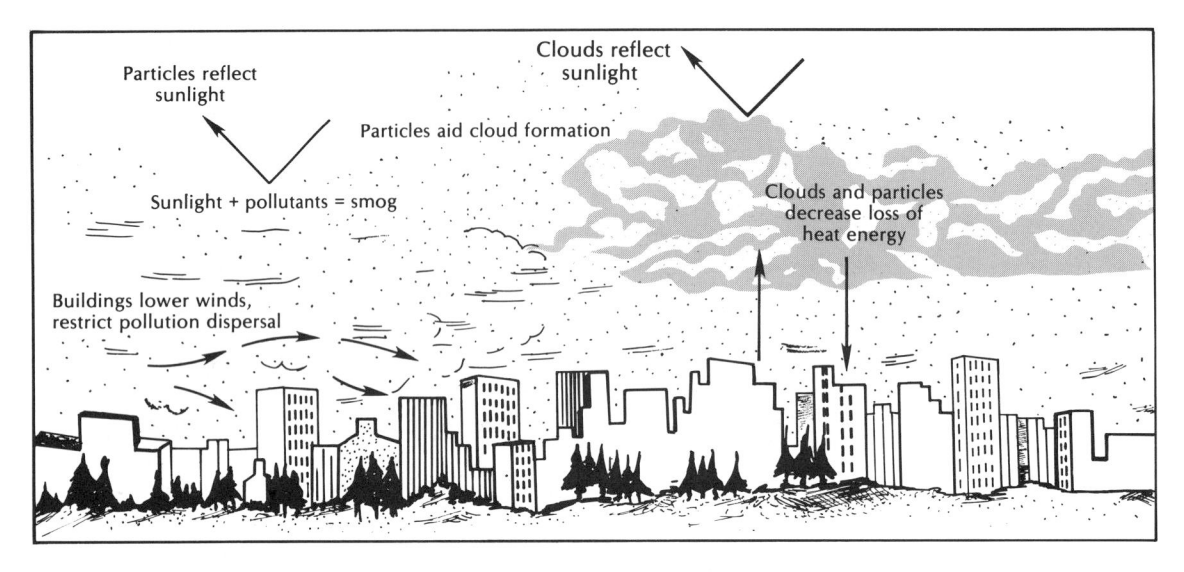

FIGURE 15-8 Schematic diagram showing the modified flows of energy in an urban environment.

little energy absorption occurs on flat land, vertical city buildings feel the full impact of the sun's rays. In the early morning and late evening, the city is absorbing more energy than surrounding rural areas.

The different surfaces that make up the city modify energy availability by changing the water balance (figure 15-9[b]). Rain falling on urban and nonurban areas is disposed of quite differently. On building-free rural surfaces some of the water is retained as soil moisture. Plants draw upon this source for their needs and eventually return the moisture to that air through transpiration. At the same time, standing water and soil moisture are evaporated. Solar energy is required for both transpiration and evaporation. In the city, pavements and buildings prohibit entry of water into the soil; with limited green areas, transpiration is minimal. Most of the rain water drains off quickly and passes to storm sewers with the result that water availability for evaporation is greatly diminished. Since both evaporation and transpiration amounts are decreased, the solar energy available for this process provides additional surface heating.

The ratio of energy used for *sensible heat flow* (which heats the atmosphere) to that of the *latent heat flux* (the latent energy in water vapor) is given by the Bowen ratio H/LE. In areas such as deserts the lack of water means that a high Bowen ratio, in the order of 20, occurs. In most vegetated areas the value is usually less than 1, indicating more transfer by latent than sensible heat. In cities, the ratio is about 4.0, showing that there is a large decrease in latent energy transfer; an increase in sensible heat has resulted. Again, this results in more available energy to heat the city atmosphere.

As areas of concentrated activities, cities are high consumers of energy; and enormous amounts of energy are imported to maintain their functions. In New York

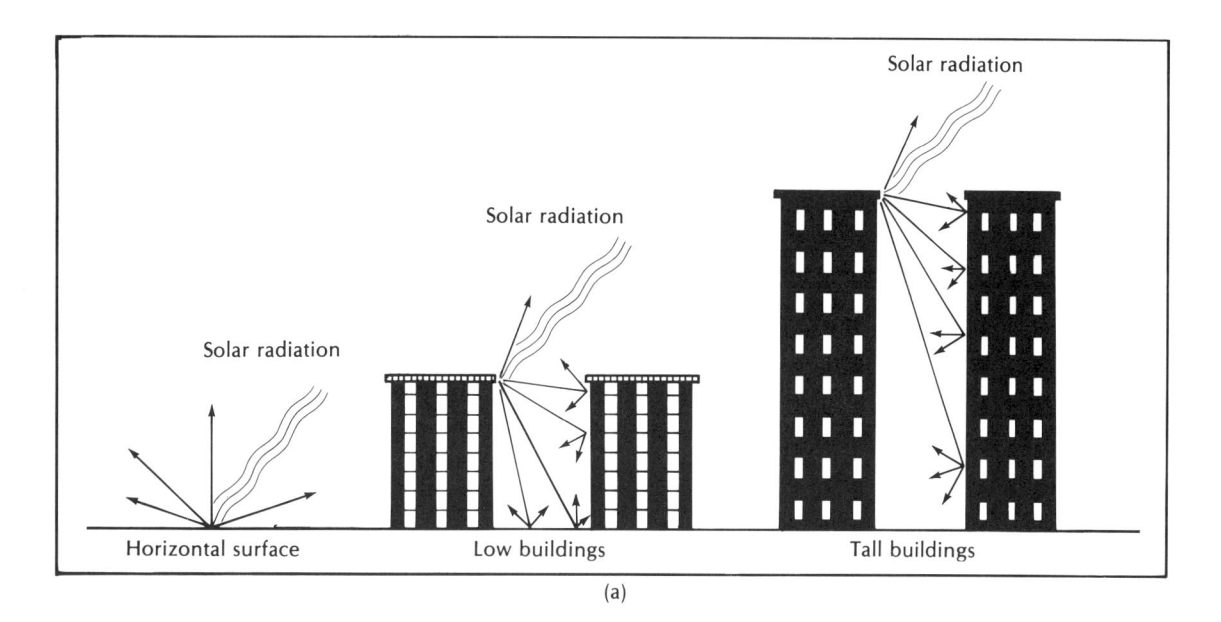

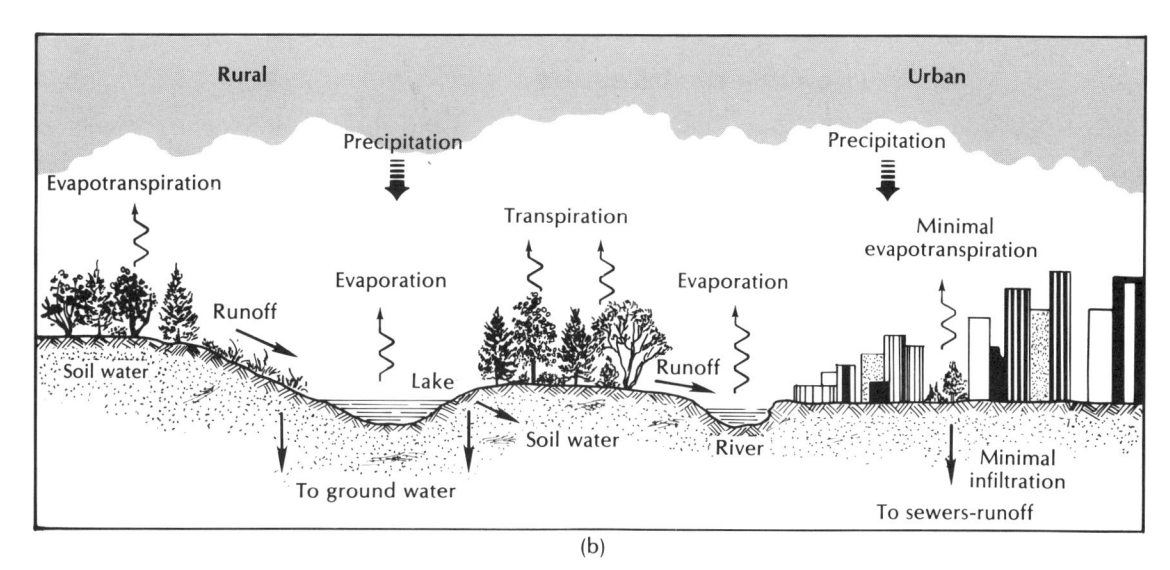

FIGURE 15-9 Modified climatic processes associated with urbanization: (a) the effects of horizontal and vertical surfaces on incoming solar radiation; (b) the disposal of precipitation on rural and urban surfaces (From *Climate and Man's Environment*, John E. Oliver, Copyright © 1973, John Wiley & Sons. Reprinted by permission of John Wiley & Sons, Inc.).

City, for example, the amount of heat generated through the burning of fossil fuels in winter is two and one-half times the amount of heat energy derived from solar radiation for the same period. This high use of energy means that much "waste" heat from factories, buildings, and transportation systems passes to the atmosphere to add to the warmth of the city.

The actual combustion of fossil fuels results in cities having higher air pollution levels than their rural counterparts. Visible evidence of air pollution over cities is the dust dome that is often produced. This pall of smoke and smog acts in a number of ways to modify the urban climate. Some incoming solar radiation strikes particles in the air and is reflected back to space. Other particles act as a nucleus onto which water vapor condenses to form clouds. That cities tend to be cloudier places than rural areas is illustrated by data from London, England: A comparison of bright sunshine hours showed that the city received 270 hours less than that of the surrounding countryside over the period of a year.

The lower input of solar energy because of increased reflection from particles and clouds results in lower temperatures in cities; this lack of energy, however, is more than counterbalanced by other effects. Some of the pollutants absorb rather than reflect energy, and the increased cloudiness reduces loss of long-wave energy to space.

The Observed Results

Of the many changes that occur in the climate of cities, the modified temperature regime is the best known and most closely studied. Cities tend to be warmer than the surrounding nonurban environment. The higher temperatures are best developed at night, when, under stable conditions, a heat island is formed. In passing from the center of the city toward the surrounding countryside, temperatures decrease slowly until, in rural areas, they remain about the same. The city is an island of warmth surrounded by cooler air. Figure 15-10(a) provides a schematic summary of this feature. It is possible to visualize the formation of an urban dome in which the normal lapse rate occurs because of the warmth of the city. This effect is in contrast to rural areas, where, during the night, a low level inversion often forms. If a slight regional wind is blowing, the warmer air of the city is swept downwind as an urban plume, modifying the rural lapse rate conditions (figure 15-10[b]).

The heat island effect has been measured in many cities around the world. Places of diverse size—from large cities like Tokyo, London, and St. Louis to smaller ones like Corvallis, Oregon, and San Jose, California—show similar patterns. Generally, when the heat island is best developed, temperature differences between city and country are as much as 5.5°C (10°F).

Wind speeds in a city are, on average, lower than those in open surrounding areas because of the increased roughness of the urban fabric. A comparison of highest expected wind speeds in the city and its airport (usually located outside the city in flat, unobstructed land) shows this to be true. In Boston, Massachusetts, the highest airport wind speed is 46 m/s (102 mph); the value for the city is 32 m/s (71 mph). The same effect is found in Chicago (city 25 m/s, 56 mph; airport 31 m/s, 70 mph) and Spokane, Washington (city 23 m/s, 52 mph; airport 35 m/s, 78 mph).

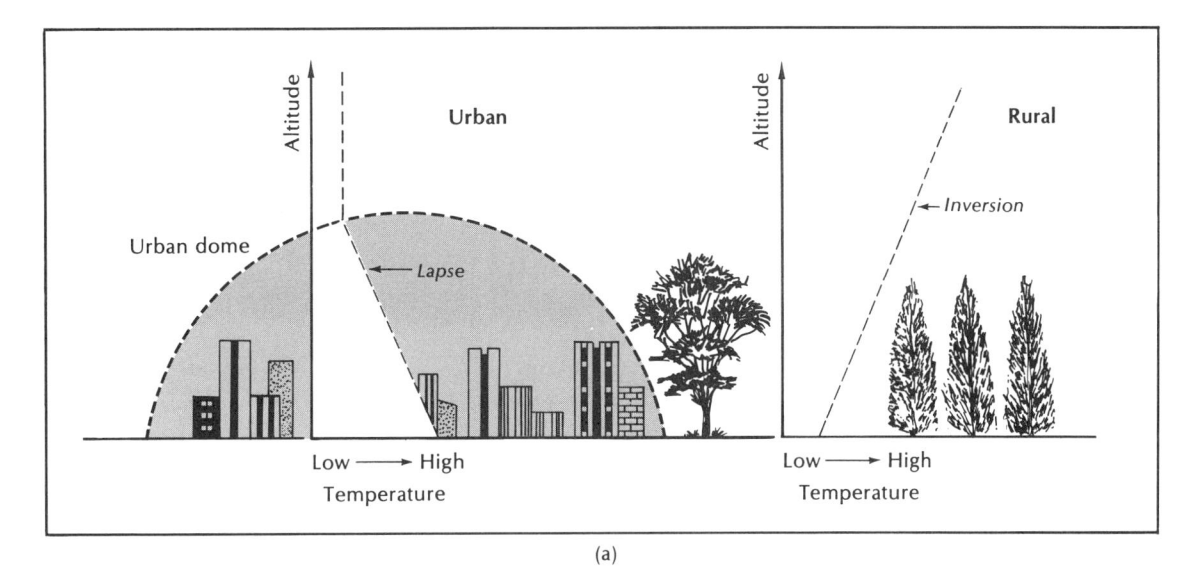

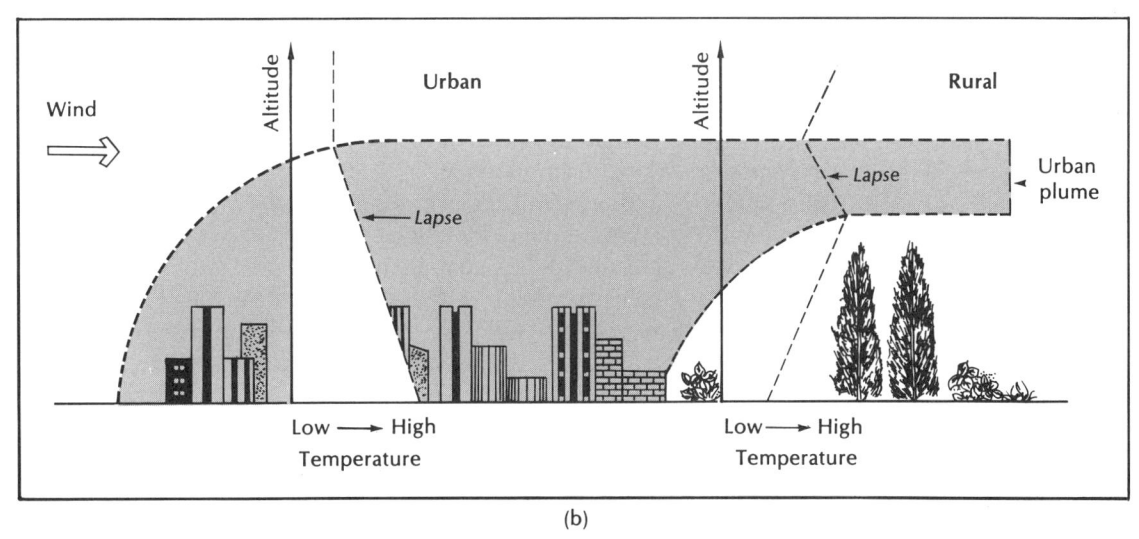

FIGURE 15-10 Comparison of lapse rates over rural and urban environments: (a) on a still night, lapse rate conditions occur in the city and an inversion in rural areas; (b) with a slight regional wind the rural conditions are modified by the creation of urban plume (From *Climate and Man's Environment*, John E. Oliver, Copyright © 1973, John Wiley & Sons. Reprinted by permission of John Wiley & Sons, Inc.).

The wind in cities may also be modified by the urban heat island. As noted, the heat island is best formed under calm conditions; in such instances, the city may create its own wind pattern. A small pressure gradient results when the warm city air rises and air is replaced by that from surrounding areas. The uplift of air is not high,

TABLE 15-3 Average changes in climatic elements caused by urbanization.

Element	Parameter	Urban vs. Rural Areas
Radiation	On horizontal surface	−15%
	Ultraviolet	−30% (winter); −5% (summer)
Temperature	Annual mean	+0.7°C
	Winter maximum	+1.5°C
	Length of freeze-free season	+2 to 3 weeks (possible)
Wind speed	Annual mean	−20 to −30%
	Extreme gusts	−10 to −20%
	Frequency of calms	+5 to +20%
Humidity	Relative—Annual mean	−6%
	Seasonal mean	−2% (winter); −8% (summer)
Cloudiness	Cloud frequency and amount	+5 to 10%
	Fogs	+100% (winter); +30% (summer)
Precipitation	Total annual	+5 to 10%
	Days (with less than 0.2 in.)	+10%
	Snow days	−14%

Source: H. E. Landsberg, "Climates and Urban Planning." In *Urban Climates*, Technical Note 10 p. 372. Geneva: World Meteorological Organization, 1970.

and a convective cycle results. This process means that the city air is recycled. If pollutants are being added to the air under these conditions, pollution levels can rise dramatically and give rise to a dust dome over the city.

The influence of cities on rainfall is still not completely understood. Generally, however, it is thought that a city gets about 10 percent more rainfall than surrounding rural areas. Intensive studies of this aspect of urban climates are part of the research project called METROMEX, an in-depth study of the conditions in St. Louis. Many significant findings resulted from this project (see the example in figure 15-11), but the results cannot always be applied to other urban centers in different climatic regimes.

Table 15-3 lists some of the basic climatic modifications that occur within cities. Examples of changes associated with the elements outlined are given, together with other parameters. It is an imposing list; it points to the amount of change that can inadvertently occur through modification of the environment by people. As the urban population of the world continues to expand, even more widespread results might be anticipated.

SUMMARY

What causes climates to change is a fascinating question. Many theories have been postulated, but no single one can totally satisfy all necessary requirements. Although there is little doubt that changes in earth-sun relationships may be a basic cause of long-term climatic change on earth, it appears that the effects must be considered in

FIGURE 15-11 Examples of results of the METROMEX study of St. Louis: (a) Pattern of average summer thunder days (1971–1975). St. Louis metropolitan area is shaded, and outline of circular raingage network is shown. (b) The 1972–1975 summer pattern of lightning-caused power outages, based on township frequencies, and the average thunder day pattern (From Stanley A. Changnon, Floyd A. Huff, Paul T. Schickedanz, and John Voger, *Summary of METROMEX*, Vol. 1. Weather Anomalies and Impacts, 1977, Illinois State Water Survey. Used by permission.).

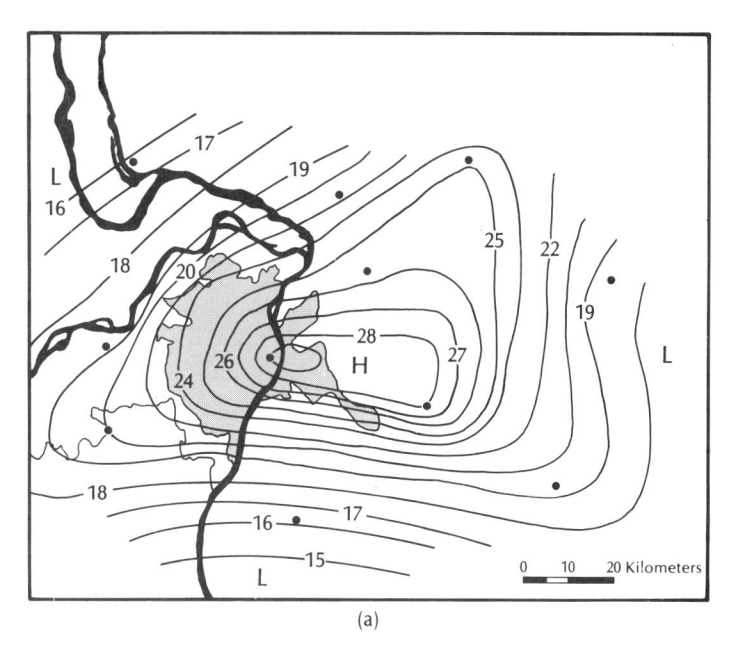

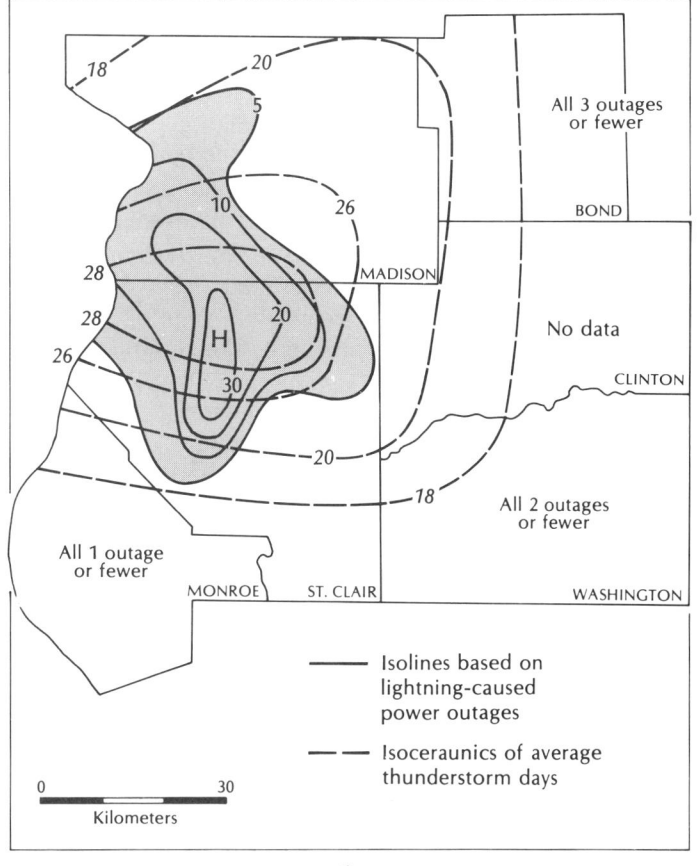

CLIMATES OF THE PAST AND FUTURE

conjunction with other factors. The problem of explaining the change is further complicated in that human activity is now a major factor in the modification of climate. To explain what is happening, the so-called natural causes must be considered together with the impacts of humans. That such is the case is well demonstrated by the extensive surface changes created by human activity and the addition of carbon dioxide to the atmosphere. In cities, the human impact is so great that a new set of climatic conditions is created.

The quest for the causes of climatic change is no longer an academic exercise. The close relationship between food production and the well-being of people is, in part, conditioned by climate. To understand and perhaps to predict future conditions to be faced by the human population is a highly significant task.

APPLIED STUDY 15

Acid Rainfall: A Change in Quality

The human impact on climate is not limited to changing the quantity of energy available or the amount of temperature variation; it also concerns the quality of certain atmospheric parameters. Of particular importance in this quality change is the problem of acid rainfall.

The relative acidity (or alkalinity) of a liquid is measured by the pH scale (figure 15-12). Note that the scale is *logarithmic*; a pH of 5 is 10 times less acidic than a pH of 4, and 100 times that of a substance with a pH of 3. Normally, the pH of rainfall, which might be considered as pure water contaminated by atmospheric carbon dioxide, is 5.6. This is acidic, so that the problem of acid rainfall really means that rainfall is becoming more acidic. Just how much the values are changing is shown in figure 15-13, which compares the acidity of rainfall in the periods 1955-1956 and 1972-1973 in the eastern United States. It can be seen that in some instances, the pH decreases by more than 1, indicating a tenfold increase in acidity.

Two questions immediately occur when viewing such data: What is the source of the increasing acidity and what are the potential impacts of such a change?

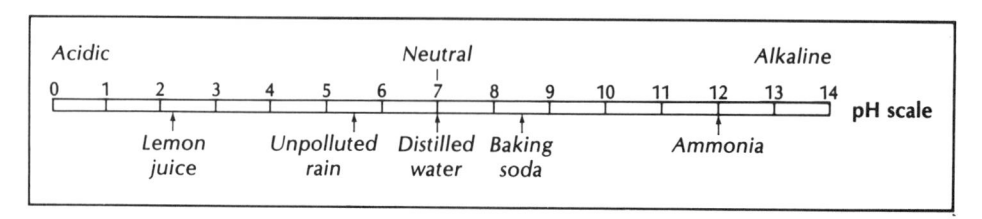

FIGURE 15-12 The scale of acidity with relative acidity of well-known items.

CAUSES OF CLIMATIC CHANGE

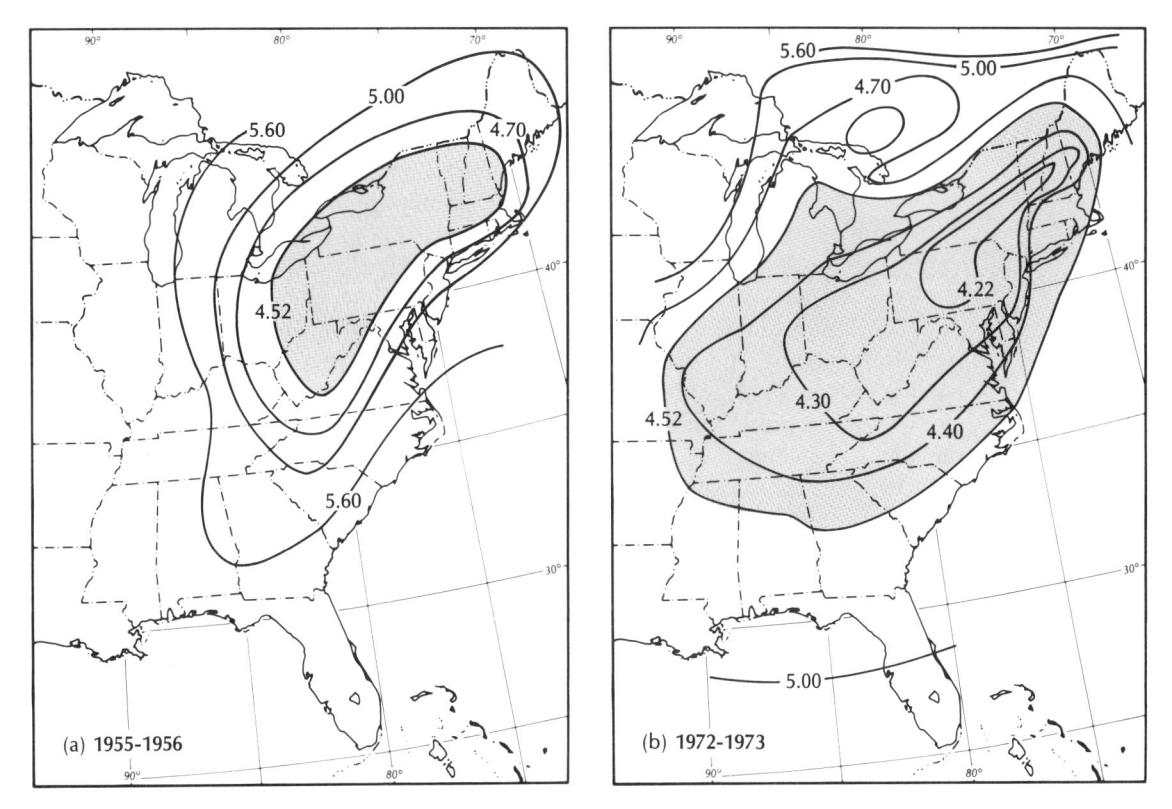

FIGURE 15–13 A comparison of maps reveals the rapid spread and intensification of the acid precipitation problem (From U.S. Environmental Protection Agency).

Acid rain originates mainly from the release of sulfur oxides and nitrogen oxides into the atmosphere through industrial and transportation sources. These chemicals are transformed into sulfuric acid and nitric acid by oxidation and hydrolysis in the atmosphere. As part of the circulating air, they are transported by the atmosphere, eventually to be washed out when precipitation occurs. This latter process may not occur until the pollutants have moved hundreds of miles from the source (particularly when they are emitted from very tall stacks), so that the problem has become global. Indeed, one of the few areas of conflict between the United States and Canada has concerned the "export" of pollutants that give rise to acid rain.

The potential effects of increasingly acid rainfall are wide-ranging. Included in such effects are the following:

1. The acidification of aquatic environments: A study of lakes in New York's Adirondack Mountains showed that more than one-third of 214 mountain lakes have pH values so acidic that no fish were present. It has also been reported that in Nova Scotia, Canada, the pH of some rivers has fallen to a point where salmon cannot live.

337

2. Accelerated weathering of buildings and structures: Acid is corrosive and results in a rapid decay of rock, cement, and metal. The geomorphological literature on the process of weathering stresses the role of acidity; clearly, the more acid the rain, the greater its negative impact.

3. The acidification and demineralization of soils: Again research indicates that the pH of soils plays an important part in its evolution and its properties. In some instances increased acidity can result in the exclusion of plants that would be present under nonacidic conditions.

These examples most often result from long-term, cumulative exposure to increasing acidity. There are times, however, when short-term highly acid rainfall has immediate effects upon a given environment.

At the present time many researchers are looking into the problem of acid rainfall. This interest will continue because there is reason to believe the acidity is increasing. In some ways this is a paradox, for in recent years the imposed controls on air polluton have led to a general rise in air quality. Some have suggested that antipollution measures may have enhanced the problem of acid rain, since much of the success has been in getting rid of flyash from the air. These grime particles are alkaline and may, in part, have acted to neutralize the acidic portion of the effluent. Although there is no conclusive evidence that this process actually occurs, it may well be an important factor.

16 Future Climates

THE NEED TO KNOW
Climate and Energy
World Food Production
Global Ecology
Other Impacts

MODELING FUTURE CLIMATES
Analog and Digital Models
Atmosphere-Ocean Relationships
Scenarios

THE CURRENT VIEW OF FUTURE CLIMATES

SUMMARY

APPLIED STUDY 16
Climatic Change and Food Production

CLIMATES OF THE PAST AND FUTURE

Thus far, the content of this book has emphasized present and past climates. There remains, obviously, a vital question: what of the climate of the future? It is not possible to answer because all of the processes that produce changes, and especially the way in which they interact, are not totally understood. It is necessary, however, to have some insight into the climate of the years ahead and its impact on human activity; this relationship is discussed in the first part of this chapter. Once the need to know is established, another question occurs. How can some idea of future climate be attained? To answer this, it is necessary to look into some of the modern techniques of climatic modeling and creating scenarios of the future.

THE NEED TO KNOW

The relationships between climate, environment, and society are extremely complex; to deal with all of them is far beyond the scope of this text. Nonetheless, some ideas of the important impact of climate can be obtained through consideration of some of the better-known relationships.

Climate and Energy

Fossil fuels provide the energy base for most modern societies. In recent years, much has been written about future sources of energy; and, given the nonrenewable nature of fossil fuels, alternative strategies have been investigated. There remains, however, an energy problem for future generations.

Problems of energy needs in the future could be heightened by climatic change. The potential impacts can be demonstrated, using energy demand by residential and commercial properties as an example. Currently, the United States uses 18 percent of its total energy demand for space heating and cooling. The amount of energy used depends largely upon the prevailing outdoor temperatures and the type of structure being heated or cooled. If, over the long term, the outside temperature changes, then the amount of energy required is altered.

Just how serious this can be is illustrated by data applicable to the recent cold winters in the United States. For the period November 1976 to May 1977, a 1.8°C decrease in normal temperatures occurred. During this period, the heating season had 22 percent more heating degree-days than the previous years. This increase led to a natural-gas and oil shortage in the Northeast, the closing of businesses, and temporary unemployment. Many of these effects were due to the lack of preparation by utility companies, but the enormous demand for energy during this period is indicative of what a general decline in temperature will accomplish. Clearly if temperatures in the years ahead do fall, the production and availability of energy will be a problem.

It might be assumed that a global warming would result in an overall decrease in energy demands. That such is not the case is seen in summer "brownouts" that occur in cities during hot summer periods. Air conditioning is a high energy user, and increased summer temperatures lead to increased energy use for cooling purposes.

Problems related to energy use in the future also concern the potential for the production of waste heat and the climatic influences of air pollution. The creation of the urban heat island is, in itself, an example of the way in which concentrated energy production and utilization can modify climate. If such effects are combined with natural climatic changes, the problem becomes much more complex and more difficult to estimate with any accuracy.

World Food Production

The production of food and the stability of an agricultural system are related closely to climate. That this is so is seen in the distribution of different types of commercial crops throughout the world. Should a change in climate occur, food production will be greatly affected. It is important to note that although the following examples use temperature as the controlling factor, a change in global temperatures will result in a change of circulation patterns, which will modify precipitation patterns.

In terms of food production, consider the fact that the last 40 years have seen a rapid growth in world population as well as a corresponding rise in the ability of farmers to feed the great mass of people. Although relying heavily upon fertilizers and improvements in crop species, the highly productive food-producing areas are delicately adjusted to prevailing climatic conditions. The difference between a bumper crop and a meager one often rests upon only slight weather differences. The significance of vagaries of weather in food production is seen in the highly variable year-to-year grain production in the U.S.S.R. Weather-induced grain losses require that massive amounts be imported from other countries. Applied Study 16 considers the potential impact of temperature change in the corn and wheat areas of North America.

Changing temperature regimes over the earth's surface would lead to a modification of the general circulation pattern, which would further result in a change in storm paths and precipitation distribution. Given the basic need for water in agriculture production, there is little doubt that areas in marginally moist regions would experience a change. The direction of the change would depend, of course, upon the nature of the temperature variation. If cooling were to take place, an equatorward movement of pressure and wind systems would result. Such a change would have dire results in many of the monsoon climates of the world and especially in the fringe areas of deserts such as the Sahel. Were a warming to occur, a poleward shift of wind belts would take place. Areas with a Mediterranean climate, which now receive precipitation from the equatorward movement of winter storms, would become increasingly more arid. Figure 16-1 illustrates the shifts schematically.

Closely associated to the influence of climatic change upon food production is the problem of the frequency and severity of pest outbreaks. It has been noted that insect pest populations will generally increase as temperature increases. With warmer and longer growing seasons, some insect pests, which produce 500–2000 offspring per female and go through a generation in 2–4 weeks, may pass through an

CLIMATES OF THE PAST AND FUTURE

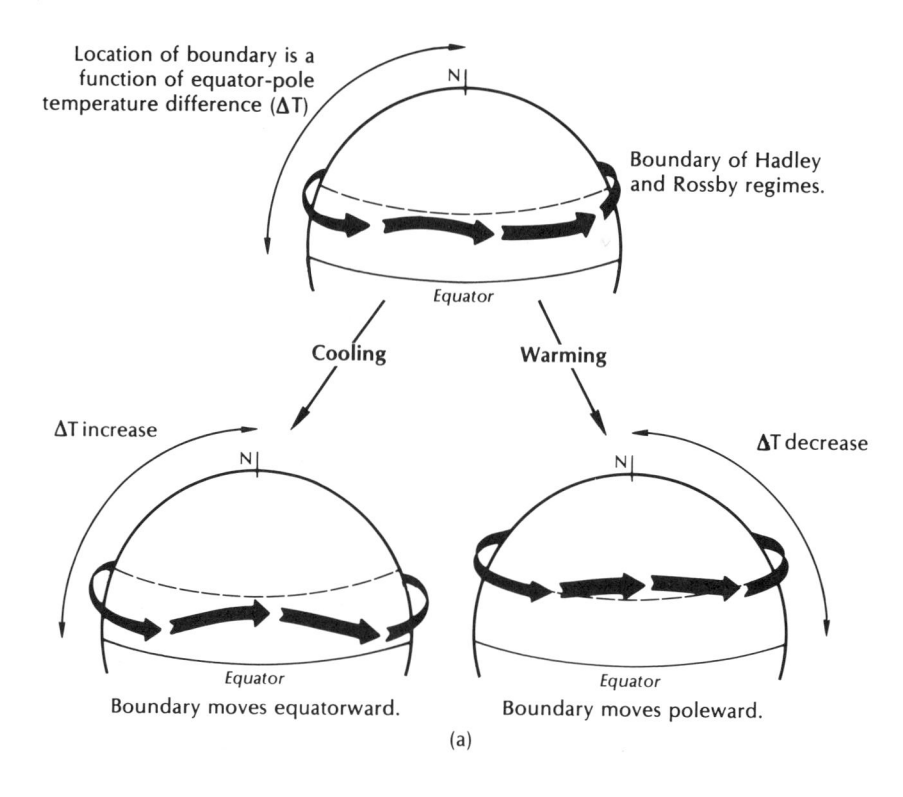

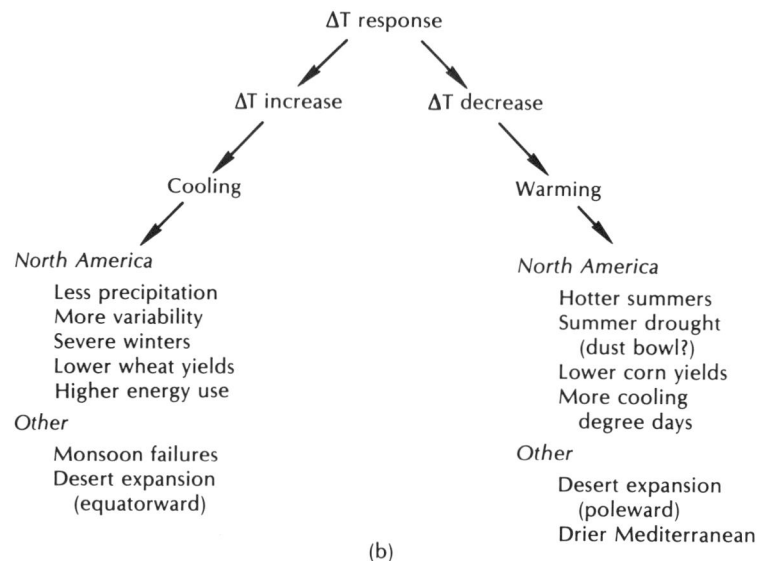

FIGURE 16-1 Changes in equator-pole temperature gradient result in modified circulation: (a) location of jet stream, indicating equatorward limit of Rossby regime as a function of cooling and warming trends; (b) some potential impacts of a modified temperature gradient and resulting circulation changes.

additional one to three generations. Since loss to pests by agriculture is currently in the order of 37 percent, such an increase could lead to very large losses.

Global Ecology

Unlike agricultural systems, in which a limited number of plant species are managed by people, natural ecosystems consist of a diversity of animals and plants that exist in balance with the prevailing environmental conditions. Most of the world's large natural systems (represented, for example, by world biomes) have been modified through human interference. However, when human activity ceases, the natural system reestablishes itself over time. When the system is disturbed by climatic change, however, the modifications that occur are permanent; a new set of ecological relationships is established.

Evidence of massive changes in world biomes exists in the changes of vegetation/climate types that have occurred in the period since the retreat of the ice of the Pleistocene. During the height of the Pleistocene, tundra and subarctic climates extended across much of the United States from the east to the west coasts, reaching as far south as a line from New Jersey through Missouri to California. These areas obviously have a very different climate today. In the same way, areas to the south of this cold regime also experienced a cooler climate. Florida, for example, had a cool temperate climate, very different from the tropical conditions that are now found.

These changes are essentially long-term ecosystem responses to climate. Ecosystems also respond to shorter-term changes and, if abnormal climate prevails over decades, to an intermediate ecosystem response in which the "climax" characteristics are modified to approach a new equilibrium.

If a long-term change in climate did occur, the effects on world biomes would be seen most clearly in those areas that are not extensively cultivated today. In the tundra, for example, a cooling of climate would result in an equatorward migration of tundra conditions and the development of more extensive permafrost. A warming would result in the poleward extension of trees, melting of permafrost, and an increase in the oxidation of surface organic material that is now either frozen or permanently saturated with water. This latter impact would cause a release of carbon dioxide to the air and an enhancement of the warming effect.

The desert world is another area that would be greatly influenced by a changing climate. Essentially, the desert boundaries would expand poleward or equatorward, depending upon the nature of the climatic change. Irrespective of the direction in which the expansion occurred, the process of desertification would result at the expanding desert boundary.

Mountainous areas would experience marked modification of ecosystem distribution. Mention has already been made of the changing snow-line levels and tree lines. Again, the pattern of mountain ecosystems would reflect the direction of change that occurs.

Other Impacts

There are many other realms of human activity that would be influenced by a change in the current climatic regime. The role of water in agriculture has already been described; but it should be remembered that as a resource, water is used in almost every aspect of human activity. Should a climatic change result in an increase or decrease in precipitation, the effects would be felt in industry, community needs, and settlement patterns.

The fishing industry, an important food source, is highly sensitive to changes in climate. It has been shown, for example, that long-term temperature changes have played a role in the success of cod fishing off West Greenland. Good catches occurred up to about 1950, but the recent temperature decline has led to a reduction in the catch. Probably the best known climatic impact upon fishing concerns the highly productive coastal waters off Peru. In this area upwelling cold water brings nutrients from the depths to provide the base of the food chain for countless fish. Periodically, the upwelling is stopped by a layer of warm water flowing south along the coast. This water movement is referred to as *El Niño* ("the child"), a name derived from its appearance at about Christmas time. El Niño totally disrupts the food chain, and fish disappear or die. Peru's fishing industry, the world's largest, is devastated. This effect, however, is but one component of the potential impact of El Niño. Later in this chapter its climatic significance is examined in more detail.

Health and disease could also be influenced by a changing climate. Of particular note in this respect is the potential for the spread of disease now largely confined to rural tropical and subtropical areas, a threat that would result from a warmer earth. Such diseases as schistosomiasis (resulting from a parasitic invasion of the body), dysentery, hookworm, and yaws (a contagious disease characterized by bone lesions) have been identified as types of diseases that could become more widespread.

Should a climatic change of major proportions occur, a major change in settlement patterns would result. The historical evidence relating migration of peoples to climatic change is well documented and provides examples of the types of response that can occur. Intimately related to this phenomenon is the impact of climatic change on current leisure and recreation patterns. A modified climate may make some areas now popular for outdoor recreation much less attractive. Given the significance of tourism in many countries, this is a factor that would cause concern.

MODELING FUTURE CLIMATES

The advent of high-speed computers has had an important effect upon research on future climates because they provide the ability to handle vast amounts of data and very complex equations. Both of the research approaches outlined in the following pages rely heavily upon computer modeling.

Analog and Digital Models

Climatologists use various types of models in their efforts to predict future climates. One type, the *analog model*, constructs a set model in which conditions are analogous or similar to those of the real world. A good illustration is the electrical analog model, which uses voltages, variable resistors, and amplifiers to represent components of energy flow in the atmosphere. Accordingly, the flow of energy from the sun can be represented by a constant electrical flow that can be modified by variable resistors representing the various atmospheric components that modify the flow of energy to the earth.

More widely employed are *digital* (numeric) computer models that use mathematical equations. Some utilize fundamental thermodynamic and hydrodynamic equations that describe the behavior of gases and forces involved in molecular and turbulent flows of fluids. The equations developed—called *primitive equation models*—are extremely complex; on a global basis even the largest digital computers cannot easily handle them. Thus a common approach to this numeric modeling is to apply equation values to a limited number of points spaced out over a grid. This procedure greatly simplifies the computations, of course, but at the same time it decreases the validity of the model.

The types of problems being examined by computer modeling are numerous, but perhaps the most significant are those that concern the general circulation models (GCM) of the atmosphere, the generation of statistical-dynamical models, and those concerning coupled ocean-atmosphere models. As the name implies, GCMs attempt to simulate the atmospheric circulation patterns using primitive equations. Although important advances have been made, even the most refined model produces simulations that differ from the actual behavior of the natural atmosphere.

One important advance in the use of computer models has been developed by researchers at the National Center for Atmospheric Research (NCAR). A Community Forecast Model (CFM) now in use is based upon an early GCM developed in Australia. The CFM represents a compromise between highly sophisticated models that use vast amounts of computer time and less costly models that provide only a first approximation of realism. This model includes systems dealing with seasonal change, atmosphere-ocean coupling, and treatment of soil moisture and snow cover. It has been used for a number of research projects ranging from paleoclimatic reconstruction of the Cretaceous period to the effects of vegetation upon regional climate.

Statistical-dynamical models use some components (transfer of momentum, heat, and moisture) as statistical effects for use in a simulation model. This feature makes them simpler than the GCM's, for in modeling the atmosphere they do not need to follow the life history of every disturbance that occurs. This limitation does not negate the validity of the model if the time span being considered is longer than the lifetime of the exchange process.

Because it is impossible to simulate all aspects of global climate without consideration of the role of the ocean, computer models that couple the ocean and atmosphere have been developed. A complete understanding of the interactions that occur requires intensive study of the climatology of the oceans; and, at the present time, the required data bank is incomplete. It is thus fortunate that a number of major research programs are underway to supply missing information. Of particular significance in this respect is the Global Atmospheric Research Program (GARP), which came about after the world community responded positively to a proposal presented by President Kennedy to the United Nations in 1961. The first research effort concerned the tropics, and, as a result, the GARP Atlantic Tropical Experiment (GATE) was launched in 1974. On a global scale, data accumulation by the First GARP Global Experiment (FGGE) in 1979 has proved enormously valuable.

Atmosphere-Ocean Relationships

Perhaps one of the most interesting avenues of current research deals with the ocean-atmosphere interface in a way that differs from that described above. Important contributions to this mode of research have been made by the International Decade of Ocean Exploration (IDOE) in which the major thrust has been multidisciplinary work on the oceans. As part of this, the Environmental Forecasting Program has as one of its own goals "improved environmental prediction through better understanding of historical climate changes, the influence of oceans on the atmosphere, and the role of ocean circulation in shaping weather and climate." An example of the type of work involved is given below. Note that in this example the research may not appear to be climatic, being based more upon individual happenings than long-term trends. The point is, however, that these seemingly short-term events are intimately related to the whole environmental complex that ultimately makes up future climates.

In 1957, immense pools of unusually warm and cold water in the North Pacific Ocean were discovered. The pools are 1000–2000 km wide, 200–300 m deep, and 1–2°C colder or warmer than is usual for that time of year. A temperature difference of this amount in such a vast pool of water means a net gain or loss of huge amounts of energy. It is thought that the pools may originate when atmospheric disturbances modify currents that border the Pacific, causing "cut-offs" in which a mass of water in a meandering current breaks away to create the drifting pool.

The significance of these pools on climate and weather is being investigated, notably by Jerome Namias and his colleagues at the Scripps Institute of Oceanography. Namias suggests that the pools act as a steering mechanism for air moving over them. For example, winds blowing from the west over a warm pool might be forced to rise or shift to the north. It is in relation to the steering effect that the pools have been considered as a cause of the recent cold winters experienced in the central and eastern United States and the anomalously warm winters that preceded them.

According to the theory, in the 1960s a cold pool in the Central Pacific caused a southward shift in the prevailing westerly winds (figure 16-2[a]). This shifting led to a dominance of cold, dry air in the eastern half of the United States, air that replaced normally warmer prevailing air masses. The same pattern was repeated in 1976–1977, leading to a severely cold winter in many parts of the United States.

In contrast, the winters of 1971–1975 were quite mild. In these years a large warm pool had displaced the cold in the Central Pacific (figure 16-2[b]). This displacement could potentially cause a northward shift of the westerlies, resulting in a dominance of warm, moist air from the Gulf and Atlantic over the eastern half of the United States.

The exact mechanism involved in the formation and effect of the ocean pools has yet to be totally explained. In fact, some researchers claim that the pools have little or no effect upon air circulation and winter conditions. Clearly, however,

FUTURE CLIMATES

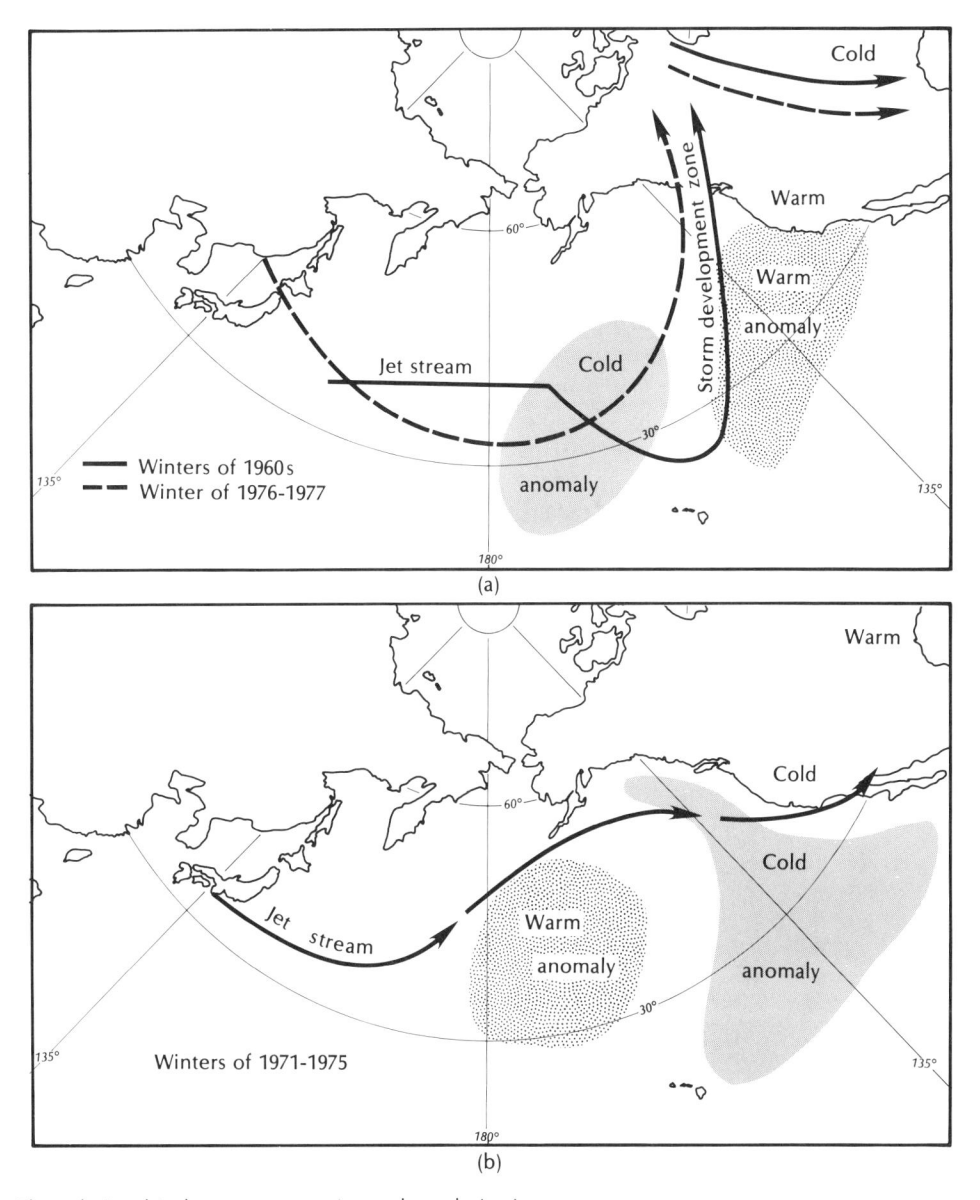

FIGURE 16-2 The relationship between oceanic pools and air circulation: (a) two bad winters and (b) a mild winter (Source: NSF, *The Dynamic Ocean*, 1978, Government Printing Office).

the research offers enough evidence to be worth pursuing. If it is shown that the pools play a critical role in winter weather conditions, then their formation can be carefully monitored; at the very least, some indication of the nature of the coming winter can be forecast. If this proves possible, explanation of the mechanisms involved should provide some insight into coming conditions over a period of years.

347

Of particular significance in recent atmosphere-ocean research is the role of El Niño, the ocean current that has already been mentioned in relation to Peru's fishing industry. Representing a warming of what is usually a cool area of the Pacific Ocean, El Niño is actually a component of a system that has a global climatic influence.

The cold waters off Peru and Ecuador result from a normal circulation pattern in which the eastern part of the equatorial Pacific Ocean is a zone of high atmospheric pressure and the western part one of lower pressure. This division causes the general flow of air to be from the east (off South America) to the west (toward Australia). The movement of air and water in this direction causes an upwelling of cold water off South America and the accumulation of warmer waters in the western part of the equatorial Pacific Ocean.

The modification of the cold El Niño current occurs when this normal pressure system is reversed, that is, when the pressure over the eastern region is lower than that in the west. This swing in the circulation is referred to as the *Southern Oscillation* and the coupled results as an *El Niño-Southern Oscillation (ENSO)* event.

The effects of an ENSO event are numerous. Locally, the warm water replacing the cold El Niño evaporates rapidly and results in high rainfall along the South American coast. On the other side of the Pacific Ocean, the rainfall diminishes and Australia experiences drought. The most wide-ranging influence, however, is caused by the heat that extends as a tongue from the coast of South America into the equatorial North Pacific Ocean. The heat source modifies the equator-pole temperature gradient so that the positions of both the subtropical and polar jets are changed. Because, as noted in earlier chapters, the seasonal climate of middle-latitude regions is determined by the upper air circulation patterns, changes in the jet streams indicate a modified pattern and result in unusual weather and climate.

Scenarios

Beyond the basic research that attempts to unravel the atmospheric intricacies that will influence climates of the future, there is also a method that uses projections based upon current events to estimate future conditions. This method consists of establishing *scenarios*. It answers the question "What would happen if . . . ?" with the scenario establishing the "if" part of the question.

A good example of the use of the scenario approach has been given in a National Science Foundation report. The study is based upon the assumption that in the foreseeable future, human activities will swamp natural forces in terms of impacts upon climate. For example, a continued increase in the use of fossil fuels at the rate of 4 percent per year would result in heat released to air and water at an alarming rate. At the 4 percent rate, by 2170 the amount of waste heat generated would be 10 percent of solar radiation reaching the earth. Such an increase would raise average earth temperature by 5–8°C. This assumption is in itself a scenario suggesting what would happen if current energy usage continues at the present state. Of course, the current rate is not necessarily that which will occur in the future, since the amount used depends upon many factors, including the growth of world popula-

TABLE 16-1 Projected range of total world energy use.

World Population	Average use* (per capita)			
	100	300	500	1000
6 billion	0.6 Q**	1.8 Q	3 Q	6 Q
10 billion	1 Q	3 Q	5 Q	10 Q
20 billion	2 Q	6 Q	10 Q	20 Q

*Million BTU per year
**Q = 10^{18}

tion, industrialization, and the types of fuels used to generate energy. These are unknown and, accordingly, a number of scenarios must be established.

The first consideration is to estimate the growth of population by the year 2100. The estimates available range from a high of 20 billion to a low of 6 billion people. It then must be ascertained how much energy this population will use; as shown in table 16-1, a considerable range exists. The data given in the table show the average per capita consumption extending from 100 to 1000 million BTU's per year. Given these ranges of data, the future sources of energy must vary; therefore, three scenarios can be postulated:

Scenario 1

World energy use reaches 20 Q BTU per year by the year 2000 (population of 20 billion at 1000 million BTU per capita per year).

Option A
All fossil fuel reserves are exhausted during the twenty-first century. Nuclear energy becomes the major energy source late in the twenty-first century and the dominant source by the year 2100.

Option B
The use of fossil fuels is curtailed by the year 2050 so that CO_2 in the atmosphere does not exceed 420 parts per million (ppm). Nuclear energy becomes the major source for an interim period in the middle part of the twenty-first century, but then solar energy takes over by the year 2100.

Scenario II

World energy use reaches 3 Q BTU per year by the year 2100 (population of 10 billion at 300 million BTU per capita per year).

Option A
Use of fossil fuels is curtailed by the year 2050 so as not to exceed 420 ppm CO_2 in the atmosphere. Nuclear energy becomes the major energy source before the middle of the 21st century and the dominant source by the year 2050.

Option B
Similar to Option A except solar energy takes over from nuclear by about the year 2075.

Scenario III

World energy use reaches 0.6 Q BTU per year by the year 2100 (population of 6 billion at 100 million BTU per capita per year).

Option A
Use of fossil fuels is curtailed by the year 2000 so as not to exceed 400 ppm CO_2 in the atmosphere. Nuclear energy becomes the major energy source early in the twenty-first century and the dominant source by the year 2025.

Option B
Similar to option A except that solar energy takes over from nuclear energy by about the year 2050.

In considering the climatic impacts of these scenarios, two major variables are considered: the release of heat in the energy conversion process and the role of carbon dioxide in enhancing the greenhouse effect. Computer modeling allows the relative impacts of these two to be related to each of the scenarios. The conclusions are as follows:

Average temperature increase (°C).

	Global	At Poles
Scenario I, Option A	2–3	≥10
Scenario I, Option B	1*	2–3*
Scenario II, Option A	1.2–1.3*	2–5*
Scenario II, Option B	1*	2–3*
Scenario III, Option A	½*	1*
Scenario III, Option B	½*	1*

*The predominant cause of the temperature increase is the CO_2 concentration in the atmosphere due to combustion of fossil fuels.

The results clearly indicate the combined roles of continued use of fossil fuels and per capita use of energy in influencing climates of the future. The greatest warming (Scenario I, Option A) occurs when fossil fuels are used throughout the twenty-first century with a world population reflecting the higher estimates; the least change occurs (Scenario III) at a low world-population estimate and when nuclear and solar power replace the fossil fuels.

In establishing and evaluating these scenarios, it must be noted that all results are tentative. Clearly, from the standpoint of dynamic climatology, the results are even more tentative because of the use of only two factors that influence climate. Beyond waste heat and carbon dioxide such things as dust, air pollution, and natural climatic change muct be considered. However, this work does show the nature of the scenario approach to climatic forecasting.

THE CURRENT VIEW OF FUTURE CLIMATES

At the present time many researchers are attempting to provide a method whereby climate in the coming century might be forecast. It is a very difficult task, and quite opposing viewpoints have been expressed. Given the shortcomings in understanding both past and present climates, it is not surprising that there is no easily obtained answer. Unfortunately, one of the most difficult things in science to establish is a unanimous consensus; it is doubtful that until a complete understanding of the atmosphere is attained there will be total agreement on what lies ahead. However, to derive some idea of future climates we can draw upon the findings of a recent conference held in Berlin. The general areas of agreement among the scientists present were the following:

- World climates have changed, they are changing now, and they will continue to change in the future.
- Although there has been a general cooling trend since about 1950, it is probable that by the year 2000 a slight warming that will be distinguishable from the normal variations in climate will occur. This increase will be largely caused by human activities that modify the climatic environment.
- If unusual occurrences (such as widespread volcanic activity, a change in mean cloudiness, Antarctic ice surges) do not occur, the global mean temperature of the middle of the next century will be greater than at any time over the last 1000 years.
- The changes will not be evenly distributed over the globe, and regional climatic changes will differ. It is expected that polar regions will have changes two or three times greater than the global average. This phenomenon will influence regional wind systems and precipitation distribution.

Thus it appears that a warming of climates is the general consensus. It is further agreed that emission of carbon dioxide into the atmosphere is a major contribution to this warming. Clearly, it would be most satisfying to assume that this is the answer; but there are areas of uncertainty that could negate this scenario. The role of aerosols, changes in polar ice cover, and the like could totally modify this forecast. Much remains to be known before climatic forecasting is an exact science.

SUMMARY

Like meteorologists, climatologists are involved in trying to predict the future. For the climatologist the time periods involved are much longer and the forecasts more speculative. Despite this fact, the effort is necessary because climate plays an integral role in the needs of people throughout the world. The need to know about climates of the future concerns the potential impacts on energy requirements, food supplies, water needs, and ecological environments. The array provides the applied climatologist with a wealth of research problems.

Climatic modeling forms a significant part of forecasting future climates. The methods of investigation vary, with a lot of attention being paid to ocean-atmosphere relationships at the current time. To account for the many possible outcomes, scenarios are established; these provide a range of potential impacts of any change that might occur in future climates.

What actually lies ahead has not been completely resolved. The general interpretation, however, suggests that climates are warming; by the middle part of the next century, the prevailing temperatures will be higher than they are today.

APPLIED STUDY 16

Climatic Change and Food Production

In recent years, the devastating impacts of crop failure have been widely covered by the communications media. Scenes showing the suffering of peoples in Ethiopia, northeast Brazil, and India have graphically displayed the results of weather-induced crop failure. Consider, however, that these disasters are relatively short-term; eventually agricultural production tends to recover. Such might not be the case if the abnormal were to become the normal because of a change in climate.

There is little doubt that a change in world climate would create massive food-production problems in the marginal areas where crop failures now occur periodically. Beyond such areas a change would also influence agriculture in regions now considered highly productive. As an example of this, corn and wheat production in North America can be considered.

In recent years there have been enormous increases in agricultural yields in North America. The increase is almost entirely attributable to improved farming techniques, particularly the scientific application of fertilizers. Figure 16-3 shows how the increase in corn yields in the Corn Belt states almost parallels the application of nitrogen fertilizer. The peaks and valleys in the left-hand diagram illustrate the relative influence of the climatic factor in corn production, with higher production occurring in years with good growing conditions. Using such data, it is possible to evaluate the relative income losses that result from deviations from normal temperature and precipitation. Such a study was completed for the Spring Wheat Belt of the United States. The data in table 16-2 were derived from that study. The findings indicate that variations from the average climatic conditions have significant monetary effect on production, irrespective of the role of fertilizers. Given the fact that the climatic factor can be assessed through evaluation of historic data, it becomes possible to estimate how productivity will be modified by a changed climate.

The evaluation of the impact of climatic change on the Corn Belt states (here assumed to be Ohio, Indiana, Illinois, and Iowa) suggests an interesting result. The long-term mean summer month temperature of these areas is June,

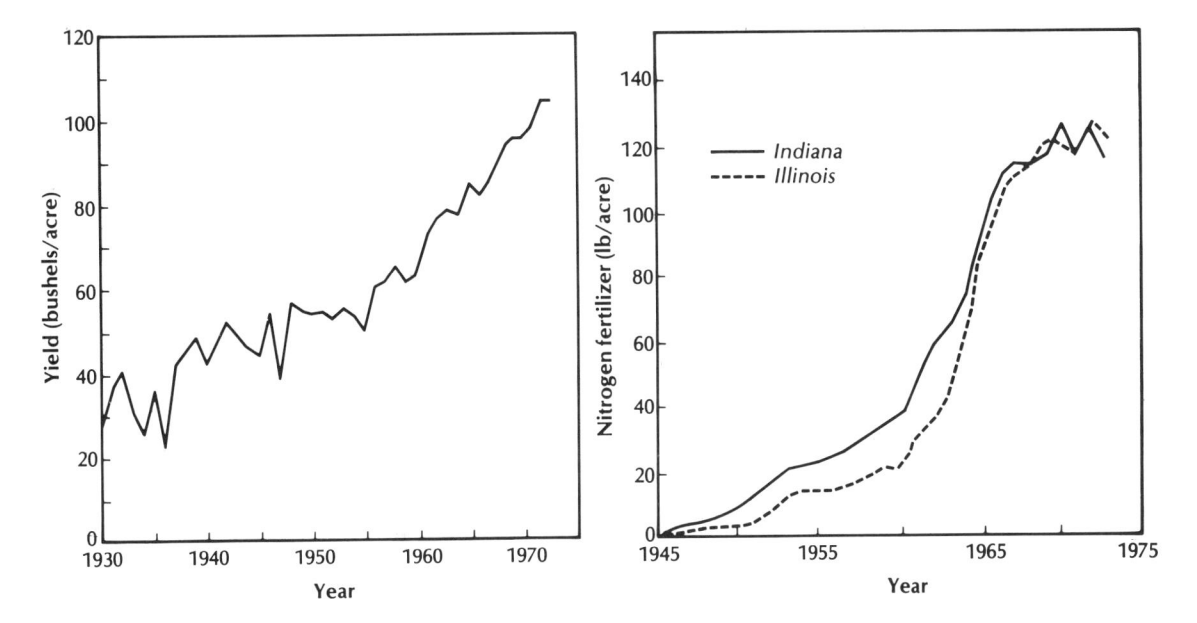

FIGURE 16-3 The effects of climate on corn production.

22.2°C (72°F); July, 24.4°C (76°F); August, 23.3°C (74°F). Research has shown, however, that highest yields are obtained at a mean of 22.2°C (72°F). This finding means that, given sufficient rainfall, highest yields are obtained when temperatures are slightly below normal. Thus, a slight cooling of the summer would suggest an increase in corn yields. Conversely, a warming would lead to lower corn production.

Such a scenario can only be assumed correct if nothing else changes. This assumption is somewhat misleading because a probable effect of a modified climatic regime is an increase in the variability of conditions. This probability is particularly true of the precipitation regime. To assess future agricultural yields correctly requires consideration of all the climatic elements.

The problem is further complicated by the influence of temperature change on the growing season of crops. Thus although the winter wheat production in the United States may increase because of a decline in temperature,

TABLE 16-2 U.S. income losses due to temperature and precipitation deviations from normal (millions of dollars).

	Temperature		Precipitation	
	+1°C	−1°C	+20%	−20%
Preseason conditions	—	—	+21	−30
April	+40	−40	+22	−25
May	−22	+13	+4	−37
June	−70	+70	+37	−44
July	−78	+70	−2	−2
Whole season	−131	+136	+82	−139

wheat-growing areas to the north may well be negatively influenced by a shortening of the growing season.

Climatic change will also influence animal husbandry. As an example, a study was carried out on the relative impact of temperature and precipitation changes on sheep production in Australia. At Esperance, a station in the Mediterranean climate of southwest Australia, the normal stocking rate is 10 sheep per hectare. If this rate is maintained, the profit from sheep farming is modified as follows:

Percentage changes of profit as a function of climatic change.

Temperature		Precipitation	
+1°C	−1°C	+10%	−10%
+6%	−9%	+14%	−1%

This example shows that an increase in both temperature and precipitation leads to increased productivity. Lower temperatures and precipitation reduce productivity.

The attempt to deduce potential impacts of climatic change upon agriculture cannot include all of the feedback mechanisms that follow the change. One need only consider the extended effects of drought in grazing areas where the limited precipitation reduces available animal food and competition leads to overgrazing. This response then results in modification of the surface characteristics, which then may lead to dust-bowl conditions, surface albedo modification, and a modified climate. There is no simple answer to exactly how agricultural productivity may be influenced by climatic change.

Appendix 1
Background Notes

1. THE GAS LAWS
2. RADIATION LAWS
3. ENERGY FLOW REPRESENTATION
4. THE HEAT BUDGET
5. LAPSE RATES AND STABILITY
6. THORNTHWAITE'S WATER BALANCE
7. VORTICITY AND ANGULAR MOMENTUM

APPENDIX 1

BACKGROUND NOTE 1: THE GAS LAWS

1. *Boyle's laws* describe relationships among pressure, volume, and density of gases. In each case the equations assume temperature to be held constant.
   The first law states

   $$P_0 V_0 = P_1 V_1 = K$$

   where P_0 and P_1 and V_0 and V_1 are the pressure and volume at two times, 0 and 1, respectively. K is a constant. The relationship indicates that an increase in pressure results in a decrease in volume and vice versa.
   The second law relates pressure (P) to density (D)

   $$\frac{P}{D} = K \text{ (at a constant temperature)}$$

   As pressure increases so does density.

2. If pressure is considered constant, the volume of a gas can be related to temperature by *Gay-Lussac's law*:

   $$V_t = V_0 \left(1 + \frac{t}{273}\right)$$

   where V_t is the volume at a temperature t (in °C) and V_0 is the volume at 0°C.

3. *Charles' law* provides the relationship between pressure and temperature when volume is held constant.

   $$P_t = P_0 \left(1 + \frac{t}{273}\right)$$

   Here P_t is pressure at a given temperature, t (in °C); and P_0 is pressure at 0°C.

4. The *combined gas law* uses a gas constant, r, to combine the relationships of gas pressure, volume, and temperature.

   $$PV = rT$$

   This *equation of state* can be written in a form that draws upon the molecular weight of a mass to derive a universal gas constant, R, and the density of the gas

   $$P = \rho RT$$

   in which P is pressure, ρ is density, R is the universal gas constant (2.87×10^6 erg/g° K), and T is temperature. This equation allows derivation of any one variable if the other two are known.

BACKGROUND NOTE 2: RADIATION LAWS

1. Radiant energy is characterized by wavelength (λ). The product of wavelength and frequency of the wave is a constant, the speed of light (c).

$$c = \lambda f$$

where λ is wavelength, f is frequency, and c the speed of light (3×10^{10} cm/s). The direct relationship indicates that as λ increases, f will decrease. Short waves are characterized by high frequencies, long waves by low frequencies.

2. Basic energy laws apply to a black body; black bodies absorb all incoming energy and radiate at a rate determined by their temperature. The rate of radiation (F) is given by the *Stefan-Boltzmann law*:

$$F = \sigma T^4$$

where F is the flux of radiation (ly/min), T is temperature in degrees Kelvin, and σ is a constant (0.813×10^{-10} cal/cm^2/min/deg^4). This equation shows that the flux of energy from a black body is related to its temperature. The higher the temperature, the greater the flux.

3. Radiators other than black bodies will not be perfect radiators. These gray bodies emit less radiation; the emissivity (ϵ) is thus less than 1 and is given by

$$F = \epsilon \sigma T^4$$

where F, σ, T^4 are components of the Stefan-Boltzmann equation and ϵ is the emissivity of the gray body at a particular wavelength.

4. *Kirchoff's law* relates emissivity to absorption (or absorptivity) of radiation at a particular wavelength.

$$\alpha\lambda = \epsilon\lambda$$

$\alpha\lambda$ is the fractional amount of energy that is absorbed (compared to that of a black body) at a given wavelength while $\epsilon\lambda$ is emissivity at that wavelength. This law shows that strong absorbers of radiation are also strong emitters.

5. If the temperature of a black body is known, its wavelength of maximum emission (λ_{max}) is given by *Wien's law*.

$$\lambda_{max} = \frac{K}{T}$$

in which K is a constant (2897) and T is temperature in degrees Kelvin. The inverse relationship means that the higher the temperature of a black body, the shorter the wavelength of maximum emission.

APPENDIX 1

BACKGROUND NOTE 3: ENERGY FLOW REPRESENTATION

In attempting to standardize the representation of energy flows, the WMO suggests the following symbols.

$Q\downarrow$ $(Q\downarrow = K\downarrow + L\downarrow)$	Downward radiation
$K\downarrow$ $(K\downarrow = S + D)$	Global solar radiation
S	Vertical component of direct solar radiation
D	Diffuse solar radiation
$L\downarrow$ $(A\downarrow)$ $(L\downarrow = A\downarrow)$	Downward atmospheric radiation
$Q\uparrow$ $(Q\uparrow = K\uparrow + L\uparrow)$	Upward radiation
$K\uparrow$	Reflected solar radiation
R	Upward solar radiation reflected by earth's surface alone
$L\uparrow$	Upward terrestrial radiation
Lg	Upward terrestrial radiation—surface
r	Reflected atmospheric radiation
$A\uparrow$	Upward atmospheric radiation
Q^*	Net radiation
K^*	Net solar radiation
L^*	Net terrestrial radiation

Given these symbols, components of the earth's energy budget can be represented by equations. Of particular importance is net radiation:

$$\text{Net radiation} = \text{Incoming} - \text{Outgoing}$$
$$Q^* = Q\downarrow - Q\uparrow$$

or
$$Q^* = K^* + L^*$$

Since
$$K^* = K\downarrow - K\uparrow \quad \text{and} \quad L^* = L\downarrow - L\uparrow$$
$$Q^* = (K\downarrow - K\uparrow) + (L\downarrow - L\uparrow)$$

Alternatively since
$$Q\downarrow = K\downarrow + L\downarrow \quad \text{and} \quad Q\uparrow = K\uparrow + L\uparrow$$

then
$$Q^* = (K\downarrow + L\downarrow) - (K\uparrow + L\uparrow)$$

Source: World Meteorological Organization (1971), *Guide to Meteorological Instrument and Observing Practices,* 4th ed., World Meteorological Organization No. 8TP.3, Geneva.

BACKGROUND NOTE 4: THE HEAT BUDGET

Consider a column of the earth's surface extending down to where vertical heat exchange no longer occurs (figure A1). The net rate (G) at which heat in this column changes depends upon the following:

- ☐ Net radiation ($K\downarrow - K\uparrow$) + ($L\downarrow - L\uparrow$)
- ☐ Latent heat transfer (LE)
- ☐ Sensible heat transfer (H)
- ☐ Horizontal heat transfer (S)

In symbolic form:

$$G = (K\downarrow - K\uparrow) + (L\downarrow - L\uparrow) - LE - H \pm S$$

Since

$$(K\downarrow - K\uparrow) + (L\downarrow - L\uparrow) = Q^*$$

then

$$G = Q^* - LE - H \pm S$$

in terms of Q^*

$$Q^* = G + LE + H \pm S$$

The column will not experience a net change in temperature over an annual period; that is, it is neither gaining nor losing heat over that time, so $G = 0$ and can be dropped from the equation.

$$Q^* = LE + H \pm S$$

This equation will apply to a mobile column, such as the oceans. On land where subsurface flow of heat is negligible, S will be unimportant. The land heat budget becomes

$$Q^* = LE + H$$

Note that a complete expression of the heat budget at a location must consider other items. For example, Flohn gives an equation that includes H (as T_L) and LE (as T_v) but also itemizes the variables listed on the following page.

FIGURE A1 The heat budget.

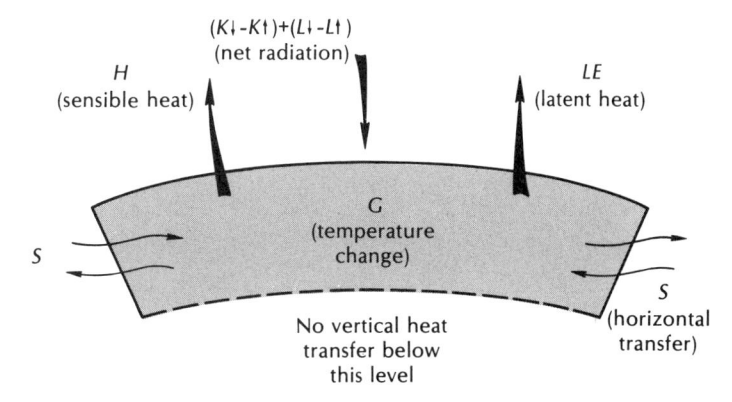

APPENDIX 1

- ☐ Heat transmission into land (T_B) or sea (T_M) masses
- ☐ Heat used in melting ice and snow (T_S)
- ☐ Heat used in warming falling precipitation (T_N)
- ☐ Heat generated by wind friction with the ground (T_R)
- ☐ Heat energy used in biologic processes (T_{BIOL})

Flohn's representation is

$$Q^* = T_B + T_L + T_v + T_S + T_N + T_R + T_{BIOL}$$

Source: Herman Flohn, *Climate and Weather*, World Universal Library, New York, 1969, p. 23.

BACKGROUND NOTE 5: LAPSE RATES AND STABILITY

Stable atmospheric conditions resist upward or downward displacement of air; unstable conditions permit an upward or downward disturbance to develop into extensive vertical motion. Cloudforms, precipitation probability, and air-pollution levels are determined in part by atmospheric stability.

Analysis of stability involves adiabatic processes. Air that is compressed or expanded experiences a change in internal energy without loss or gain of heat to the surrounding environment. The change in temperature in a parcel of air can be easily calculated by means of Poisson's equation:

$$\frac{T}{T_0} = \left(\frac{P}{P_0}\right)^{0.286}$$

If we know the initial temperature (T_0) and pressure (P_0), then the temperature (T) at any other pressure level (P) can be calculated. Development of equations allows the adiabatic rate for dry air in the troposphere to be calculated. This *dry adiabatic lapse rate* (DALR) is 5.5°F/1000 ft, or 1°C/100 m.

Comparison of the dry adiabatic lapse rate with the prevailing environmental lapse rate, explained in the text, determines the stability of air. Should lifting eventually produce saturation of a rising parcel of air, a new process becomes involved. The release of latent heat from the condensing water vapor reduces the rate of adiabatic cooling. A moist, or wet, adiabatic rate prevails. This rate decrease is variable, but the moist adiabatic lapse rate (MALR) is commonly on the order of 3.2°F/1000 ft, or 0.5°C/100 m.

The relationships between adiabatic and environmental lapse rates produce a number of conditions of stability (note that the schematic diagrams in figure A2 illustrate relative displacement upward; downward displacement follows the same principles):

1. *Stable equilibrium*   The temperature decrease with height is less than the adiabatic rates (Figure A2[a]). Any displaced parcel of air will return to its original position. Such a condition resists vertical motion.

FIGURE A2 Relative upward displacement of air: (a) stable equilibrium; (b) conditionally stable; (c) neutral equilibrium; (d) unstable equilibrium.

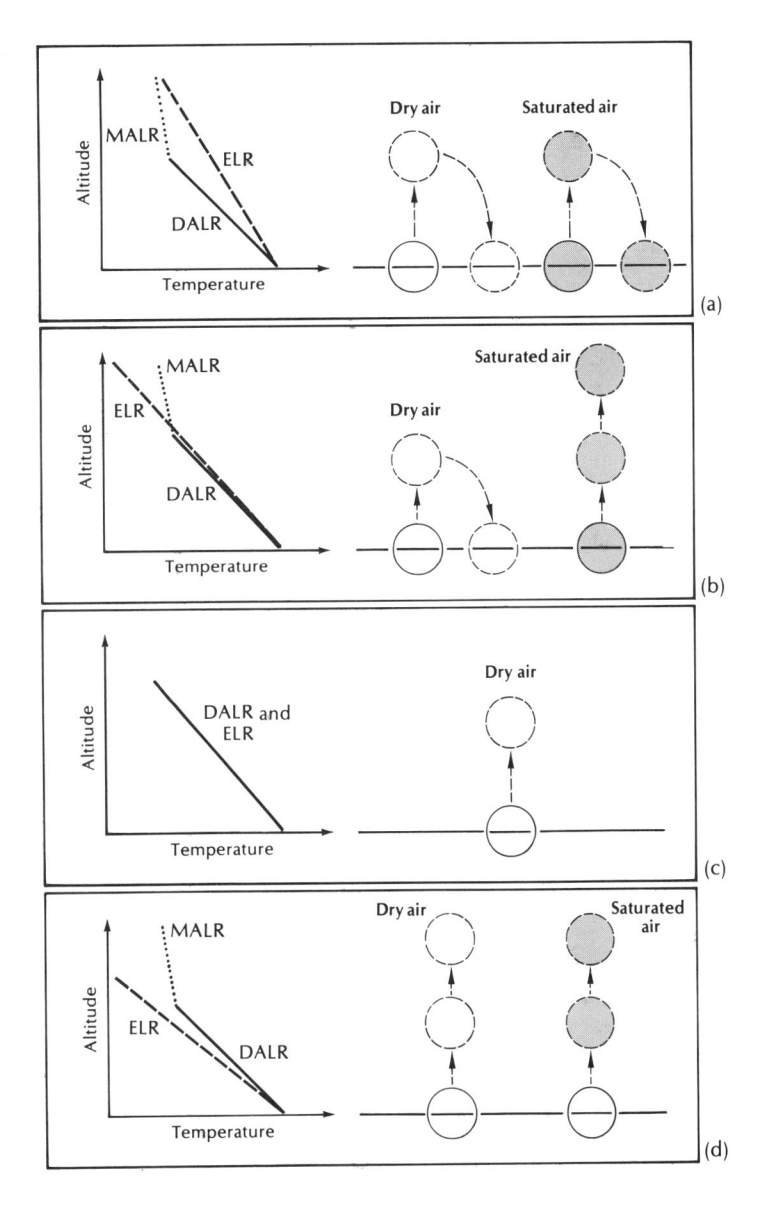

2. *Conditionally stable equilibrium* In this case, the environmental lapse rate is between the values for dry and saturated adiabatic rates (figure A2[b]). A dry parcel of air will return to its original position on uplift *unless* saturation is attained. Should the moist adiabatic lapse rate prevail, the air parcel will continue to be displaced.

3. *Neutral equilibrium* Neutral equilibrium is a rare situation in which the dry adiabatic lapse rate and the environmental lapse rate are the same (figure A2[c]). No matter to which height a dry parcel of air is displaced, it

APPENDIX 1

will remain in place. If, in an even rarer situation, such an air parcel becomes saturated, it then becomes unstable.

4. *Unstable equilibrium* If the prevailing environmental lapse rate is greater than the dry adiabatic lapse rate, a parcel of air will continue to move upon displacement (figure A2[*d*]). Such a situation leads to rapid vertical motion.

BACKGROUND NOTE 6: THORNTHWAITE'S WATER BALANCE

It will appear evident, from all that has been said thus far, that a balance between incoming and outgoing moisture is attained and that the balance will reflect the climatic regime that exists. A number of writers have made use of the budget approach to evaluate the climatic and hydrologic differences that exist over the earth's surface. Probably the best known is the Thornthwaite method; table A1 illustrates Thornthwaite's bookkeeping method.

The potential evapotranspiration, shown in row 1, is the adjusted potential evapotranspiration that is derived by substituting monthly temperature (°C) into an empiric formula derived by Thornthwaite. The heat index, i, is calculated at the same time. The results obtained provide the unadjusted PE, which is corrected through the use of a prepared nomograph.

Row 2 provides values for monthly precipitation. The difference between precipitation and potential evapotranspiration, row 3, provides the amount of moisture available after evapotranspiration requirements have been satisfied. The excess above the difference will either go to soil water or occur as runoff. The amount of water stored by the soil is highly variable and will depend upon the nature of the soil. In his initial work, Thornthwaite used a value of 4 in. (10 cm) as a general value to be applied to all water balance studies. Although this does enable comparison of the balance for different stations, it is not realistic in terms of precise water balance

TABLE A1 Sample water balance computation.

	Jan.	Feb.	Mar.	Apr.	May	June	July	Aug.	Sept.	Oct.	Nov.	Dec.	Yr.
1 PE	0.5	0.7	1.2	2.0	3.1	3.9	4.8	4.4	3.4	2.0	0.9	0.4	27.3
2 P	5.6	4.4	4.0	2.3	2.0	1.9	0.5	0.7	1.6	3.6	5.6	6.7	38.9
3 P−PE	5.1	3.7	2.8	0.3	−1.1	−2.0	−4.3	−3.7	−1.8	1.6	4.7	6.3	
4 ΔST	0	0	0	0	−1.1	−2.0	−0.9	0	0	+1.6	+2.4	0	
5 ST	4.0	4.0	4.0	4.0	2.9	0.9	0	0	0	1.6	4.0	4.0	
6 AE	0.5	0.7	1.2	2.0	3.1	3.9	1.4	0.7	1.6	2.0	0.9	0.4	18.4
7 D	0	0	0	0	0	0	3.4	3.7	1.8	0		0	8.9
8 S	5.1	3.7	2.8	0.3	0	0	0	0	0	0	2.3	6.3	20.5

*PE—potential evapotranspiration; P—precipitation; P−PE—precipitation minus evapotranspiration; ΔST—change in soil moisture storage since previous month; ST—soil moisture storage at the end of the month; AE—actual evapotranspiration; D—water deficit; S—water surplus. Data in inches.
Source: After Carter (1965)

studies. One of the most significant modifications of the Thornthwaite budget is the introduction of the varying soil moisture-holding capacities. However, for demonstration purposes, the 4 in. value is retained here.

Rows 4 and 5 give the amount of moisture stored in the soil and the change in storage (ΔST) since the previous month, using the 4 in. value. As long as precipitation is greater than potential evapotranspiration the value remains 4 in. However, as soon as the relationship is reversed, potential evapotranspiration is greater than precipitation, plants draw upon the available soil moisture, and the soil storage falls below capacity. As soon as the cumulative excess of potential evapotranspiration over potential is greater than 4 in., soil moisture is assumed to be totally utilized. A deficit period then exists. As row 6 shows, actual evaporation is lower than the potential at this time. Obviously, at the end of the deficit period, when P is greater than potential evapotranspiration (PE), the 4 in. must be restored to the soil before any surplus occurs. Rows 7 and 8 provide the balance of water in terms of deficit and surplus, and it will be noted that the deficit does not occur as soon as potential evapotranspiration is greater than precipitation because of the period of soil moisture utilization.

The preceding data can be shown graphically, as in figure A3. Notice that the graph is divided into four areas: the surplus and deficit periods, the period of utilization, and the period of recharge.

FIGURE A3 Water balance for Vancouver, Washington (From Douglas B. Carter, *Fresh Water Resources*, Association of American Geographers, 1965. Used with permission.).

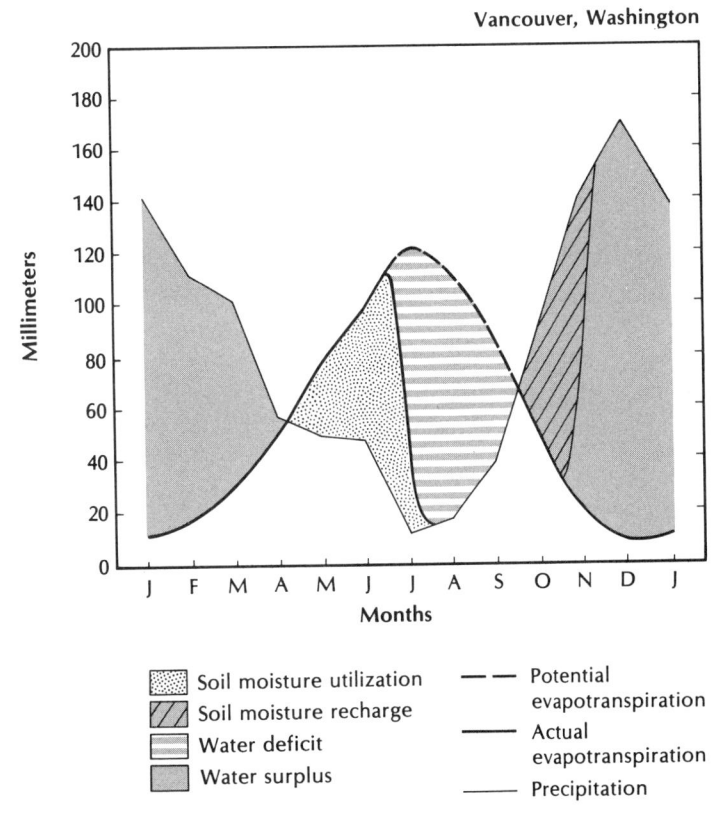

APPENDIX 1

BACKGROUND NOTE 7: VORTICITY AND ANGULAR MOMENTUM

Figure A4 shows a cyclonic circulation. The rate of spin of an air particle is given by its vorticity (Q). In a constant circular motion, such as that shown in figure A4, vorticity is computed by the circulation path (C) divided by the area enclosed by that path (A). The value of C is the circumference of the circle ($C = 2\pi R$) multiplied by the rotational velocity V. Thus:

$$Q \text{ (Vorticity)} = C \text{ (Circulation)}/A \text{ (Area)}$$
$$= 2\pi RV/\pi R^2$$
$$= 2V/R$$

The rate of spin depends upon the radius of curvature of the circulating air and its velocity.

The same variables, V and R, are used to derive the *angular momentum*, the amount of rotational spin. To obtain this, the mass of the rotating body is incorporated. Of significance in respect to the atmosphere is the fact that angular momentum is conserved when rotating air moves from one place to another. This *conservation of angular momentum* is given by:

$$\text{mass} \times \text{velocity} \times \text{radius of curvature} = \text{constant}$$

or

$$mVR = K$$

This relationship shows that if a unit mass of air ($m = 1$) changes its radius of curvature, then its velocity must also change. An example of this, using quite arbitrary and nondimensioned values, is given in figure A5.

A frequently seen example of the conservation of angular momentum at work occurs when an ice skater changes from a slow to a rapid whirl. This effect

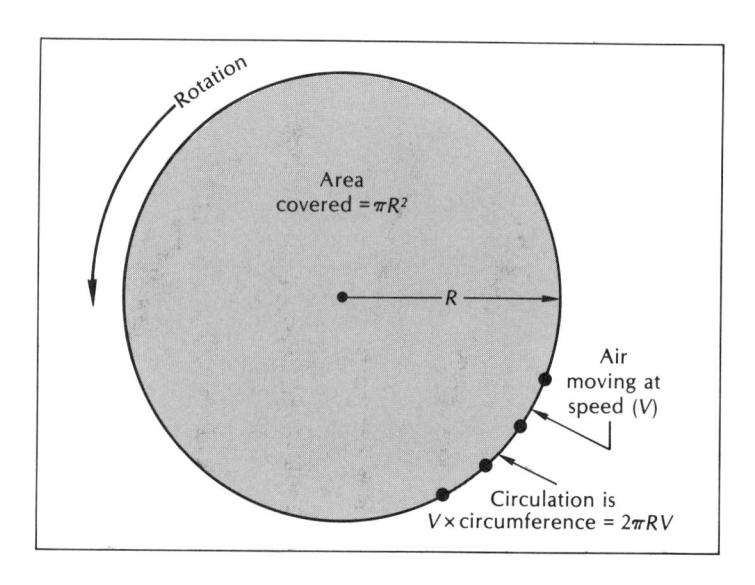

FIGURE A4 Cyclonic circulation around a center.

occurs because after gaining momentum, the skater stretches upward to reduce the radius of the spin. The stretching-up motion that enables air to increase in velocity probably occurs in a tornado. Figure A6 provides an example of how this occurs.

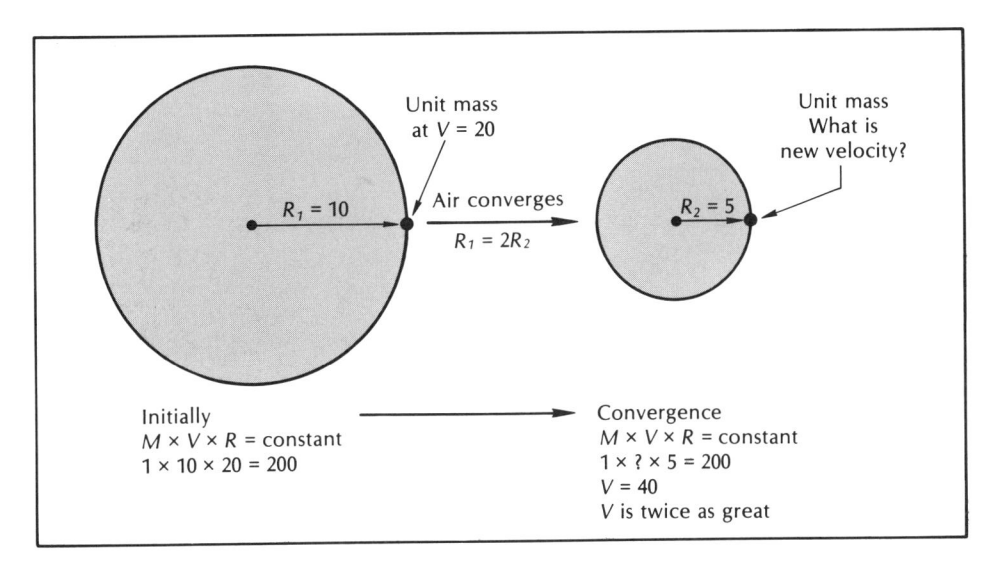

FIGURE A5 Following the conservation of angular momentum, reduction of the radius of a mass in circular motion will lead to an increase in its velocity.

Unit mass at $V = 20$
$R_1 = 10$
Air converges
$R_1 = 2R_2$
Unit mass What is new velocity?
$R_2 = 5$

Initially
$M \times V \times R$ = constant
$1 \times 10 \times 20 = 200$

Convergence
$M \times V \times R$ = constant
$1 \times ? \times 5 = 200$
$V = 40$
V is twice as great

FIGURE A6 Stretching of a column of air will lead to an increase in velocity of the circulating winds.

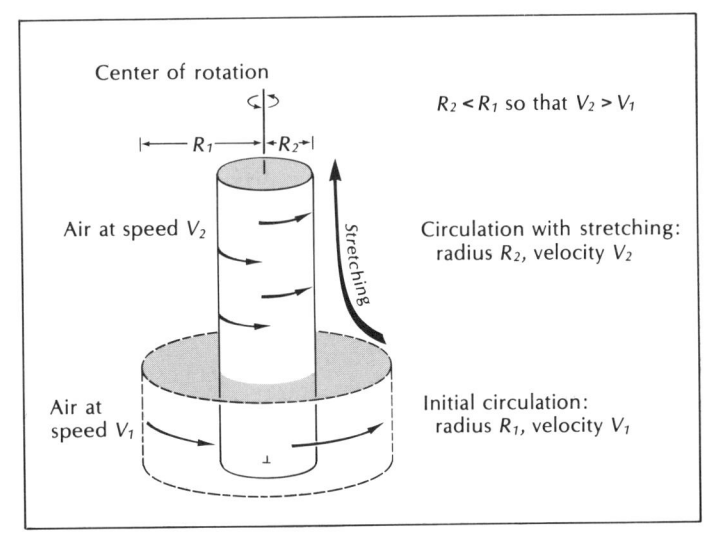

Center of rotation
R_1 R_2
Air at speed V_2
Stretching
Air at speed V_1

$R_2 < R_1$ so that $V_2 > V_1$

Circulation with stretching: radius R_2, velocity V_2

Initial circulation: radius R_1, velocity V_1

Appendix 2
Glossary

Absolute humidity   The ratio of the mass or weight of water vapor per unit volume of air; for example, grams per cubic meter.
Adiabatic   Heating or cooling in gases due strictly to the expansion and contraction of the gases.
Advection   Mass motion in the atmosphere; in general, horizontal movement of the air.
Air mass   A large body of air that is characterized by homogeneous physical properties at any given altitude.
Albedo   The reflectivity of the earth environment, generally measured in percentage of incoming radiation.
Angstrom   A unit of length equal to 10^{-8} cm used in measuring electromagnetic waves.
Anticyclone   An area of above-average atmospheric pressure characterized by a generally outward flow of air at the surface.
Atmosphere   The mixture of gases that surrounds the earth.
Atmospheric circulation   The motion within the atmosphere that results from inequalities in pressure over the earth's surface. When the average of the entire globe is considered, it is referred to as the *general circulation* of the atmosphere.
Aurora   A display of colored light seen in the polar skies; called *aurora borealis* in the Northern Hemisphere and *aurora australis* in the Southern Hemisphere.

Black body   A substance or body that is a perfect absorber and a perfect radiator.
Blizzard   High winds accompanied by blowing snow usually associated with winter cold fronts in mid-latitudes.
Bora   A cold, dry winter blowing down off the highlands of Yugoslavia and affecting the Adriatic coast.
Buoyant   Less dense than the surrounding medium and thus able to float.

Calorie A measurement equal to the amount of heat needed to raise 1 g water 1°C; equal to 4.19 Joules.
Centrifugal force The apparent force exerted outward on a rotating body or on an object traveling on a curved path.
Chinook A warm, dry wind blowing down off the Rocky Mountains of western North America.
Climate All of the types of weather that occur at a given place over time.
Climatic regime The annual cycles associated with various climatic elements; for example, the thermal regime is the seasonal patterns of temperature, and the moisture regime is the seasonal patterns of precipitation.
Cloud streets Long, thin lines of clouds forming in the trade winds when winds are steady and of low velocity over helical currents.
Cold front The leading edge of a cold air mass where it displaces a warmer air mass.
Cold wave An unseasonably cold spell that can occur at any time of year.
Condensation The change of state from a gas to a liquid.
Conduction Energy transfer directly from molecule to molecule. It takes place most readily in solids in which molecules are tightly packed.
Convection Mass movement in a fluid or vertical movements in the atmosphere.
Convergent Moving toward a central point of area; coming together.
Coriolis force An apparent force caused by the earth's rotation. It is responsible for deflecting winds clockwise in the Northern Hemisphere and counter clockwise in the Southern Hemisphere.
Cyclone Any rotating low-pressure system.

Deflation The lifting and removal of earth particles by wind action.
Dewpoint The temperature at which saturation would be reached if the air mass were cooled at constant pressure without altering the amount of water vapor present.
Diurnal Occurring in the daytime or having a daily cycle.
Divergence The condition that exists when the distribution of winds within a given area results in a net horizontal outflow of air from the region. In divergence at lower levels, the resulting deficit is compensated for by a downward movement of air from aloft; hence, areas of divergent winds are unfavorable to cloud formation and precipitation.
Doldrums An area near the equator of very ill-defined surface winds associated with the equatorial convergence zone.
Dust devil A small cyclonic circulation, or dust swirl, produced by intense surface heating. They are most common in arid regions and resemble miniature tornadoes in appearance.

Easterly wave A weak large-scale convergence system that is part of the secondary circulation of the tropics.
Electromagnetic spectrum The range of energy that is transferred as wave motions and that does not require any intervening matter to make the transfer. The waves travel at the speed of light, 186,000 mi/s.
Electromagnetic waves Waves characterized by variations of electric and magnetic fields.
Evapotranspiration The total water loss from land by the combined processes of evaporation and transpiration.
Exosphere The outermost region of the atmosphere from which particles may escape to space. The first interaction of solar radiation with the atmosphere occurs here.

GLOSSARY

Fluid A substance capable of flowing easily.
Foehn A wind occurring in central Europe which is the same wind as the chinook wind of North America; also called a *leste wind*.

Gas law The pressure exerted by a gas is proportional to its density and absolute temperature.
Geostrophic wind A wind aloft flowing parallel to the pressure gradient, with the pressure gradient and the Coriolis force in balance.
Greenhouse effect The process by which the heating of the atmosphere is compared to a common greenhouse. Sunlight (shortwave radiation) passes through the atmosphere to reach the earth. The energy reradiated by the earth is at a longer wavelength, and its return to space is inhibited by atmospheric carbon dioxide and water vapor. This process acts to increase the temperature of the lower atmosphere.

Hadley cell A convectional cell operating as part of the general circulation located approximately between the Tropic of Cancer or the Tropic of Capricorn and the equator.
Harmattan A dry, dust-laden wind blowing south from the Sahara Desert.
Heat wave Any unseasonably warm spell, which can occur any time of the year.
Hectare A metric unit of area equal to 2.47 acres.
Hurricane A tropical cyclone that develops in the Atlantic Ocean.

Infrared radiation Radiation in the range longer than red. Most sensible heat radiated by the earth and other terrestrial objects is in the form of infrared waves.
Intertropical convergence zone The seasonally migrating, low-pressure zone located approximately at the equator, where the northeast and southeast trade winds converge. Comprised largely of moist and unstable air, it provides copious precipitation. Also referred to as the Intertropical Front (ITF).
Inversion A reversal of the normal atmospheric regime; for example, a temperature increase with height.
Ionosphere A zone of the upper atmosphere characterized by gases that have been ionized by solar radiation.
Isobar A line on a map or chart connecting points of equal barometric pressure.

Jet stream A high-speed flow of air that occurs in narrow bands of the upper-air westerlies.

Katabatic wind Any air blowing downslope as a result of the force of gravity.

Langley A measure of radiation intensity equal to 1 g-cal/cm^2.
Latent energy Energy temporarily stored or concealed, such as the heat contained in water vapor.
Leveche A dry, dust-laden wind blowing from the Sahara Desert into Spain.

Mesosphere The layer of the atmosphere above the stratosphere where temperatures drop fairly rapidly with increasing height.
Metabolism The chemical processes that sustain organisms.
Microclimate The climate of a small area, such as a forest floor or small valley.

Millibar A unit of pressure equal to 1000 dynes/cm².
Mistral A cold, dry gravity wind blowing down off the Alps and affecting the French and Italian Riviera.

Occluded front A warm mass of air trapped when a cold front overtakes a warm front.
Opaqueness The degree to which light will not pass through a substance.
Ozone Oxygen in the triatomic form (O_3); highly corrosive and poisonous.
Ozone layer The layer of ozone, 25 km above the earth's surface, that absorbs ultraviolet radiation from the sun.

Periodic Occurring or appearing at regular intervals, as the sun's rising and setting.
Photochemical A chemical change that either releases or absorbs radiation.
Photoperiod The period of each day when direct solar radiation reaches the earth's surface, approximately sunrise to sunset.
Pluvial Pertaining to precipitation.
Polar front The storm frontal zone separating air masses of polar origin from air masses of tropical origin.
Pressure gradient The amount of pressure change occurring over a given distance.

Rayleigh scattering The scattering of solar radiation by particles in the earth's atmosphere.
Relative humidity The ratio of the amount of water present in the air to the amount of water vapor the air can hold, multiplied by 100.
Rossby waves Upper-air waves in the middle and upper troposphere of the middle latitudes with wavelengths of 4000–6000 km; named for C. G. Rossby, the meteorologist who developed the equations for parameters governing the waves.

Santa Ana A chinook wind occurring in southern California and northern Mexico.
Sastrugi Ripples produced in snow by persistent gravity winds in Antarctica.
Saturation vapor pressure The maximum amount of water vapor that the atmosphere can hold at a given temperature.
Sensible temperature The sensation of temperature the human body feels in contrast to the actual heat content of the air recorded by a thermometer.
Sirocco A hot, dry wind blowing north across the Mediterranean Sea.
Solar constant The mean rate at which solar radiation reaches the earth.
Specific gravity The ratio of a unit mass of a substance to a unit mass of water.
Specific heat The amount of heat needed to raise 1 g of a substance 1°C.
Specific humidity The ratio of the mass or weight of water vapor in the air to a unit of air including the water vapor, such as grams of water vapor per kilogram of wet air.
Stationary front A cold or warm front that has ceased to move; the boundary between two stagnant air masses.
Stefan Boltzmann's law A law pertaining to radiation that states the amount of radiant energy emitted by a black body is proportional to the fourth power of the absolute temperature of the body.
Stratopause The upper boundary of the stratosphere.
Sublimation The transition of water directly from the solid state to the gaseous state, without passing through the liquid state; or vice versa.
Subsidence Descending or settling, as in the air.
Synoptic climatology A study of climatology that relates local and regional climates to atmospheric circulation patterns.

Terrestrial Pertaining to the land, as distinguished from the sea or air.
Terminal fall velocity The rate at which a particle will fall through a fluid when the acceleration due to gravity is balanced by friction.
Thermodynamics The science of the relationship between heat and mechanical work.
Thermosphere The zone including the ionosphere and exosphere.
Thunderstorm A convective cell characterized by vertical cumuliform clouds.
Tornado An intense vortex in the atmosphere with abnormally low pressure in the center and a converging spiral of high velocity wind.
Trade winds Two belts of winds that blow almost constantly from easterly directions and are located on the equatorward sides of the subtropical highs.
Transpiration The process by which water leaves a plant and changes to vapor in the air.
Tropical cyclone A large rotating low-pressure storm that develops over tropical oceans, called a *hurricane* in the Atlantic and a *typhoon* in the Pacific Ocean.
Tropopause The upper boundary zone of the troposphere, marked by a discontinuity of temperature and moisture.
Troposphere The lower layer of the atmosphere marked by decreasing temperature, pressure, and moisture with height; the layer in which most day-to-day weather changes occur.
Twilight The period before sunrise and after sunset in which refracted sunlight reaches the earth.
Typhoon See *tropical cyclone*.

Ultraviolet radiation Radiation of a wavelength shorter than violet. The invisible ultraviolet radiation is largely responsible for sunburn.

Van Allen belts Two zones of charged particles existing around the earth at very high altitudes, associated with the earth's magnetic field.
Vapor pressure The partial pressure of the total atmospheric gaseous mixture that is due to water vapor, also called *vapor tension*.
Virga A thin veil of rain seen hanging from a thunderstorm but not reaching to the ground. The droplets are evaporated before they reach the ground.
Viscosity The internal friction in fluids that offers resistance to flow.
Vortex A whirling or rotating fluid with low pressure in the center.

Warm front A zone along which a warm air mass is displacing a colder one.
Waterspout A tornado occurring at sea that touches the surface and picks up water.
Wavelength The linear distance between the crests or the troughs in a successive wave pattern.
Weather The state of the atmosphere at any one point in time and space.
Willywaw A bora wind in Alaska, Greenland, and coastal Antarctica.

Zenith A point in space directly above a person's head; a point on a line passing through the point of observation and the center of the earth.

REFERENCES CITED

Barry, R. G. The World Hydrologic Cycle. In R. J. Chorley (Editor), *Water, Earth and Man.* London: Methuen, 1969.
Broecker, W. S. Climatic Change: Are We on the Brink of a Pronounced Global Warming? *Science* 1975, 189, 460–463.
Calder, N. *The Weather Machine.* New York: Viking Press, 1974.
Flint, R. F. *Glacial and Quaternary Geology.* New York: John Wiley & Sons, 1970.
Geiger, R. *The Climate Near the Ground*, 4th ed. Cambridge, Mass: Harvard University Press, 1965.
Giles, B. D. On Isobars, Isohypses and Isopachs or Pressure, Contour and Thickness Charts. *Weather*, 1976, *31*, 113–121.
Hershfield, D. M., Brakesiek, D. L., and Comer, G. H. Some Measures of Agricultural Drought. In E. F. Schultz (Editor), *Floods and Droughts: Proceedings of the International Hydrology Symposium.* Ft. Collins, Colo.: Colorado State University Press, 1972.
Hidore, J. J. *A Geography of the Atmosphere.* Dubuque, Iowa: W. C. Brown, 1969.
Hidore, J. J. *Physical Geography: Earth Systems.* Glenview, Ill: Scott, Foresman, 1974.
Mather, J. R. *Climatology: Fundamentals and Applications.* New York: McGraw-Hill Book Company, 1979.
Munn, R. E. *Descriptive Micrometeorology.* New York: Academic Press, 1966.
Ooi, Jin-bee. Rural Development in Tropical Areas. *Journal of Tropical Geography*, 1965, *12*(1).
Oke, T. R. *Boundary Layer Climates.* New York: Halsted Press, 1978.
Oliver, J. E. *Climate and Man's Environment.* New York: John Wiley & Sons, 1973.
Oliver, J. E. *Perspectives on Applied Physical Geography.* Belmont, CA: Wadsworth Publishing Company, 1977.
Oliver, J. E. *Climatology: Selected Applications.* New York: V. H. Winston & Sons, A Halsted Press Book in the Scripta Series in Geography, 1981.
Palmer, W. C. *Meteorological Drought.* U.S. Weather Bureau Research Paper No. 45. U. S. Department of Commerce. Washington, D. C.: Government Printing Office, 1965.
Reed, J. W. Wind Power Climatology, *Weatherwise*, 1974, *27*, 237–242.
Stringer, E. T. *Foundations of Climatology.* San Francisco: W. H. Freeman, 1975.
Yoshino, M. M. *Climate in a Small Area.* Tokyo: University of Tokyo Press, 1975.

Index

A

Absolute humidity, 85
Absorption, 26–27
Acclimatization, 289
Acid rainfall, 336–38
Acidity, 336
Active solar systems, 40
Adiabatic lapse rates, 81, 360–61
Adirondack Mountains, 337
Advection fog, 81
Aerology, 4
Africa, 209
Agassiz, Louis, 294
Age of Discovery, 7
Age of fishes, 299
Agricultural drought, 103
Agriculture, 306, 326
Air mass source regions, 230
Air mass system, 192–96
Air mass thunderstorms, 136
Air masses, 73–74, 120
Air masses, North America, 252–53
Air pollution, 177–80, 327, 341
Alaska, 277
Albedo, 51, 178
Albedo changes, 326

Alfisols, 248
Alkalinity, 336
Alpine glaciation, 296
Amazon basin, 203
American Revolution, 311
Amniote egg, 300
Amphibian development, 299
Amundsen-Scott Station, 279
Angle of incidence, 29
Angular momentum, 113, 364
Angular velocity, 113
Animal husbandry, 354
Antarctic Basin, 277–78
Antarctic extremes, 280
Anticyclones, 122
Applied climatology, 5, 18–20
Arctic air, 233
Arctic Basin, 273–77
Arctic Ocean, 276
Arid regions, 97
Aridisols, 201
Arizona, 246
Art and paintings, 314
Asia, 215
Asian monsoon, 129
Aspect, 57
Aswan, 218

Atmosphere, composition, 9–13
Atmosphere, layers, 14
Atmosphere-Ocean Relationships, 346
Atmospheric changes, 327–29
Atmospheric circulation, 115–16, 345
Atmospheric gases, 13
Atmospheric modification, 320–21
Atmospheric Science, 4
ATS-1, 7
Australia, 354

B

Baja California, 219
Bergeron-Findeisen process, 88
Biomes, 343
Biosphere, 4
Black body, 357
Blizzard, 261, 280
Bolivia, 283
Bowen ratio, 330
Boyle's law, 356
Brownouts, 340
Building codes, 157
Byrd Station, 278

C

California, 263–64
Canopy layer, 168
Cape Horn, 282
Carbon cycle, 327
Carbon dioxide, 9, 327–28
Carbon 14, 299
Carboniferous, 299, 301
Carrying capacity, 222–24
Chaco Canyon, 304
Chamonix, 311
Change in state, 69
Charles' law, 356
Cherrapunji, 126, 215
Chile, 237
Chinook, 242
Chromosphere, 25
Cilaos, 213
City climates, 329–34
Classification, air mass, 192–96
Classification, approaches to, 183

Classification, empiric, 184
Classification, genetic, 184, 192–96
Classification, Köppen, 185–90
Classification, rationale, 182
Classification, Thornthwaite, 190
CLIMAP, 298
Climate reconstructed, 307
Climatic sheath, 168
Climatography, 5
Climatology, defined, 4
Climatology, history, 6–9
Clouds, 80–83, 88
Cloud seeding, 156
Cloud streets, 121
Coastal deserts, 219
Coastal development hazards, 156
Coastal stations, 265
Coast ranges, 266
Cold front, 91, 137
Cold pole, 123
Cold waves, 233
Collision-coalescence, 88
Combined gas law, 356
Computer depictions, 300
Condensation, 70, 79–80
Conductivity, water, 68
Constant pressure charts, 171
Contemporary literature, 304
Continental air masses, 73
Continental distribution, 321–22
Continental divide, 254
Continental glaciation, 296
Continentality, 240
Continental uplift, 321
Continents, water balance, 104
Convectional rainfall, 238
Convective lifting, 90
Convergence, 95, 112
Core analysis, 299
Coriolis effect, 114
Corn Belt, 352–53
Corn production, 353
cP air masses, 229. See also Air masses
Cretaceous, 300
Crop failure, 352
Cyclones, 122
Cyclonic lifting, 91
Cyclonic precipitation, 238
Cyclonic storms, 236

D

Danube River, 233
Day length, 33
Daylight, North America, 232
Death Valley, 218, 246
Deforestation, 324
Degree days, 65–66, 340
Dendrochronology, 303
Desertification, 221–25, 326
Deserts, 118, 199–200, 246
Desert storms, 220
Devonian, 299
Dew, 80
Dewfall, 166
Dewpoint, 79
Dinosaurs, 300
Diurnal rainfall, 213
Divergence, 95, 112
Doldrums, 121
Donora, 177
Doppler radar, 144
Drainage basin, 99
Drizzle, 81
Drought, 102–4, 225, 256, 304
Dry climates, 186
Dryness, 102
Dust storms, 220

E

Earth rotation, 113
Earth's axis, 32
Earth-sun relationships, 29–31, 317
Easterly wave, 121
Ecosystem, 343
Ecuador, 285
Egypt, 217
El Azizia, 218
El Niño, 176, 344, 348
Electromagnetic spectrum, 24
Empiric systems, 185–92
Energy, alternate sources, 40
Energy, defined, 22
Energy balance, polar, 274
Energy balance, planetary, 35–38
Energy equivalents, 137
Energy flow, representation, 358

Energy receipts, global, 31–33
Energy transfer, 39
Energy transformation, 22
ENSO event, 348
Equable temperatures, 56
Equatorial climate, development, 203
Equinox, 31, 272
Eric the Red, 306
Erosion, 205
Erratics, 294
Eurasia, 245
Evaporation, 70, 74–79
Evaporites, 297
Evapotranspiration, 75–76, 190
Evapotranspirometer, 77
Exotic river, 102
Extreme temperatures, 56

F

Fahrenheit scale, 47
Fire, 263, 324
Fishing, 344
Floods, 99–104, 303
Floral evidence, 301–3
Florida, 238
Foehn wind, 285
Fog, 80–81, 266, 283, 285
Food production, 18, 341–43, 352–54
Forest climates, 169
Fossil fuels, 332, 340
Fossil plants, 301
Freezing point, 69
French Revolution, 313
Friction, 114
Frost, 80
Frostbite, 288
Frost-free season, 64, 282
F-scale, 147
Fujita, 147
Future Climates, 351

G

Galileo, 7
GARP, 8, 345
Gas constant, 13

Gas laws, 13, 51, 356
GATE, 9
Gay-Lussac's law, 356
General circulation model, 345
Genetic systems, 192–96
Geography, 5
Geologic time, 12
Geosphere, 4
Geostationary satellites, 7
Glacial advance, 296
Glacial enlargement, 311
Glacial retreat, 296
Glacial times, 307
Glaciation theory, 295
Gleissberg cycle, 320
Global ecology, 343
Global solar radiation, 28
GOES-SMS, 7
Grasslands, 200
Gravity winds, 123, 279
Great Lakes, 260
Great Plains, 242, 243, 256
Greenhouse Effect, 28, 33–34
Greenland, 306
Growing season, 6, 266

H

Hadley, 115
Hadley regime, 116
Hail, 88, 141–43
Hail insurance, 155
Hail regions, 142
Hawaii, 214
Health, 19, 205
Heat, defined, 46
Heat balance, 38–39
Heat budget, 359
Heat capacity, 55
Heat conductivity, 55
Heat island, 332–33
Heat waves, 254–55, 263
Heterosphere, 17
Heyerdahl, Thor, 120
Highland climate extremes, 284
Highland climates, 283–86
Historical period, 303
Homosphere, 17
Horse latitudes, 119
Humidity, 72, 238

Humidity, measures of, 72
Hurricane Eloise, damage, 156
Hurricanes, 147–54
Hydrologic cycle, 69–71
Hydrosphere, 4
Hygroscopic nuclei, 80
Hypothermia, 288
Hypoxia, 289

I

Ice, evidence of, 294
Ice erosion, 295
Ice pack, 273
Ice pellets, 88
Ice sheets, existing, 308
Ice sheets, maximum, 308
Icelandic records, 305
Ideal gas law, 13
Illinois, 107
Index of continentality, 240
Indicator fossils, 301
Industry, 20
Infrared energy, 33
Instrumental period, 307
Interglacial times, 307
Intermittent dryness, 102
Intertropical convergence, 95
Intertropical convergence zone, 121. See also ITCZ
Inverse square law, 25
Inversion, 51, 178
Inversion frequency, 179
Invertebrate fossils, 298
Ionosphere, 17
Irradiance, 28
Irrigation, 247
Isanomalous temperatures, 52
Isolines, 5
Isonomals, 52
Isostatic readjustment, 298
Isothermal layer, 14, 50
Isotopes, 298
ITCZ, 95, 121, 126–27, 211

J

Java, 213
Jet streams, 116, 146, 230

K

Katabatic wind, 168, 279
Katmai, 321
Keweenaw Peninsula, 262
Khartoum, 222, 223
Kirchoff's law, 357
Köppen classification, 185-90
Krakatoa, 321

L

Lake Bonneville, 297
Lake breeze, 130
Lake Lahontan, 297
Laminar flow, 14, 165
Land and sea breezes, 129
Land use, 326
Lapse rate, 50, 333, 360-61
Latent heat, 330
Latent heat transfer, 26
Latitude, 52
Lee of Lakes Effect, 261
Leveche, 220
Lightning, 140-42
Little Ice Age, 309, 311-13
Littoral zone, 266
Local climates, 162, 265
Local climatology, 169-70
London, England, 238
Lord Rayleigh, 26
Lysimeter, 77

M

Macroclimates, 162, 171
Macroclimatology, 6
Maritime air masses, 73
Maunder minimum, 320
Means, 48
Medieval chronicles, 304
Mediterranean climate, 233, 244-45
Mesa Verde, 304
Mesoclimates, 162, 170-71
Mesosphere, 14
Metabolism, 286
Meteorological drought, 103
Meteorology, defined, 4
METROMEX, 334

MET unit, 286
Microclimates, 162, 164-68, 283
Microclimatology, 6
Middle latitude summers, 229
Middle latitude winters, 230
Mid-latitude circulation, 122, 228-29
Mid-latitude cyclone, 91-93
Mid-latitude deserts, 245-47
Mid-latitude wet climates, 236-40
Mid-latitude wet-dry climates, 240-43
Mie scattering, 26
Migration, 304
Mill Creek culture, 304
Minerals, 205
Models, 344-46
Monsoon, 126, 129, 242
Mortality rates, 20
Mountain breeze, 130
Mountain building, 321
Mountain effects, 265
Mount Kilimanjaro, 322
Mount Rainier, 264
Mount St. Helens, 320
Mount Waialeale, 214
Mount Washington, 283
mP air masses, 238. *See also* Air masses

N

Namib, 219
National Center for Health, 233
National Climate Program Act, 9
National Hurricane Center, 152
NCAR, 345
Net radiation, 38, 358
Nile, 303
Nitrogen, discovery, 10
Normal temperature, 48
North America, 251-66
North Atlantic Drift, 239

O

Obliquity of the ecliptic, 317
Occluded front, 92
Ocean currents, 60, 275
Oceanic circulation, 323
Ocean levels, 322
Oceans, water balance, 104

Ocean temperatures, 119, 346
Odeillo, 40
Offshore wind, 126
Olympic Peninsula, 239
Onshore winds, 283
Optical air mass, 28
Orbital eccentricity, 318
Orographic lifting, 93–94
Orographic precipitation, 93
Oxisols, 202
Oxygen, discovery, 10
Oxygen, evolution, 11–12
Oxygen isotopes, 298
Ozone, 11, 27

P

Pacific Ocean pools, 346
Palmer Index, 103
Particulates, 328
Passive solar systems, 40
Past life, 298–303
Peat bogs, 302
Penman, 77
Perennial dryness, 102
Periglacial evidence, 297
Perihelion, 29
Periodic drought, 248
Periodic freezing, 234
Permafrost, 243, 307
Permanent snow, 50
Phoenix, Arizona, 247
Photosphere, 25
Photosynthesis, 11
Physical climatology, 5
Physiological effects, 285, 286–89
Planetary energy balance, 35
Planetary frontal zone, 122
Plant physiology, 301
Plate tectonics, 321
Pleistocene, 307
Pleistocene ice, 296
Pluvial lakes, 297
Podosols, 202
Poisson's equation, 360
Polar circulation, 122
Polar climates, 272–78
Polar desert, 279
Polar dry climate, 278–81

Polar front, 233
Polar wet and dry climate, 276–77
Polar wet climates, 281–82
Pole of inaccessibility, 278
Pollen analysis, 302–3
Potential evapotranspiration, 77, 190
Power outages, 335
Precession of equinoxes, 318
Precipitable water, 72, 259
Precipitation, frequency, 98
Precipitation, record extremes, 95
Precipitation, seasonal, 98
Precipitation, spatial variation, 94
Precipitation effectiveness, 191
Precipitation process, 88–89
Pressure, atmospheric, 14
Pressure belts, migration, 126
Pressure gradient, 112
Pressure surfaces, 171
Primitive atmosphere, 11
Primitive equations, 345
Probability, 267–69

R

Radiation, 36
Radiation, nature of, 23
Radiation fog, 81
Radiation laws, 23
Rain forests, 199
Rain-shadow, 97
Rayleigh scattering, 26
Record low temperature, 242
Redwoods, 239
Reflectance, 28
Relative humidity, 80
Resort areas, 245
Ross Ice Shelf, 278
Rossby regime, 116, 173, 228
Rossby waves, 116
Runoff, 86

S

Saffir/Simpson scale, 150
Sahara, 217, 218, 224
Sahelian drought, 225
St. Louis, 334

Salinity, 75, 275
San Diego, 264
San Francisco, 266
Sand storms, 220
Sastrugi, 280
Satellites, 7
Saturation level, 72
Saturation mixing ratio, 73
Saudi Arabia, 220
Savanna, 200
Scales of Climate, 162-63
Scattering, 26, 35
Scenarios, 348-50
Sea level changes, 297-98
Sea surface temperatures, 322
Seasonal dryness, 102
Seasonal forests, 199
Seasonal precipitation, 98
Sediments, 297
Sensible heat, 14, 37, 56, 330
Settlements, 306
Siberia, 243
Siberian high pressure, 243
Sirocco, 220
Sky color, 26
Slope orientation, 283
Snow, 88, 106, 238
Snow, water content, 84
Snow belts, 262
Snowfall, 259, 264
Snowfall, impacts, 84-85, 106-10
Snowline, 264, 285
Snowmelt, 85, 99
Soil classification, 201-2
Soil temperature, 165
Soils, 338
Solar beam, 31
Solar constant, 26
Solar energy, direct use, 40
Solar equator, 29
Solar furnace, 40
Solar homes, 42
Solar output variation, 319-20
Solar radiation intensity, 29
Solar wind, 25
Solstice, 272
Source regions, 253
Southern oscillation, 176, 348
South Pole, 279
Specific heat, 55, 68

Spring floods, 236
Spring wheat belt, 352
Sprinklers, 248
Squall lines, 137
Stability, 81, 177, 360-61
Stationary front, 92
Stefan-Boltzmann law, 357
Stored energy, 38
Storms, severe, 136-55
Storms and Planning, 155-57
Strand lines, 297
Stratopause, 14
Stratosphere, 14
Sublimation, 75
Subsidence, 119
Subtropical high, 262
Subtropical jet, 229
Sudan, 222
Summer-dry climates, 244-45
Sun, 25, 319
Sun, apparent migration, 55, 215
Sunspot cycles, 25
Sunspots, 25, 230
Supercells, 139
Supercooled water, 88
Supplemental irrigation, 247-49
Surface changes, 324-26
Surface properties, 55
Surface tension, 68
Surface winds, 123-26
Swiss Alps, 294-95
Synoptic climatology, 6, 171-76

T

Taiga, 201
Tambora, 321
Temperature, annual range, 60
Temperature, daily minimum, 48
Temperature, defined, 46
Temperature, distribution, 59-60
Temperature, diurnal, 51
Temperature, diurnal range, 84
Temperature, mean global, 35
Temperature, units, 47-48
Temperature, vertical, 49
Temperature change, 310
Temperature gradient, 60, 342
Temperature inversion, 51

Temperature lag, 47
Temperature range, 49
Temperature regimes, 55
Temperature thresholds, 64–66
Temperature units, 47–48
Terminal velocity, 88
Terrestrial radiation, 33–34
Thames river, 304
Thermal stress, 288
Thermoregulation, 288
Thermosphere, 14
Thornthwaite Classification, 190–92
Thornthwaite method, 77, 362–63
Thunder, 139
Thunder days, 335
Thunderstorm development, 90
Thunderstorms, 136–39
Timber, 205
Topography, 168
Tornadoes, 143–47
Tornado hazard, 146
Torricelli, 7
Trade wind inversion, 149
Trade winds, 119
Transamazon highway, 203–5
Transmissivity, 28
Transpiration, 7, 78–79
Tree line, 58
Tree ring analysis, 303
Tropical circulation, 116–21
Tropical climates, 208–10
Tropical cyclones, 147–54
Tropical deserts, 216–19
Tropical easterlies, 121
Tropical wet climate, 210–14
Tropical wet-dry climate, 214–16
Tropopause, 14, 51
Troposphere, 14
Tundra, 201, 277
Turbulent flow, 14, 165
Twilight, 273

U

Ultraviolet radiation, 27
Upper air circulation, 174, 265
Upper Mississippi Valley, 247
Urban climatology, 49, 329–34
Urbanization, 331, 334

U-shaped valleys, 294
U.S.S.R., 341

V

Valley breeze, 130
Valley inversions, 178
Vapor pressure, 75
Vegetation and climate, 196–201
Vegetation distribution, 196
Verkoyansk, 242
Vertebrate animals, 299
Vikings, 306
Volcanic activity, 320
Vorticity, 364
Vostok, 278

W

Warm front, 92
Waste heat, 332, 341
Water, density, 68
Water, phase changes, 70
Water, physical properties, 68–69
Water, yield, 86
Water balance, 104–5, 233–36, 362–63
Water balance, urban, 330–31
Water resources, 19
Watershed, 99
Water vapor, 72, 78, 89
Water vapor transfer, 89–93
Water year, 236
Wavelength, 23
Weathering, 338
Westerlies, 123
Westerly circulation, 229
White Christmas, 267
Wind, energy from, 132–33
Wind, global patterns, 115
Wind chill, 233
Wind chill equivalent, 287
Wind erosion, 225, 258
Wind power, 132
Winds, diurnal, 129
Winds, seasonal changes, 126
Winds, surface, 123–26
Windward slopes, 283
Winter dry climate, 243

Winter storms, 260
Winter storms, impacts, 108–10
WMO, 8

Y

Year without a summer, 312

Z

Zonal climates, 52
Zonal winds, 122